Bioimaging

Bioimaging

Imaging by Light and Electromagnetics in Medicine and Biology

Edited by

Shoogo Ueno

CRC Press
Taylor & Francis Group
Boca Raton London New York

CRC Press is an imprint of the
Taylor & Francis Group, an **informa** business

First edition published 2020

by CRC Press
6000 Broken Sound Parkway NW, Suite 300, Boca Raton, FL 33487-2742
and by CRC Press
2 Park Square, Milton Park, Abingdon, Oxon, OX14 4RN

ISBN: 978-0-367-20304-7 (hbk)
ISBN: 978-0-367-49043-0 (pbk)
ISBN: 978-0-429-26097-1 (ebk)

Typeset in Times
by Deanta Global Publishing Services, Chennai, India

Contents

Preface vii

Acknowledgments xiii

Editor xv

List of Contributors xvii

Chapter 1 Introduction 1

Shoogo Ueno

Chapter 2 Molecular Imaging of Viable Cancer Cells 15

Mako Kamiya and Yasuteru Urano

Chapter 3 Molecular Vibrational Imaging by Coherent Raman Scattering 37

Yasuyuki Ozeki, Hideaki Kano, and Naoki Fukutake

Chapter 4 Magnetic Resonance Imaging: Principles and Applications 75

Masaki Sekino and Shoogo Ueno

Chapter 5 Chemical Exchange Saturation Transfer and Amide Proton Transfer Imaging 101

Takashi Yoshiura

Chapter 6 Diffusion Magnetic Resonance Imaging in the Central Nervous System 121

Kouhei Kamiya, Yuichi Suzuki, and Osamu Abe

Chapter 7 Magnetic Particle Imaging 155

Keiji Enpuku and Takashi Yoshida

Chapter 8 Sensing of Magnetic Nanoparticles for Sentinel Lymph Nodes Biopsy 185

Masaki Sekino and Moriaki Kusakabe

Chapter 9 Optimizing Reporter Gene Expression for Molecular Magnetic Resonance Imaging: Lessons from the Magnetosome 201

Qin Sun, Frank S. Prato, and Donna E. Goldhawk

Chapter 10 Magnetic Control of Biogenic Micro-Mirror..215

   Masakazu Iwasaka

Chapter 11 Non-Invasive Techniques in Brain Activity Measurement Using Light or Static
   Magnetic Fields Passing Through the Brain 233

   Osamu Hiwaki

Index 249

Preface

Biomedical imaging and sensing technologies have been rapidly developing and expanding in a variety of fields in medicine and biology. This book, titled *Bioimaging: Imaging by Light and Electromagnetics in Medicine and Biology*, explores new horizons in bioimaging from the molecular and cellular level to the human brain. Bioimaging is applied in a variety of fields with numerous techniques in different spectrums. For this book, we limit the focus to imaging by light and electromagnetics for application in medicine and biology.

In 2015, we edited a book titled *Biomagnetics: Principles and Applications of Biomagnetic Stimulation and Imaging*. Also published by CRC Press Taylor & Francis Group, one-third of the book was related to bioimaging, but molecular imaging was not included. However, imaging and sensing technologies have rapidly accelerated since that time.

I received an inquiry from the editor in physics at CRC Press Taylor & Francis Group in August 2018, asking me about the possibility of a new book. It was a good opportunity for me to start making a new framework. I, fortunately, have coworkers and scientists in different institutions, colleagues at the University of Tokyo, and friends overseas, who are experts in biomedical imaging. I planned to make the book consist of more than ten chapters from the molecular level to the human brain. It is fortunate that 11 chapters arrived by the deadline of manuscript submission, although a few chapters did not arrive by the cutoff date. The rest will be hopefully available in an online readable version in parallel to this published book.

A brief outline of each chapter is as follows:

CHAPTER 1. INTRODUCTION (SHOOGO UENO)

A history of the discovery of electromagnetic fields is briefly described in the opening chapter. After classification of electromagnetic fields from lower frequencies to higher frequencies, medical devices and biomedical imaging systems used in different frequency bands are discussed. The vacuum ultraviolet ray with a wavelength of around 200 nm is the boundary of ionizing radiation and non-ionizing radiation. It is emphasized that medical devices and biomedical imaging tools are used in different frequency bands either in the ionizing radiation region or in the non-ionizing radiation region.

Then, the history and overview of imaging tools are described. In advances in biomedical imaging and stimulation, these include computed tomography (CT), magnetic resonance imaging (MRI), magnetoencephalography (MEG), transcranial magnetic stimulation (TMS), magnetic particle imaging (MPI), and near-infrared spectroscopic (NIRS) imaging. In advances in molecular and cellular imaging, green fluorescent protein (GFP), optical fluorescence and cancer therapy, optogenetics and studies of neuronal circuit dynamics in the brain, Raman scattering and coherent Raman scattering (CRS) microscopy, molecular imaging based on MRI, and magnetic orientation of living systems and biogenic micromirrors are introduced.

CHAPTER 2. MOLECULAR IMAGING OF VIABLE CANCER CELLS (MAKO KAMIYA AND YASUTERU URANO)

Molecular imaging that visualizes cellular functions or molecular processes inside the body is described in Chapter 2. Probes used for molecular imaging are targeted to biomarkers for specific visualization of targets or pathways. Optical imaging based on fluorescence emission from fluorophores is introduced as a promising technique.

To achieve a reliable molecular imaging for the detection of viable cancer cells, two categories of unique optical fluorescence imaging probes are developed: always-on probes and activatable probes. Always-on probes consist of fluorescent reporting units (fluorophores) linked to tumor-targeting moieties such as antibodies, affibodies, or small-molecular ligands. Activatable probes of which fluorescence is initially suppressed is turned on by a molecular switch at tumor sites, providing a cancer-specific signal with high sensitivity and a high tumor-to-background signal ratio.

In this chapter, first, optical fluorescence imaging probes available for visualizing cancer cells are reviewed. Second, activatable probes, which offer particular advantages in terms of providing a cancer-specific signal with high sensitivity and a high tumor-to-background signal ratio, are explored. Third, various switching mechanisms and design strategies for activatable bioimaging probes are discussed. Finally, some applications demonstrating the usefulness of the designed probes are presented.

CHAPTER 3. MOLECULAR VIBRATIONAL IMAGING WITH COHERENT RAMAN SCATTERING (YASUYUKI OZEKI, HIDEAKI KANO, AND NAOKI FUKUTAKE)

Raman scattering is a light scattering phenomenon involving molecular vibration of substances that interact with light, providing the vibrational spectrum of molecules. The Raman scattering technique has the potential to observe the behavior of molecules, however, the intensity of Raman scattering is quite weak, and it takes a long acquisition time.

Coherent Raman scattering (CRS) microscopy using two-color and/or broadband laser pulses is introduced to overcome the problems in the original Raman scattering technique. CRS enables a dramatic improvement in the acquisition time and sensitivity of Raman scattering. Optical microscopy based on CRS allows for biological imaging with molecular vibrational contrast. CRS microscopy is opening up a variety of applications which include label-free molecular analysis of cells and tissues, monitoring of metabolic activities, and supermultiplex imaging with more than ten colors.

In this chapter, the general concepts and recent applications of CRS microscopy are described. Specifically, the basic principles of CARS and stimulated Raman scattering (SRS) microscopy are explained, showing practical instrumentations. The applications to label-free imaging, metabolic imaging, and supermultiplex or ultramultiplex imaging are introduced. In addition, the classical and quantum-mechanical pictures for various CRS processes as well as the theory of image formation in CRS microscopy are explained in detail.

CHAPTER 4. MAGNETIC RESONANCE IMAGING: PRINCIPLES AND APPLICATIONS (MASAKI SEKINO AND SHOOGO UENO)

Magnetic resonance imaging (MRI) provides a wealth of anatomical and physiological information. The advantages of MRI over other modalities are high spatial resolution and no radiation exposure. This chapter gives a basic introduction to MRI consisting of physical principles and a few biomedical applications such as functional imaging of the brain, mapping of electric properties of tissues, and MRI-guided focused ultrasound therapy.

An MRI system is equipped with a strong magnet which induces nuclear magnetization in the target body. The MRI signals originate from the rotation of the nuclear magnetization under the magnetic fields. The application of a gradient magnetic field causes a spatial variation of signal frequency and an image is reconstructed by means of Fourier transformation. Details of these phenomena are comprehensively presented with schematic illustrations, sample images, and equations. Functional MRI visualizes a local change in blood oxygenation associated with neuronal electrical activities. The magnetic susceptibility of blood depends on the oxygenation level, which affects the signal intensity of the MRI. Functional MRI is applicable to both evoked and spontaneous brain activities. The neuronal activities are relevant to electrical properties of tissues.

Applications of MRI include mapping of electrical properties such as conductivity, permittivity, and weak magnetic fields arising from neuronal currents. Pattern recognition and various artificial intelligence techniques are attracting attention because of the ability to extract information from MRI. State-of-the-art imaging processing techniques for MRI are also introduced in this chapter.

CHAPTER 5. CHEMICAL EXCHANGE SATURATION TRANSFER AND AMIDE PROTON TRANSFER IMAGING (TAKASHI YOSHIURA)

Molecular imaging based on MRI is an attractive and challenging theme to be studied for both basic research and clinical diagnosis.

Chemical exchange saturation transfer (CEST) is a novel molecular magnetic resonance (MR) contrast based on saturation transfer from solutes to bulk water molecules via proton exchanges. CEST helps detect low concentration molecules *in vivo*. A variety of diamagnetic and paramagnetic compounds have been proposed as exogenous CEST agents for specific purposes such as reporting pH or monitoring metabolism, although only a few are clinically applicable.

On the other hand, recent studies have shown the feasibility and possible clinical impact of CEST imaging of endogenous mobile molecules with amide, amine, and hydroxyl protons. Amide proton transfer (APT) imaging is a specific type of CEST imaging which detects endogenous mobile proteins and peptides in tissue. Currently, APT imaging is considered to be the most clinically relevant CEST imaging, and it has actually started to show promising results, especially in oncologic imaging.

In this chapter, first, the theory of CEST and APT imaging are explained with illustrated figures. Next, the proposed clinical utilities of CEST and APT imaging and their pitfalls are described. Finally, potentials for future development are discussed.

CHAPTER 6. DIFFUSION MAGNETIC RESONANCE IMAGING IN THE CENTRAL NERVOUS SYSTEM (KOUHEI KAMIYA, YUICHI SUZUKI, AND OSAMU ABE)

Diffusion MRI measures the diffusion of water molecules along different directions. Such information provides images sensitive to the underlying tissue structure on the order of micrometers, with orders of magnitude below the nominal image voxel size. Because of this unique advantage, diffusion MRI gained popularity in both clinical and scientific communities. Its power to probe tissue microstructure non-invasively made a significant impact on clinical medicine, as represented by the discovery of a profound decrease of apparent diffusivity in acute stroke in 1990. Also, diffusion anisotropy observed in the nervous tissue led to the development of diffusion tensor imaging (DTI), as well as a methodology for tracing macroscopic fiber tracts (tractography) that is now widely applied for presurgical planning. Since these founding works, significant improvements have been made in both scanner hardware and acquisition schemes, pushing the limit of the available data range for clinical study. These have led to the development of several newer analytic frameworks suitable for extended acquisition, as well as advanced tractography methods, enhancing our understanding of diseases and brain anatomy. Moreover, a biologically inspired compartmental model has been proposed in an effort to derive metrics that are ready for naïve interpretation and translation to histology such as axon diameter, intra-cellular water fraction, and compartmental diffusivities.

This chapter describes what has been achieved so far with diffusion MRI, focusing on applications in clinical scanners and a discussion about the problems currently under debate. The first part of chapter is devoted to basic concepts of conventional DTI and tractography, along with a few examples which showcase the contribution of diffusion imaging to clinical medicine. The second part introduces technical advances that followed DTI. Applications of tractography for quantification of the brain network, i.e., connectome, are also briefly described. Finally, limitations, pitfalls, and challenges in modern diffusion imaging are discussed.

CHAPTER 7. MAGNETIC PARTICLE IMAGING (KEIJI ENPUKU AND TAKASHI YOSHIDA)

Magnetic particle imaging (MPI) is a new imaging technique for *in vivo* medical diagnosis. MPI uses magnetic nanoparticles (MNPs) as tracers. In medical applications, MNPs accumulated in the human or animal body are magnetized using an excitation field, and the resulting signal from the MNPs can be obtained. The three-dimensional (3D) positions and numbers of the MNPs can then be reconstructed from the measured data through the solution of an inverse problem.

The MPI method has high sensitivity and high spatial resolution for MNP detection. It can detect MNPs with concentration as low as $1\sim2$ ng/μL with mm or sub-mm spatial resolution. Furthermore, it can directly visualize and quantify MNPs without being affected by surrounding materials. Prototype MPI systems for small animals have already been developed, and a lot of pre-clinical imaging results have been demonstrated, confirming the usefulness of the MPI method.

The following are described in this chapter: (1) the operating principle of MPI. (2) Signal strength and spatial resolution in MPI. These values are determined by dynamic and non-linear magnetization of MNPs under AC and DC fields. (3) MPI hardware. System configuration to achieve high spatial resolution using field-free point (FFP) or field-free line (FFL). (4) Imaging techniques for 3D detection of MNPs. Reconstruction of MNP distribution from the measured data through the solution of an inverse problem.

CHAPTER 8. SENSING OF MAGNETIC NANOPARTICLES FOR SENTINEL NODES BIOPSY (MASAKI SEKINO AND MORIAKI KUSAKABE)

Magnetic nanoparticles (MNPs) are used as medical tracers as well as MRI contrast agents. These applications are particularly useful for the diagnosis of cancer. Metastasis to lymph nodes occurs when cancer cells reach the nodes through lymphatic vessels. When MNPs are injected in the proximity of a cancer lesion, the MNPs enter the lymphatic vessels and follow the path of metastasis. Accumulation of the MNPs in the nodes can be detected using magnetic sensors or magnetic imaging modalities. This technique is useful for identifying the sentinel lymph nodes to which cancer cells enter first following the lymphatic flow from the cancer lesion. Biopsy of the sentinel lymph node is necessary for staging the cancer. The present gold standard technique for identifying the sentinel lymph node uses radioisotope tracers and a gamma-ray detector. While the radioisotope method exhibits a high identification rate, this technique is available only in hospitals with nuclear medicine capabilities. The use of MNPs should be extended as an alternative to the radioisotope method because it can be used in any hospital.

This chapter introduces physical characteristics and pharmacokinetics of MNPs, and the development of a dedicated handheld magnetometer for sensing MNPs.

CHAPTER 9. OPTIMIZING REPORTER GENE EXPRESSION FOR MOLECULAR MAGNETIC RESONANCE IMAGING (QIN SUN, FRANK S. PRATO, AND DONNA E. GOLDHAWK)

Iron perturbs the magnetic resonance (MR) signal and thus can influence measured MR parameters. All cells and tissues require an iron co-factor for various functions. Therefore, iron-related MR measures inherently target integral cellular activities. In addition, differential regulation of iron enhances the distinction between tissue and cell types, which enables MRI to non-invasively monitor changes in iron-related organ function.

The chapter authors have drawn on the best-known example of effective, gene-based iron-labeling in single cells – the magnetosome found in magnetotactic bacteria (MTB) – for the most efficient expression system associated with the utility of reporter genes for *in vivo* molecular imaging by MRI.

This chapter reviews concepts in reporter gene expression for molecular MRI. This includes the regulation of iron compartmentalization and metabolism in mammalian cells, and the influence it has on MR detection of iron contrast. In addition, the chapter authors examine the synthesis of magnetosomes and propose strategies, inspired by this bacterial structure, for improving the detection of reporter gene expression by MRI.

CHAPTER 10. MAGNETIC CONTROL OF BIOGENIC MICRO-MIRROR (MASAKAZU IWASAKA)

Most materials, such as proteins and DNA, show orientation under strong magnetic fields. This interesting phenomenon, called magnetic orientation, has been studied by many groups. The critical threshold for the magnetic field in magnetic orientation, however, has not been clarified. Masakazu Iwasaka demonstrates that a thin micro-mirror from a fish scale with high reflectivity exhibits a distinct magnetic response at 100 mT. A dramatic event under a magnetic field is the decrease in light scattering from guanine crystals as well as the rapid rotation against the applied magnetic field. Enhancement of light scattering intensity is also observed when the three vectors of light incidence, magnetic field, and observation are orthogonally directed. The results indicate that biogenic guanine crystals have a large diamagnetic anisotropy along the surface in parallel and normal directions. The thin biogenic plates, with a size of micro-meter to submicron-meter, can act as a non-invasive and magnetically controlled micro-mirror for light irradiation control.

The first section in the chapter describes a method for detecting the alignment direction of micro-particles floating in water via magnetic orientation. The second section discusses optical and magnetic properties of biogenic micro-mirror particles.

CHAPTER 11. NON-INVASIVE TECHNIQUES IN BRAIN ACTIVITY MEASUREMENT USING LIGHT OR STATIC MAGNETIC FIELDS PASSING THROUGH THE BRAIN (OSAMU HIWAKI)

This chapter deals with a novel optical imaging technique to obtain imaging of brain activities based on near-infrared spectroscopy (NIRS).

The signals detected using NIRS originate from changes in the concentration of oxyhemoglobin and deoxy-hemoglobin due to vascular changes in the cortex. However, following an electrophysiological response, the hemodynamic response is delayed by several seconds. Furthermore, in conventional NIRS, the distance between a light source and a detector located on the scalp ranges from 20 to 40 mm. As the light diffuses in all directions inside the head, the distribution of the light is rather complex. This means that it is difficult to obtain brain imaging with a high spatial resolution. To improve these problems, a new method of NIRS is proposed by the chapter author.

The first part of this chapter introduces a non-invasive brain function measurement using light including NIRS. The latter part discusses new techniques which enable the non-invasive measurement of brain activity with a high spatial and a high temporal resolution using near-infrared light or static magnetic fields.

This book provides an explanation of the physical, biophysical, and biochemical principles of biomedical imaging, as well as applications of these techniques in neuroscience, clinical medicine, and healthcare. The book aims to educate graduate and undergraduate students, young scientists, medical doctors, and engineers on how these techniques are used in hospitals and why they are so promising. It is a brief overview of the recent research trends in bioimaging and the book acts as an informative guide for researchers and engineers to further study this field.

Acknowledgments

I thank the chapter authors for their intellectual contributions. It is due to the great efforts of chapter authors that this book has been published. Further, I thank the staff of CRC Press Taylor & Francis Group, in particular, Rebecca Davies, editor, physics, for contacting me to discuss a book on bioimaging in August 2018, and her consistent support and valuable advice regarding the book proposal and all the editing processes; Kirsten Barr, editorial assistant, for her assistance and support during the editing processes for the whole year; Damanpreet Kaur, editorial assistant, for her assistance; and the staff of the editorial project development team.

Finally, but not least, I greatly appreciate the anonymous reviewers who gave valuable comments for the publication of this book.

Shoogo Ueno

Editor

Shoogo Ueno, PhD, received his BS, MS, and PhD (Dr Eng) degrees in electronic engineering from Kyushu University, Fukuoka, Japan, in 1966, 1968, and 1972, respectively. Dr. Ueno was an Associate Professor with the Department of Electronics, Kyushu University, from 1976 to 1986. From 1979 to 1981, he spent his sabbatical with the Department of Biomedical Engineering, Linkoping University, Linkoping, Sweden, as a Guest Scientist. He served as a Professor in the Department of Electronics, Graduate School of Engineering, Kyushu University from 1986 to 1994 and subsequently served as a Professor in the Department of Biomedical Engineering, Graduate School of Medicine, the University of Tokyo, Tokyo, Japan, from 1994 to 2006. During this time, he also served as a Professor in the Department of Electronic Engineering, Graduate School of Engineering, the University of Tokyo. In 2006, he retired from the University of Tokyo as Professor Emeritus. Since then, he has been a Professor with the Department of Applied Quantum Physics, Graduate School of Engineering, Kyushu University, and he served as the Dean of the Faculty of Medical Technology, Teikyo University, Fukuoka, Japan from 2009 through 2012.

Dr. Ueno's research interests include biomedical imaging, bioelectromagnetics, bioengineering, and brain research, in particular, neuroimaging, transcranial magnetic stimulation (TMS), electro- and magnetoencephalography (EEG and MEG), electric impedance and current imaging based on magnetic resonance imaging (MRI), magnetic control of biological cell growth and cell destruction, and bioelectromagnetic approaches to the treatments of brain diseases.

Dr. Ueno is a Life Fellow (2011) and a Fellow of the Institute of Electrical and Electronics Engineers, IEEE (2001), a Fellow of the American Institute for Medical and Biological Engineering, AIMBE (2001), and a Fellow of the International Academy for Medical and Biological Engineering, IAMBE (2006). He was a founding member of the International Advisory Board of International Conferences on Biomagnetism in 1987. He was the President of the Bioelectromagnetics Society, BEMS (2003–2004), the Chairman of the Commission K on Electromagnetics in Biology and Medicine of the International Union of Radio Science, URSI (2000–2003), the President of the Japan Biomagnetism and Bioelectromagnetics Society (1999–2001), the President of Magnetics Society of Japan (2001–2003), and the President of the Japanese Society for Medical and Biological Engineering (2002-2004). He received the *Doctor Honoris Causa* from Linkoping University (1998). He was the 150th Anniversary Jubilee Visiting Professor at Chalmers University of Technology, Gothenburg, Sweden (2006), and a Visiting Professor at Simon Frasier University, Burnaby, Canada (1994) and Swinburne University of Technology, Hawthorn, Australia (2008). He received the IEEE Magnetics Society Distinguished Lecturer during 2010 and *d'Arsonval* Medal from the Bioelectromagnetics Society (2010). He is a Fellow of the URSI (2017).

On the personal side, he was born in Arao, Kumamoto, Japan, on 1 October 1943, and has lived mostly in Fukuoka since 1962. He and his wife, Haruko, have two sons and four grandchildren.

List of Contributors

Osamu Abe
Department of Radiology
Graduate School of Medicine
The University of Tokyo
Tokyo, Japan

Keiji Enpuku
Department of Electrical and Electronic
   Engineering
Graduate School of Information Science and
   Electrical Engineering
Kyushu University
Fukuoka, Japan

Naoki Fukutake
Nikon Corporation
Kanagawa, Japan

Donna E. Goldhawk
Imaging, Lawson Health Research Institute
and
Medical Biophysics
Western University
and
Collaborative Graduate Program in Molecular
   Imaging
Western University
London, Ontario, Canada

Osamu Hiwaki
Department of Biomedical Information Sciences
Graduate School of Information Sciences
Hiroshima City University
Hiroshima, Japan

Masakazu Iwasaka
Molecular Bioinformation Research Division
Research Institute for Nanodevice and Bio
   Systems
Hiroshima University
Higashi Hiroshima, Japan

Kouhei Kamiya
Department of Radiology
Graduate School of Medicine
The University of Tokyo
Tokyo, Japan

Mako Kamiya
Department of Biomedical Engineering
Graduate School of Medicine
The University of Tokyo
Tokyo, Japan

Hideaki Kano
Department of Applied Physics
Graduate School of Pure and Applied
   Sciences
University of Tsukuba
Tsukuba, Ibaraki, Japan
and
Tsukuba Research Center for Energy Materials
   Science (TREMS)
University of Tsukuba
Tsukuba, Ibaraki, Japan

Moriaki Kusakabe
Research Center for Food Safety
Graduate School of Agricultural and Life
   Science
The University of Tokyo
Tokyo, Japan

Yasuyuki Ozeki
Department of Electrical Engineering and
   Information Systems
Graduate School of Engineering
The University of Tokyo
Tokyo, Japan

Frank S. Prato
Imaging, Lawson Health Research Institute
and
Medical Biophysics
Western University
and
Collaborative Graduate Program in Molecular
   Imaging
Medical Imaging
Physics and Astronomy
Western University
London, Ontario, Canada

Masaki Sekino
Department of Electrical Engineering and
    Information Systems
Graduate School of Engineering
The University of Tokyo
Tokyo, Japan

Qin Sun
Imaging, Lawson Health Research Institute
and
Medical Biophysics
Western University
and
Collaborative Graduate Program in Molecular
    Imaging
Western University
London, Ontario, Canada

Yuichi Suzuki
Department of Electrical Engineering and
    Information Systems
Graduate School of Engineering
The University of Tokyo
Tokyo, Japan

Shoogo Ueno
Department of Biomedical Engineering
Graduate School of Medicine
The University of Tokyo
Tokyo, Japan

Yasuteru Urano
Department of Biomedical Engineering
Graduate School of Medicine
The University of Tokyo
Tokyo, Japan
and
Graduate School of Pharmaceutical Sciences
The University of Tokyo
Tokyo, Japan
and
CREST, Japan Agency for Medical Research
    and Development
Tokyo, Japan

Takashi Yoshida
Department of Electrical and Electronic
    Engineering
Graduate School of Information Science and
    Electrical Engineering
Kyushu University
Fukuoka, Japan

Takashi Yoshiura
Department of Radiology
Graduate School of Medical and Dental
    Sciences
Kagoshima University
Kagoshima, Japan

1 Introduction

Shoogo Ueno

CONTENTS

1.1 Introduction 1
1.2 Light and Electromagnetic Fields in Medicine and Biology 2
1.3 Advances in Biomedical Imaging and Stimulation..4
    1.3.1 Computed Tomography 4
    1.3.2 Magnetic Resonance Imaging 5
    1.3.3 Magnetoencephalography...5
    1.3.4 Transcranial Magnetic Stimulation 6
    1.3.5 Magnetic Particle Imaging 6
    1.3.6 Near-Infrared Spectroscopic Imaging 7
1.4 Advances in Molecular and Cellular Imaging..7
    1.4.1 Green Fluorescent Protein 7
    1.4.2 Optical Fluorescence and Cancer Therapy...8
    1.4.3 Optogenetics and Studies of Neuronal Circuit Dynamics in the Brain...........8
    1.4.4 Raman Scattering and Coherent Raman Scattering Microscopy......................8
    1.4.5 Molecular Imaging Based on Magnetic Resonance Imaging............................9
    1.4.6 Magnetic Orientation of Living Systems and Biogenic Micromirrors..............9
1.5 Summary 10
References...10

1.1 INTRODUCTION

Imaging techniques have been rapidly developing and expanding in a variety of fields in medicine and biology. This chapter gives an overview of a range of imaging techniques from molecular and cellular levels to the human brain, focusing on the imaging techniques using light and electromagnetics. We start with a brief history of the discovery of electromagnetic fields and review the imaging techniques using light, electromagnetic fields, or electromagnetic techniques in medicine and biology.

The overview includes green fluorescent protein (GFP) and its applications to bioimaging, molecular imaging of viable cancer cells, optogenetics and studies of neuronal circuit dynamics in the brain, molecular vibrational imaging by coherent Raman scattering, principles and applications of magnetic resonance imaging (MRI) such as functional MRI (fMRI), diffusion MRI of the human nervous system, and chemical exchange saturation transfer (CEST) and amide proton transfer (APT) imaging. Other important and new technologies include magnetic particle imaging (MPI) using magnetic nanoparticles and magnetic sensors, usage of the reporter gene for molecular MRI, magnetic control of the biogenic micromirror, and brain imaging by near-infrared spectroscopy (NIRS) and static magnetic fields. Transcranial magnetic stimulation (TMS) is also reviewed to discuss the functional mapping and imaging of the human brain associated with fMRI, magnetoencephalography (MEG), electroencephalography (EEG), and systems neuroscience based on optogenetics.

Historically, computed tomography (CT) and positron emission tomography (PET), as well as ultrasonic imaging systems, have been introduced in the medical world as biomedical imaging tools, and these imaging tools have changed the medical world dramatically in modern medicine. In this chapter, we do not describe ultrasonic imaging, but we focus our discussion on recent advances in electromagnetic-based and light-based biomedical imaging and stimulation techniques, which has brought a significant impact in medicine and biology in recent decades.

1.2 LIGHT AND ELECTROMAGNETIC FIELDS IN MEDICINE AND BIOLOGY

Electromagnetic fields or electromagnetic waves were theoretically predicted by James Clerk Maxwell (1831–1879) in the UK in 1864, and the existence of electromagnetic waves was experimentally verified by Heinrich Rudolf Hertz (1857–1894) in Germany in 1888. Guglielmo Marches Marconi (1874–1937) in Italy applied electromagnetic waves for telecommunication over the English Channel between France and England in 1899. As Maxwell pointed out, light is one of the electromagnetic waves in a specific frequency band.

Maxwell derived the so-called four fundamental electromagnetic equations, and he obtained an equation to show the propagation velocity v of electromagnetic waves in free space or in vacuum space as follows.

$$v = \frac{1}{\left(\varepsilon_0 \mu_0\right)^{1/2}} \tag{1.1}$$

where ε_0 is the dielectric constant or permittivity and μ_0 is the magnetic permeability in vacuum space.

Applying $\varepsilon_0 = 8.8542 \times 10^{-12}$ F/m and $\mu_0 = 4\pi \times 10^{-7}$ H/m,
we obtain

$$v = 2.998 \times 10^8 \text{ m/s} \tag{1.2}$$

If we use $\varepsilon_0 = 1/(36\pi) \times 10^{-9}$ F/m and $\mu_0 = 4\pi \times 10^{-7}$ H/m,
we obtain

$$v = 3.00 \times 10^8 \text{ m/s.} \tag{1.3}$$

These values coincide with the light velocity c = 2.99792458×10^8 m/s.

Here, the relationship between frequency and wavelength is described for further discussion.

Since the electromagnetic waves propagate at the light velocity in vacuum space, multiplication of wavelength λ and frequency f of electromagnetic waves are equal to the light velocity c as shown by

$$c = \lambda \times f \tag{1.4}$$

Therefore, the wavelength λ is given by

$$\lambda = c / f \tag{1.5}$$

The wavelength of 300 Hz electromagnetic fields is 1000 km.
The wavelength of 300 kHz (300×10^3 Hz) electromagnetic fields is 1 km
The wavelength of 300 MHz (300×10^6 Hz) electromagnetic fields is 1 m
The wavelength of 300 GHz (300×10^9 Hz) electromagnetic fields is 1 mm
The wavelength of 300 THz (300×10^{12} Hz) electromagnetic fields is 1 μm
The wavelength of 300 PHz (300×10^{15} Hz) electromagnetic fields is 1 nm

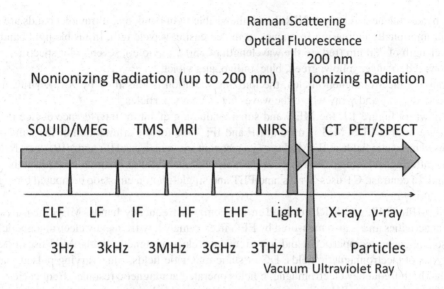

FIGURE 1.1 Classification of electromagnetic fields. Medical devices and biomedical imaging systems used at different frequency bands are marked along the frequency axis. The electromagnetic fields or electromagnetic waves are legally handled up to 3 THz or 3000 GHz. Electromagnetic fields lower than a frequency at the vacuum ultraviolet ray of its wavelength around 200 nm are called non-ionizing radiation. The electromagnetic fields higher than a frequency at the vacuum ultraviolet ray are called ionizing radiation. The electromagnetic fields are classified from DC to higher frequencies by ELF (extremely low frequency), LF (low frequency), IF (intermediate frequency), HF (high frequency), and EHF (extremely high frequency) electromagnetic fields. Visible light is in the frequency range from red (780 nm) to violet (380 nm). X-ray and γ-ray are in the ionizing radiation range. In non-ionizing radiation or non-ionizing electromagnetic fields, SQUID (superconducting quantum interference device) and MEG (magnetoencephalography) are used at ELF; MPI (magnetic particle imaging) is used at LF and IF; TMS (transcranial magnetic stimulation) uses pulsed magnetic fields at IF; MRI (magnetic resonance imaging) uses its resonant frequencies at HF; and NIRS (near-infrared spectroscopy) uses a near-infrared (NIR) ray. Raman scattering and optical fluorescence are used at the visible light band. In contrast, CT (computed tomography) uses X-rays, and PET (positron emission tomography) and SPECT (single-photon emission computed tomography) use γ-rays.

Figure 1.1 shows the classification of electromagnetic fields used in telecommunication and medical applications. Medical devices and biomedical imaging systems used at different frequency bands are marked along the frequency axis.

The usage of electromagnetic fields is legally law-handled up to 3 THz or 3000 GHz. The electromagnetic fields below a point of the vacuum ultraviolet ray at around 200 nm in wavelength are called non-ionizing radiation or non-ionizing electromagnetic fields. In contrast, the electromagnetic fields higher than a point at the vacuum ultraviolet ray at around 200 nm in wavelength are called ionizing radiation or ionizing electromagnetic fields. Exposure to ionizing radiation causes ionization of molecules in cells and living tissues, which may result in undesirable effects on living systems when the intensity of irradiation exceeds a threshold level. Radiation therapy for cancer diseases use ionizing radiation.

In non-ionizing electromagnetic fields, frequency bands are classified as follows:

Extremely low frequency (ELF) electromagnetic fields (DC ~ 3 kHz)
Low frequency (LF) electromagnetic fields (3 kHz ~ 10 kHz)
Intermediate frequency (IF) electromagnetic fields (10 kHz ~ 10 MHz)
High frequency (HF) electromagnetic fields (10 MHz ~ 6 GHz)
Extremely high frequency (EHF) electromagnetic fields (6 GHz ~ 3 THz)

In light bands, the near-infrared (NIR) band, the visible light band, and ultraviolet bands are classified in promotionally increasing frequency or in decreasing wavelength. In visible light band, from the wavelength of 780 nm (red) to the wavelength of 380 nm (violet), seven-color spectra exist like a rainbow; red, orange, yellow, green, blue, indigo, and violet.

In ionizing electromagnetic fields, the ionizing radiation is classified by X-ray, γ-ray, and the areas beyond X-ray and γ-ray where the waves act as heavy particles.

As shown in Figure 1.1, the MEG and superconducting quantum interference device (SQUID) systems are used at ELF; MPI is used at LF and IF; TMS for the stimulation of the human brain uses pulsed magnetic fields at IF; MRI uses its resonant frequencies at HF; and NIRS uses NIR rays. Raman scattering and optical fluorescence for molecular and cellular imaging are used at the visible light band. In contrast, CT uses X-rays, and PET and single-photon emission computed tomography (SPECT) use γ-rays.

MEG, MPI, TMS, and MRI are used in non-ionizing frequency bands. MEG measures brain electrical activities and is also measured by EEG in extremely low frequency electromagnetic fields. MPI uses magnetic nanoparticles and low frequency electromagnetic fields. MRI uses three different types of electromagnetic fields; DC or static magnetic fields, time-varying pulsed magnetic fields, and high frequency electromagnetic fields operated at magnetic resonant frequencies.

Most molecular and cellular imaging are used in the visible light band in the non-ionizing radiation region.

In contrast, CT, PET, and SPECT are used in the ionizing radiation region.

It is important to understand that these medical devices and medical imaging tools are used in deferent frequency bands either in the ionizing radiation region or in the non-ionizing radiation region.

1.3 ADVANCES IN BIOMEDICAL IMAGING AND STIMULATION

1.3.1 COMPUTED TOMOGRAPHY

CT is a great invention of imaging technology. Allan M. Cormack (1924–1998) in the USA proposed principles of image reconstruction to get transverse axial imaging of the inside of the body exposed to external irradiation of X-rays. He published the principles of computed tomography in the *Journal of Applied Physics* in 1963 and 1964 (Cormack, 1963, 1964). Prior to Cormack's papers, in 1917, Johann Karl August Radon (1887–1956) in Austria derived a transformation to reconstruct imaging by projected data obtained by projection from infinite directions (Radon, 1917). The essence is the same in the Cormack and Radon papers in solving the inverse problem of the line integral which contains the information of imaging data. Cormack worked for the invention of computed tomography without notice of the Radon transformation, but, afterwards, he referred Radon's work in his lectures. After all, the Radon transformation in 1917 was re-recognized on the occasion of the appearance of computed tomography.

Godfrey N. Hounsfield (1919–2004) in the UK invented X-ray computed tomography in 1972, and Hounsfield and his group commercialized the CT system, called EMI scan, in 1973, and he published papers in 1973 (Hounsfield, 1973, Ambrose and Hounsfield, 1973). Because of its innovative non-invasive imaging technology, the invention of CT has dramatically changed the medical world. Cormack and Hounsfield were awarded the Nobel Prize in Physiology and Medicine in 1979.

It is noteworthy that Cormack mentioned in his paper that his method is also useful for imaging in PET (Cormack, 1964). When a positron meets an electron near the positron, the positron and electron combine and vanish, which results in the emission of a pair of electromagnetic waves, i.e., the emission of γ-rays in the opposite directions. By calculating the time of flight (TOF) of the γ-rays between the γ-ray detectors positioned in the circle around the body, we can estimate the position of the source of γ-ray radiation in the body. CT and PET are very useful tools in clinical medicine today.

1.3.2 MAGNETIC RESONANCE IMAGING

The principle of MRI was proposed by Paul Christian Lauterbur (1929–2007) in the USA in 1973 (Lauterbur, 1973). The basic principle of imaging is to introduce gradient coils to produce linearly increased or decreased gradient magnetic fields in addition to the main static magnetic fields. The resonant frequency of proton in the tissues varies with the position, which gives us the spatial information in the space of the Fourier Transformation. The gradient magnetic fields are transiently applied as time-varying magnetic fields or pulsed magnetic fields which are computer-controlled in pulse sequences.

Sir Peter Mansfield (1933–2017), in the UK in 1977, introduced a new pulse sequence system, called echo planar imaging (EPI), to obtain imaging in a very short time (Mansfield, 1977, Mansfield and Maudsley, 1977).

Compared with CT and PET, MRI uses no X-rays or γ-rays, it uses only three types of electro-magnetic fields; static magnetic fields, time-varying pulsed magnetic fields, and radio frequency (RF) electromagnetic fields operated at the magnetic resonant frequencies. Therefore, MRI is safer than CT and PET from a viewpoint of ionizing radiation.

Lauterbur and Mansfield were awarded the Nobel Prize in Physiology and Medicine in 2003, 24 years after the award for the invention of CT.

Because of its variety of potential abilities, MRI has been well developed in different variational applications such as MR angiography, diffusion tensor MRI (DTI), fMRI, MRI-based impedance and current imaging, molecular MRI, and so on.

Seiji Ogawa (1934–) in Japan invented fMRI based on blood oxygenation level dependent (BOLD) effects to visualize the functional organizations in the brain in 1990 (Ogawa et al., 1990), and demonstrated the usefulness of fMRI using human subjects (Ogawa et al., 1992). fMRI has been used in a variety of fields from basic brain research to cognitive and social sciences because of its usefulness and non-invasive imaging methodology without any injection of contrast agencies into human subjects. Using the BOLD effect, fMRI is able to detect the functional signals in the brain related to the changes in magnetic susceptibility of oxyhemoglobin (diamagnetism) and deoxyhemoglobin (para-magnetism) in the blood. The BOLD effect reflects on the local spin-spin relaxation T_2^*, and the changes in T_2^* are detected as the functional signals associated with neuronal electrical activities.

A method of spin diffusion measurements was introduced by Stejskal and Tanner in 1965 (Stejskal and Tanner, 1965), and around 20 years later, diffusion-weighted imaging and DTI have been well developed for clinical use by the efforts of many investigators (Le Bihan et al., 1986, Turner and Le Bihan, 1990, Le Bihan et al., 1992, Basser et al., 1994). Recent advances in diffusion MRI and DTI have become powerful tools to visualize structures of neuronal fibers and neuronal networks in the human brain.

MRI also has potential abilities to visualize electrical properties in the brain. For example, Ueno and his group proposed methods of MRI-based impedance imaging (Ueno and Iriguchi, 1998, Yukawa et al., 1999) and conductivity imaging (Sekino et al., 2005, Sekino et al., 2009). Toward the detection of neuronal currents by MRI, studies were reported by Kamei et al. in 1999 (Kamei et al., 1999) and Sekino et al. in 2009 (Sekino et al., 2009). Further studies are needed for MRI-based neuroimaging (Ueno and Sekino, 2015).

Molecular imaging based on MRI is also very important and attractive, and this technique will be mentioned and discussed in Session 1.4.5.

1.3.3 MAGNETOENCEPHALOGRAPHY

Electrical activities of neurons in the brain generate electric fields, called the electroencephalogram (EEG), on the scalp of the head, as well as producing magnetic fields, called the magnetoencephalogram (MEG), with the signal of ~100 fT (100×10^{-15}T) over the head. The EEG was first measured by Hans Berger (1873–1941) in Germany in 1929 (Berger, 1929). The MEG was first measured by David Cohen (1927–) in the USA by using a magnetometer in 1968, and later by using an ultra-sensitive magnetic sensor, the SQUID system, in a magnetically shielded room in 1972 (Cohen 1968, Cohen, 1972). The SQUID used in the MEG measurement was made by Zimmermann et al. in 1970 (Zimmermann et al., 1970). The SQUID is based on the principle of superconductive tunneling in quantum mechanics theoretically predicted by Brian David Josephson (1940–) in the UK in 1962 (Josephson, 1972). Josephson was awarded the Nobel Prize in Physics shared by other scientists Leona Esaki (1925–) in Japan and Ivar Giaever (1929–) in the USA in 1973.

The MEG is a useful tool in measuring dynamic electrical activities in the brain with a high temporal (~m second) and a high spatial (~mm) resolution. The MEG has the advantage in temporal and spatial resolution compared with other tools such as fMRI and NIRS, but the disadvantage is solving the inverse problem in estimating the electrical sources in the brain from the data measured by the MEG systems. There are some attempts to combine MEG and MRI to get neuronal imaging without solving the inverse problems. In this case, the MEG system is operated in ultra-low field MRI, reducing the noise of electromagnetic fields from the MRI system (Clark et al., 2007, Mossle et al., 2006, Vesanen et al., 2014). Further studies are needed for better imaging systems with better spatial resolution and a shorter acquisition time.

1.3.4 Transcranial Magnetic Stimulation

TMS is a technique to stimulate the human brain transcranially by a coil positioned on the surface of the head. Barker et al. in the UK reported TMS using a round coil to stimulate the human brain in 1985 (Barker et al., 1985). The success of human brain stimulation by TMS brought a strong impact on the scientific community. By TMS with a round coil, however, it was difficult to stimulate a targeted area of the brain. Ueno et al. proposed a method of localized brain stimulation by TMS with a figure-eight coil in 1988 (Ueno et al., 1988), and succeeded in human brain stimulation within a 5-mm resolution (Ueno et al., 1989, Ueno et al., 1990). TMS with a figure-eight coil enables localized stimulation of the cerebral cortex and is now used worldwide in cognitive brain research, clinical neurophysiology, and other basic and clinical medicine (Ueno, 1994). Around 1995–2000, repetitive TMS (rTMS) was developed, and it began being used in the treatment of various intractable neurological and neuropsychological diseases including depression as reported by George and Wassermann in 1994 (George and Wassermann, 1994). Risk and safety aspects of rTMS with pulse trains of high frequencies need to be investigated (Wassermann 1998, Pascual-Leone et al., 2002). The therapeutic applications of rTMS have been accelerated in recent years, and, for example, an international symposium on rTMS treatments was organized by Youichi Saitoh in Japan in 2016, where rTMS treatments of intractable pain, Parkinson's disease, stroke recovery and rehabilitation, and depression were discussed (Saitoh, 2016).

The combination of TMS with DTI, fMRI, and EEG accelerates the studies of brain dynamics. For example, the DTI and *in vivo* fiber tractography (Mori et al., 1999) contribute to the studies on DTI-based neural trajectories in TMS (De Geeter et al., 2015). EEG measurements just after the onset of brain stimulation by TMS with a figure-eight coil are important to study dynamic neuronal connectivity in the brain. Ilmoniemi et al. (1977) successfully measured neuronal electrical responses to magnetic brain stimulation. The combination of TMS with EEG, MEG, fMRI, or DTI enables us to study dynamic intra- and interhemispheric connectivity for the deeper understanding of cortico-cortical and interhemispheric interactions, cortical inhibitory processes, cortical plasticity and oscillations, and so on.

1.3.5 Magnetic Particle Imaging

Magnetic nanoparticles (MNPs) are used as medical tracers as well as MRI contract agents. MPI is an imaging technique to visualize MNPs for biomedical applications (Gleich and Weizenecker 2005, Weizenecker et al., 2009). MPI allows us to obtain the direct quantitative imaging of the spatial distribution of MNPs with a high spatial and a high temporal resolution. MPI exploits the unique characteristics of MNPs when AC and DC gradient fields are applied. The DC gradient field produces the so-called field-free point (FFP) or field-free line (FFL) where the field becomes zero. The FFP or FFL makes it possible to detect MNPs with a high spatial resolution because the magnetic signal is selectively generated from the MNPs located near the FFP or FFL. Understanding of the dynamic magnetization of MNPs under AC and DC gradient fields is important, as MPI hardware and imaging technology are developed (Yoshida et al., 2013, Enpuku et al., 2017).

Applications of MNPs are useful for cancer detection. Metastasis to lymph nodes occurs when cancer cells reach the nodes through lymphatic vessels. When MNPs are injected to the proximity of a cancer lesion, the MNPs in the nodes can be detected using magnetic sensors or magnetic imaging modalities. The early detection of the sentinel lymph node (SLN) is essential to block the scattering of metastasis into the whole body.

1.3.6 NEAR-INFRARED SPECTROSCOPIC IMAGING

NIRS is a non-invasive method to obtain imaging of living body by applying NIR rays into the body. However, the red light and NIR rays with wavelengths around 700 nm–1500 nm are scattered and absorbed by the mediums in the living tissues and go through the tissues in various directions. For the realization of imaging by NIRS, the problems related to the varieties in scattering and absorption are to be solved.

NIRS has been studied as a method to monitor blood flow and oxygen metabolism in living tissues since Jobsis reported a paper on the oxygenation status of the heart and brain of animals using near-infrared light (Jobsis, 1977). In the 1990s, a so-called functional NIRS (fNIRS) was developed to visualize brain functional activities related to the hemoglobin changes in cerebral blood flow linked to neuronal activities in the brain (Ferrari and Quaresima, 2012). Further studies are needed for better imaging systems with a high spatial resolution.

A unique method is proposed to obtain brain imaging by light and NIRS by putting a light-emitted diode on the palate of the upper jaw (Hiwaki and Miyaguchi, 2018). This method is promising, however, a disadvantage is to put the photodiode on the palate inside the mouth.

1.4 ADVANCES IN MOLECULAR AND CELLULAR IMAGING

1.4.1 GREEN FLUORESCENT PROTEIN

Aequorin and green fluorescent protein (GFP) were discovered by Osamu Shimomura (1928–2018) in Japan in 1962 (Shimomura et al., 1962). Using the jelly fish *Aequorea victoria*, Shimomura's target was a luminescent substance, aequorin. GFP was isolated as a by-product of aequorin owing to its bright conspicuous fluorescence. Both are unusual proteins but they had no particular importance when Shimomura and his co-authors first reported them; 40 years after their discovery, they are well known and widely used, aequorin as a calcium probe and GFP as a marker protein (Shimomura, 2005).

Martin Chalfie (1947–) in the USA and his co-workers used GFP as a marker for gene expression in 1994 (Chalfie et al., 1994). Based on their experiments, Chalfie and his co-workers concluded that a complementary DNA for the *Aequorin victoria* GFP produces a fluorescent product when expressed in prokaryotic (*Escherichia coli*) or eukaryotic (*Caenorhabditis elegans*) cells. Because exogenous substrates and cofactors are not required for this fluorescence, GFP expression can be used to monitor gene expression and protein localization in living organisms.

Roger Yonchien Tsien (1952–2016) in the USA accelerated and expanded the studies of GFP for the established biomarker of gene expression and protein targeting in intact cells and organisms. Mutagenesis and engineering of GFP into chimeric proteins are opening new vistas in physiological indicators, biosensors, and photochemical memories (Tsien, 1998).

Shimomura, Chalfie, and Tsien were awarded the Nobel Prize in Chemistry in 2008.

1.4.2 Optical Fluorescence and Cancer Therapy

Molecular imaging is a type of biomedical imaging that visualizes cellular functions or molecular processes inside the body. Probes used for molecular imaging are targeted to biomarkers for specific visualization of targets or pathways. Many kinds of modalities have been employed for non-invasive molecular imaging. Among the many modalities, optical imaging is promising. Optical imaging is based on fluorescence emission from fluorophores that are targeted to, or accumulated by, cancer cells, or activated by molecular targets that are overexpressed in cancer. Fluorescence imaging of malignant and premalignant lesions can be used as a guide for diagnostic biopsy or intraoperatively to aid accurate surgical resection. Urano and his group are leading in the field of optical fluorescence imaging for cancer detection, and in his laboratory, unique probes and cancer detections have been developed, for example, the results include rapid cancer detection by topically spraying a fluorescent probe, and activatable probes which offer particular advantages in terms of providing a cancer-specific signal with high sensitivity and a high tumor-to-background signal ratio (Urano et al., 2005, 2009, 2011, Kamiya et al., 2007, 2011).

To achieve the reliable molecular imaging, designing and fabrication of reliable and selective optical fluorescence imaging probes are essential. Two major categories of optical fluorescence imaging probes for cancer detection are developed: always-on probes and activatable probes. Always-on probes consist of fluorescent reporting units (fluorophores) linked to tumor-targeting moieties such as antibodies, affibodies, peptides, or small-molecular ligands. Activatable probes of which fluorescence is initially suppressed is turned on by a molecular switch at tumor sites, providing a cancer-specific signal with high sensitivity and high tumor-to-background signal ratio.

1.4.3 Optogenetics and Studies of Neuronal Circuit Dynamics in the Brain

Karl Deisseroth (1971–) in the USA and his group published a paper on millisecond-timescale, genetically targeted optical control of neural activity, and demonstrated millisecond-timescale control of neuronal spiking, as well as control of excitatory and inhibitory synaptic transmission in 2005 (Boyden et al., 2005). They adapted the naturally occurring algal protein Channelrhodopsin-2, a rapidly gated light-sensitive cation channel, by using lentiviral gene delivery in combination with high-speed optical switching to photo-stimulate mammalian neurons. Deisseroth appeals that this technology allows the use of light to alter neural processing at the level of single spikes and synaptic events, yielding a widely applicable tool for neuroscientists and biomedical engineers. Thus, Deisseroth has spearheaded "optogenetics," a new methodological discipline in which cellular activities are controlled by light and has revolutionized systems neuroscience research.

By the use of genetic engineering, Deisseroth made it possible to study neuronal circuit dynamics in brain regions of interest *in vivo*. Optogenetics is now widely accepted as a powerful tool to address causal relationships between neuronal activity and behavior.

Optogenetics is rapidly expanding in various fields in the neuroscience world, but this technology is only used for animal studies. Studies for human subjects are not allowed because the gene delivery and optical control of neuronal activities by inserting optical fibers into human brain regions is forbidden.

Instead, the new findings obtained by optogenetics will certainly give useful information and suggestions to human brain research by using TMS, fMRI, and other tools in biomedical imaging and stimulation.

1.4.4 Raman Scattering and Coherent Raman Scattering Microscopy

Chandrasekhara Venkata Raman (1888–1970) in India discovered a new type of light scattering in 1928 (Raman and Krishnan, 1928). Raman was awarded the Nobel Prize in Physics in 1930, only two years after the discovery of the so-called Raman scattering. Raman scattering is an inelastic light scattering phenomenon involving molecular vibration of substances that interact with light, providing the vibrational spectrum of molecules. The intensity of Raman scattering is quite weak, and it takes a long acquisition time. If we introduce Coherent Raman scattering (CRS) microscopy using two-color and/or broadband laser pulses, the acquisition time and sensitivity in Raman imaging are dramatically improved.

Optical microscopy based on CRS allows for biological imaging with molecular vibrational contrast. CRS microscopy is opening up a variety of applications which include label-free molecular analysis of cells and tissues, monitoring of metabolic activities, and super-multiplex imaging with more than ten colors. Two types of CRS are mainly introduced; coherent anti-Stokes Raman scattering (CARS) microscopy and stimulated Raman scattering (SRS) microscopy. CARS microscopy is advantageous for acquiring broadband vibrational spectra, while SRS is useful for high-speed and sensitive analysis of vibrational spectra in a relatively narrow bandwidth (Ozeki et al., 2009).

1.4.5 MOLECULAR IMAGING BASED ON MAGNETIC RESONANCE IMAGING

CEST is a novel MRI-based molecular imaging demonstrated by Ward and Balaban in 2000 (Ward and Balaban, 2000). CEST allows for the imaging of low-concentration endogenous and exogenous molecules with spatial resolution comparable with those in imaging obtained by conventional MRI. In addition, CEST imaging has the potential to provide physiological information which includes pH and glucose metabolism. CEST contrast is achieved by applying a saturation pulse at the resonant frequency of a slow-intermediate exchanging proton site of endogenous or exogenous CEST agents. APT imaging is one method of CEST imaging, and APT imaging has a higher contrast effect compared with other imaging methods in CEST (Zhou et al., 2003).

CEST imaging requires a sufficiently slow exchange on the magnetic resonance time scale to allow selective irradiation of the protons of interest. As a result, magnetic labeling is not limited to radio frequency saturation but can be expanded with slower frequency-selective approaches such as inversion, gradient dephasing, and frequency labeling (van Zijl and Tadav, 2011). By solving the difficulties in CEST imaging, many approaches to clinical diagnoses are reported which include the diagnosis of brain tumors by APT imaging (Togao et al., 2014, 2016). CEST imaging has opened a new horizon in MRI-based molecular imaging.

A unique approach to molecular MRI using reporter genes has been studied. As a type of endogenous contrast, gene-based iron-labeling in single cells, i.e., the magnetosome found in magnetotactic bacteria, has been used for molecular MRI (Goldhawk et al., 2012).

1.4.6 MAGNETIC ORIENTATION OF LIVING SYSTEMS AND BIOGENIC MICROMIRRORS

Magnetic orientation is a phenomenon where biological materials align in the direction in parallel or perpendicular to magnetic fields when the materials are exposed to magnetic fields. Torbet and his co-workers in France first reported magnetic orientation of fibrin gels in strong magnetic fields in 1981 (Torbet et al., 1981). When diamagnetic materials with a high anisotropy in magnetic susceptibility are exposed to magnetic fields, the materials align in the direction either in parallel or perpendicular to the magnetic fields depending on the anisotropy of magnetic susceptibility of the materials.

Using this effect, for example, bone growth acceleration is observed when samples of a mixture of bone morphogenetic protein 2 (BMP-2) and collagen are exposed to 8-T magnetic fields for 60 hours at the beginning period in bone formation (Kotani et al., 2002). The studies of magnetic orientation of adherent cells and other biogenic samples have been reported (Ueno et al., 1993, Testorf et al., 2002, Iwasaka and Ueno, 2003, Iwasaka and Mizukawa, 2013).

Applications of magnetic orientation to optically functional materials in living systems are particularly of interest and importance in biomimicry and optic technology. For example, in the case of fish, the control of light reflection from the body is important during daylight in becoming less visible against predators. Guanine-based bio-reflectors are the most representative biogenic materials for daylight reflection control. Magnetic control of the biogenic micromirror related to guanine crystal platelets from fish and other materials has been studied (Iwasaka et al., 2015).

1.5 SUMMARY

Recent advances in bioimaging were summarized, focusing on imaging by light and electromagnetics in medicine and biology. This area is still developing and expanding from the molecular and cellular level to the human brain. This book covers the recent advances in biomedical imaging with wide spectra composed of different imaging tools. Chapter authors are experts in each field, and they wrote their chapters to explain the background, principles, and applications in each bioimaging area. We hope readers enjoy the essence of this expanding field.

REFERENCES

Ambrose, J. and Hounsfield, G. 1973. Computerized transverse axial scanning (Tomography) Part 2. *British Journal of Radiology* **46**: 1023–1047.

Barker, A. T., Jalinous, R., and Freeston, I. L. 1985. Non-invasive magnetic stimulation of human motor cortex. *The Lancet* **i**(8437): 1106–1107.

Basser, P. J., Mattiello, J., and Le Bihan, D. 1994. Estimation of the effective self-diffusion tensor from the NMR spin echo. *Journal of Magnetic Resonance Series B* **103**(3): 247–254.

Berger, H. 1929. Uber das Elektroenkephalogramm des Menschen. *Archiv fur Psychiatrie und Nervenkrankheiten* **87**(1): 527–570.

Boyden, E. S., Zhang, F., Bamberg, E., Nagel, G., and Deisseroth, K. 2005. Millisecond-timescale, genetically targeted optical control of neuronal activity. *Nature Neuroscience* **8**(9): 1263–1268.

Chalfie, M., Tu, Y., Euskirchen, G., Ward, W. W., and Prasher, D. C. 1994. Green fluorescent protein as a marker for gene expression. *Science* **263**(5148): 802–805.

Clark, J., Hatridge, M., and Mossle, M. 2007. SQUID detected magnetic resonance in microtesla fields. *Annual Reviews of Biomedical Engineering* **9**: 389–413.

Cohen, D. 1968. Magnetoencephalography: Evidence of magnetic fields produced by alpha-rhythm currents. *Science* **161**(3843): 784–786.

Cohen, D. 1972. Magnetoencephalography: Detection of the brain electrical activity with a superconducting magnetometer. *Science* **175**(4022): 664–666.

Cormack, A. M. 1963. Representation of a function by its line integrals, with some radiological applications II. *Journal of Applied Physics* **34**(9): 2722–2727.

Cormack, A. M. 1964. Representation of a function by its line integrals with some radiological applications. *Journal of Applied Physics* **35**(10): 2908–2913.

De Geeter, N., Crevecoeur, G., Leemans, A., and Dupre, L. 2015. Effective electric fields along realistic DTI-based neural trajectories for modelling the stimulation mechanisms of TMS. *Physics in Medicine and Biology* **60**(2): 453–471.

Enpuku, K., Tsujita, K., Nakamura, T., Sasayama, T., and Yoshida, T. 2017. Biosensing utilizing magnetic markers and superconducting quantum interference devices. *Superconducting Science and Technology* **30**: 053002.

Ferrari, M. and Quaresima, V. 2012. A brief review on the history of human functional near-infrared spectroscopy (fNIRS) development and fields of application. *Neuroimage* **63**(2): 921–935.

Georg, M. S. and Wassermann, E. M. 1994. Rapid-rate transcranial magnetic stimulation and ECT. *Convulsive Therapy* **10**(4): 251–254.

Gleich, B. and Weizenecker, J. 2005. Tomographic imaging using the nonlinear response of magnetic particles. *Nature* **435**(7046): 1214.

Goldhawak, D., Rohani, R., Sengupta, A., Gelman, N., and Prato, F. 2012. Using the magnetosome to model effective gene-based contrast for magnetic resonance imaging. *WIRES Nanomed Nanobiotechnology* **4**(4): 378–388.

Hiwaki, O. and Miyaguchi, H. 2018. Noninvasive measurement of dynamic brain signals using light penetrating the brain. *PLoS One* **13**(1): e0192095.

Hounsfield, G. 1973. Computerized transverse axial scanning (Tomography) Part 1 Description of system. *British Journal of Radiology* **46**(552): 1016–1022.

Ilmoniemi, R. J., Virtanen, J., Karhu, J., Aronen, H. J., Naatanen, R., and Katila, T. 1997. Neuronal responses to magnetic stimulation reveal cortical reactivity and connectivity. *Neuroreport* **8**(16): 3537–3540.

Iwasaka, M. and Ueno, S. 2003. Detection of intracellular macromolecule behavior under strong magnetic fields by linearly polarized light. *Bioelectromagnetics* **24**(8): 564–570.

Iwasaka, M. and Mizukawa, Y. 2013. Light reflection control in biogenic micro-mirror by diamagnetic orientation. *Langmuir* **29**(13): 4328–4334.

Iwasaka, M., Mizukawa, Y., and Roberts, N. W. 2015. Magnetic control of the light reflection anisotropy in a biogenic guanine microcrystal platelet. *Langmuir* **32**(1): 180–187.

Jobsis, F. F. 1977. Noninvasive, infrared monitoring of cerebral and myocardial oxygen sufficiency and circulatory parameters. *Science* **198**(4323): 1264–1267.

Josephson, B. D. 1972. Possible new effects in superconductive tunneling. *Physics Letters* **1**: 251–253.

Kamei, H., Iramina, K., Yoshikawa, K., and Ueno, S. 1999. Neuronal current distribution imaging using magnetic resonance. *IEEE Transactions on Magnetics* **35**(5): 4109–4111.

Kamiya, M., Kobayashi, H., Hama, Y., Koyama, Y., Bernardo, M., Nagano, T., Choyke, P. L., and Urano, Y. 2007. An enzymatically activated fluorescence probe for targeted tumor imaging. *Journal of the American Chemical Society* **129**(13): 3918–3929.

Kamiya, M., Asanuma, D., Kuranaga, E., Takeishi, A., Sakabe, M., Miura, M., Nagano, T., and Urano, Y. 2011. β-Galactosidase fluorescence probe with improved cellular accumulation based on a spirocyclized rhodol scaffold. *Journal of the American Chemical Society* **133**(33): 12960–12963.

Kotani, H., Kawaguchi, H., Shimoaka, T., Iwasaka, M., Ueno, S., Ozawa, H., Nakamura, K., and Hoshi, K. 2002. Strong static magnetic field stimulates bone formation to a definite orientation *in vivo* and *in vitro*. *Journal of Bone and Mineral Research* **17**(10): 1814–1821.

Lauterbur, P. C. 1973. Image formation by induced local interactions: Examples employing nuclear magnetic resonance. *Nature* **242**(5394): 190–191.

Le Bihan, D., Breton, E., Lallemand, D., Grenier, P., Cabanis, E., and Laval-Jeantet, M. 1986. MR imaging of intravoxel incoherent motions: Application to diffusion and perfusion in neurologic disorders. *Radiology* **161**(2): 401–407.

Le Bihan, D., Turner, R., Douek, P., and Patronas, N. 1992. Diffusion MR imaging: Clinical applications. *American Journal of Radiology* **159**(3): 591–599.

Mansfield, P. 1977. Multi-planar image formation using NMR spin echoes. *Journal of Physics C: Solid State Physics* **10**: L55–L58.

Mansfield, P. and Maudsley, A. A. 1977. Planar spin imaging by NMR. *Journal of Magnetic Resonance* **27**(1): 101–119.

Mori, S., Crain, B. J., Chacro, V. P., and van Zijl, P. C. 1999. Three-dimensional tracking of axonal projections in th brain by magnetic resonance imaging. *Annals of Neurology* **45**(2): 265–269.

Mossle, M., Han, S. I., Myers, W. R., Lee, S. K., Kelso, N., Pines, A., and Clarke, J. 2006. SQUID-detected microtesla MRI in the presence of metal. *Journal of Magnetic Resonance* **179**(1): 146–151.

Ogawa, S., Lee, T. M., Kay, A. R., and Tank, D. W. 1990. Brain magnetic resonance imaging with contrast dependent on blood oxygenation. *Proceedings of the National Academy of Sciences of the United States of America* **87**(24): 9868–9872.

Ogawa, S., Tank, D. W., Menon, R., Ellermann, J. M., Kim, S. G., Merkle, H., and Ugurbil, K. 1992. Intrinsic signal changes accompanying sensory stimulation: Functional brain mapping with magnetic resonance imaging. *Proceedings of the National Academy of Sciences of the United States of America* **89**(13): 5951–5955.

Ozeki, Y., Dake, F., Kajiyama, S., Fukui, K., and Itoh, K. 2009. Analysis and experimental assessment of the sensitivity of stimulated Raman scattering microscopy. *Optics Express* **17**(5): 3651–3658.

Pascual-Leone, A., Davey, N., Rothwell, J., Wassermann, E. M., and Puri, B. K. 2002. *Handbook of Transcranial Magnetic Stimulation*. Hodder Arnold, London.

Radon, J. 1917. Uber die Bestimmung von Funktionen durch ihre Integralwerte lange gewisser Mannigfaltgkeiten. Berichte uber die Verhandlungen der Koniglich-Sachsischen Akademie der Wissenschaften zu Leipzig, Mathematisch-Physische Klasse [Reports on the proceedings of the Royal Saxonian Academy of Science at Leipzig, mathematical and physical section]. *Leipzig: Teubner* **69**: 262–277. Translation: Radon J. Parks, P. C. 1986. On the determination of functions from their integral values along certain manifolds. *IEEE Transactions on Medical Imaging* **5**(4): 170–176.

Raman, C. V. and Krishnan, K. S. 1928. A new type of secondary radiation. *Nature* **121**(3048): 501–502.

Saitoh, Y. 2016. *Abstracts of the International Symposium on rTMS Treatments*. Tetsumon Memorial Hall, The University of Tokyo, Tokyo, Japan, 1–41.

Sekino, M., Inoue, Y., and Ueno, S. 2005. Magnetic resonance imaging of electrical conductivity in the human brain. *IEEE Transactions on Magnetics* **41**(10): 4203–4205.

Sekino, M., Ohsaki, H., Yamaguchi-Sekino, S., Iriguchi, N., and Ueno, S. 2009. Low-frequency conductivity tensor of rat brain tissues inferred from diffusion MRI. *Bioelectromagnetics* **30**(6): 489–499.

Sekino, M., Ohsaki, H., Yamaguchi-Sekino, S., and Ueno, S. 2009. Toward detection of transient changes in magnetic resonance signal intensity arising from neuronal electrical activities. *IEEE Transactions on Magnetics* **45**(10): 4841–4844.

Shimomura, O., Johnson, F. H., and Saiga, Y. 1962. Extraction, purification and properties of aequorin, a bioluminescent protein from the luminous hydromedusan, *Aequorea. Journal of Cellular and Comparative Physiology* **59**: 223–239.

Shimomura, O. 2005. The discovery of aequorin and green fluorescent protein. *Journal of Microscopy* **217**(1): 3–15.

Stejskal, E. O. and Tanner, J. E. 1965. Use of spin echo in pulsed magnetic field gradient to study anisotropic, restricted diffusion and flow. *The Journal of Chemical Physics* **43**: 3579–3603.

Testorf, M. F., Oberg, P. A., Iwasaka, M., and Ueno, S. 2002. Melanophore aggregation in strong static magnetic fields. *Bioelectromagnetics* **23**(6): 444–449.

Togao, O., Yoshiura, T., Keupp, J., Hiwatashi, A., Yamashita, K., Kikuchi, K., Suzuki, Y., Suzuki, S. O., Iwaki, T., Hata, N., Mizoguchi, M., Yoshimoto, K., Sagiyama, K., Takahashi, M., and Honda, H. 2014. Amide proton transfer imaging of adult diffuse gliomas: Correlation with histopathological grades. *Neuro Oncology* **16**(3): 441–448.

Togao, O., Hiwatashi, A., Keupp, J., Yamashita, K., Kikuchi, K., Yoshiura, T., Yoneyama, M., Kruiskamp, M. J., Sagiyama, K., Takahashi, M., and Honda, H. 2016. Amide proton transfer imaging of diffuse glioma: Effect of saturation pulse in parallel transmission-based technique. *PLoS One* **11**(5): e0155925.

Torbet, J., Freyssinet, J. M., and Hudry-Clergeon, G. 1981. Oriented fibrin gels formed by polymerization in strong magnetic fields. *Nature* **289**(5793): 91–93.

Tsien, R. T. 1998. The green fluorescent protein. *Annual Review of Biochemistry* **67**: 509–544.

Turner, R. and Le Bihan, D. 1990. Single-shot diffusion imaging at 2.0 Tesla. *Journal of Magnetic Resonance* **86**(3): 445–452.

Ueno, S., Tashiro, T., and Harada, K. 1988. Localized stimulation of neuronal tissues in the brain by means of a paired configuration of time-varying magnetic fields. *Journal of Applied Physics* **64**(10): 5862–5864.

Ueno, S., Matsuda, T., and Fujiki, M. 1989. Localized stimulation of the human cortex by opposing magnetic fields. In *Advances in Biomagnetism*, Williamson, S. J., Hoke, M., Stroink, G., and Kotani, M., Eds. Springer, New York, 529–532.

Ueno, S., Matsuda, T., and Fujiki, M. 1990. Functional mapping of the human motor cortex obtained by focal and vectorial magnetic stimulation of the brain. *IEEE Transactions on Magnetics* **26**(5): 1539–1544.

Ueno, S., Iwasaka, M., and Tsuda, H. 1993. Effects of magnetic fields on fibrin polymerization and fibrinolysis. *IEEE Transactions on Magnetics* **29**(6): 3352–3354.

Ueno, S. (Ed.) 1994. *Biomagnetic Stimulation*. Plenum Press, New York, 1–136.

Ueno, S. and Iriguchi, N. 1998. Impedance magnetic resonance imaging: A method for imaging of impedance distributions based on magnetic resonance imaging. *Journal of Applied Physics* **83**(11): 6450–6452.

Ueno, S. and Sekino, M. (Eds.) 2015. *Biomagnetics: Principles and Applications of Biomagnetic Stimulation and Imaging*. CRC Press Taylor & Francis Group, Boca Raton, London, New York, 1–343.

Urano, Y., Kamiya, M., Kanda, K., Ueno, T., Hirose, K., and Nagano, T. 2005. Evolution of fluorescein as a platform for finally tunable fluorescence probes. *Journal of the American Chemical Society* **127**(13): 4888–4894.

Urano, Y., Asanuma, D., Hama, Y., Koyama, Y., Barrett, T., Kamiya, M., Nagano, T., Watanabe, T., Hasegawa, A., Choyke, P. L., and Kobayashi, H. 2009. Selective molecular imaging of viable cancer cells with pH-activatable fluorescence probes. *Nature Medicine* **15**(1): 104–109.

Urano, Y., Sakabe, M., Kosaka, N., Ogawa, M., Mitsunaga, M., Asanuma, D., Kamiya, M., Young, M. R., Nagano, T., Choyke, P. L., and Kobayashi, H. 2011. Rapid cancer detection by topically spraying a γ-glutamyltranspeptidase-activated fluorescent probe. *Science Translational Medicine* **3**(110): 110–119.

van Zijl, P. C. M. and Tadav, N. N. 2011. Chemical exchange saturation transfer (CEST): What is in a name and what isn't? *Magnetic Resonance in Medicine* **65**(4): 927–948.

Vesanen, P. T., Nieminen, J. O., Zevenhoven, K. C. J., Hsu, Y. C., and Ilmoniemi, R. J. 2014. Current-density imaging using ultra-low-field MRI with zero-field encoding. *Magnetic Resonance Imaging* **32**(6): 1–5.

Ward, K. M., Aletras, A. H., and Balaban, R. S. 2000. A new class of contrast agents for MRI based on proton chemical exchange dependent saturation transfer (CEST). *Journal of Magnetic Resonance* **143**(1): 79–87.

Wassermann, E. M. 1998. Risk and safety of repetitive transcranial magnetic stimulation: Report and suggested guidelines from the International workshop on the Safety of Repetitive Transcranial Magnetic Stimulation, June 5–7, 1996. *Electroencephalography and Clinical Neurophysiology* **108**(1): 1–16.

Weizenecker, J., Gleich, B., Rahmer, J., Dahnke, H., and Borgert, J. 2009. Three-dimensional real-time *in vivo* magnetic particle imaging. *Physics in Medicine and Biology* **54**(5): L1–L10.

Yoshida, T., Othman, N. B., and Enpuku, K. 2013. Characterization of magnetically fractionated magnetic nanoparticles for magnetic particle imaging. *Journal of Applied Physics* **114**(17): 173908.

Yukawa, Y., Iriguchi, N., and Ueno, S. 1999. Impedance magnetic resonance imaging with external AC field added to main static field. *IEEE Transactions on Magnetics* **35**(5): 4121–4123.

Zimmermann, J. E., Thiene, P., and Harding, J. T. 1970. Design and operation of stable rf-biased superconducting point-contact quantum devices and a note on the properties of perfectly clean metal contacts. *Journal of Applied Physics* **41**(4): 1572–1580.

Zhou, J., Lal, B., Wilson, D. A., Latera, I., and van Zjil, P. C. 2003. Amide proton transfer (APT) contrast for imaging of brain tumors. *Magnetic Medicine in Medicine* **50**: 1120–1126.

2 Molecular Imaging of Viable Cancer Cells

Mako Kamiya and Yasuteru Urano

CONTENTS

2.1 Introduction 15
2.2 Optical Fluorescence Imaging Probes (Activatable Probes and Always-on Probes) 16
2.3 Fluorescence Switching Mechanisms for Activatable Probes 17
    2.3.1 Self-Quenching 17
    2.3.2 Förster Resonance Energy Transfer 17
    2.3.3 Photoinduced Electron Transfer 18
    2.3.4 Intramolecular Spirocyclization 18
2.4 Design Strategies of Activatable Fluorescence Probes for Bioimaging 18
    2.4.1 PeT-based Fluorescence Imaging Probes 18
    2.4.2 Spirocyclization-based Fluorescence Imaging Probes 19
2.5 Fluorescence Imaging of Cancer with Activatable Probes 24
    2.5.1 Combination of Enzyme Pre-Targeting and Activatable Probes 24
    2.5.2 Combination of Tumor-Targeting Antibody and Acidic-pH-Activatable Probes 25
    2.5.3 Activatable Fluorescence Probes Targeted to Glycosidases 26
    2.5.4 Aminopeptidase-Targeted Activatable Fluorescence Probes for Rapid and
         Sensitive Tumor Imaging 28
    2.5.5 Carboxypeptidase-Targeted Activatable Fluorescence Probes for Rapid and
         Sensitive Tumor Imaging 30
2.6 Conclusion 30
References 32

2.1 INTRODUCTION

Cancer is a global healthcare concern: its annual incidence worldwide in 2018 was estimated to be 18.1 million people, while mortality was 9.6 million people. It is said that one out of five to six people worldwide develop cancer during their lifetime, and one out of eight to ten people die from cancer. Since there is a better chance of recovery when cancer is found and treated at an early stage, as judged in terms of improved five-year survival rates, various types of cancer screening tests, such as blood tests, urine tests, cytology, endoscopic examination, and medical imaging, have been developed and are in routine clinical use. However, substantial numbers of cancer patients are still diagnosed late due to the lack of obvious cancer signs or symptoms, or for other reasons. Available types of cancer treatment include surgery, radiation therapy, chemotherapy, immunotherapy, hormone therapy, and so on. In the case of surgical treatment, accurate and complete resection of the tumor is critical to achieve a cure.

Molecular imaging is a type of medical imaging that visualizes cellular functions or molecular processes inside the body. In general, probes used for molecular imaging are targeted to biomarkers for specific visualization of targets or pathways, thus contributing to the aim of achieving precision medicine by improving diagnostic performance for early cancer detection, tumor staging, and surgical guidance. Many kinds of modalities have been employed for non-invasive molecular imaging, including positron emission tomography (PET), single-photon emission computed tomography

(SPECT), magnetic resonance imaging (MRI), and optical imaging. Among these modalities, optical imaging is based on fluorescence emission from fluorophores that are targeted to, or accumulated by, cancer cells, or activated by molecular targets that are overexpressed in cancer. Optical fluorescence imaging offers many advantages, such as high sensitivity, low cost, portability, real-time capability, and freedom from the problems associated with the use of radiolabeled probes or ionizing radiation. Further, it can be used over a wide spectral range from visible to near-infrared (NIR), thus enabling multi-color imaging, and the fluorescence signal can be activatable, thus providing a high target-to-background ratio.[1] For these reasons, there is great interest in the application of optical fluorescence imaging to guide surgery and endoscopy. For example, fluorescence imaging of malignant and pre-malignant lesions can be used to guide diagnostic biopsy or intraoperatively to aid accurate surgical resection. In this chapter, we first review optical fluorescence imaging probes available for visualizing cancer cells. We then focus on activatable probes, which offer particular advantages in terms of providing a cancer-specific signal with a high sensitivity and high tumor-to-background signal ratio. We discuss various switching mechanisms and design strategies for activatable bioimaging probes, and we present some applications demonstrating the usefulness of the designed probes.

2.2   OPTICAL FLUORESCENCE IMAGING PROBES
(ACTIVATABLE PROBES AND ALWAYS-ON PROBES)

There are two major categories of optical fluorescence imaging probes that have been used for imaging cancer.[1] The first is "always-on probes," which consist of fluorescent reporting units (fluorophores) linked to tumor-targeting moieties such as antibodies, affibodies, peptides, or small-molecular ligands (Figure 2.1a). Since these always-on probes emit a constant fluorescence signal regardless of their situation, their pharmacokinetics must be optimized so that a sufficient number of the imaging probes are accumulated at the tumor site and the excess non-target-bound probes are excreted from the body to obtain a high target-to-background ratio (TBR). In general, small-molecular probes have faster pharmacokinetics, which might allow imaging at earlier time points after administration, but can sometimes lead to too-rapid clearance before successful targeting. In contrast, when fluorophores are conjugated to macromolecules such as antibodies, the probes show slower pharmacokinetics, which gives sufficient time for targeting, but can sometimes produce a high background due to remaining non-target-bound probes.

Most of the currently available optical fluorescence imaging probes are always-on probes, and there have been several examples of successful first-in-human studies. A pioneering example is intraoperative fluorescence imaging of ovarian cancer in patients by intravenous injection of folate-FITC (also known as EC17), a conjugate of folate and fluorescein, which targets folate receptor-α (FR-α).[2] FR-α is overexpressed in 95% of ovarian cancers, while it is minimally expressed in normal tissues. In this pilot study by Ntziachristos et al., the fluorescence signal was detectable intraoperatively (2–8 h after iv injection) in all patients with malignant tumor expressing FR-α without any serious adverse event, and surgeons were able to find and resect an additional 29% of malignant lesions that were not identified under white light. Another example is fluorescence imaging of colorectal cancer in patients by fluorescence colonoscopy after intravenous administration of GE-137, a conjugate of a 26-amino-acid cyclic peptide targeted to human tyrosine kinase c-Met and a fluorescent cyanine dye.[3] In this report, Hardwick et al. demonstrated that GE-137 visualized not only neoplastic polyps that were discernible by conventional white-light imaging, but also an additional nine polyps that were missed by conventional methods and by the unaided human eye.

The second major category is "activatable probes," whose fluorescence is initially suppressed, but is turned on by a molecular switch at tumor sites, providing a cancer-specific signal with high sensitivity and high TBR (Figure 2.1b). In order to design activatable fluorescent probes, it is important to control the fluorescence emission, and various strategies have been used to quench the fluorescence of the fluorophores, including Förster resonance energy transfer (FRET), photoinduced electron transfer (PeT),

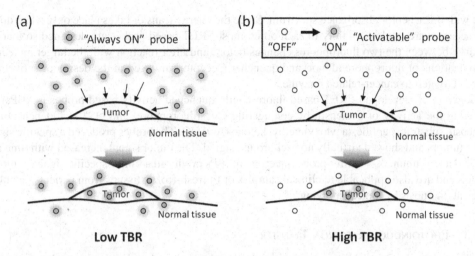

FIGURE 2.1 Two categories of fluorescent probes used for tumor imaging: (a) tumor imaging with "always-on" probes. To obtain sufficient target-to-background ratios (TBRs), it is necessary to wait until a sufficient number of probes are accumulated at the tumor site, and then excess non-targeted probes are washed out. (b) Tumor imaging with "activatable" probes. The fluorescence signal of activatable probes is activated only at tumor sites, affording high TBRs.

intramolecular spirocyclization, and self-quenching. The molecular switch at tumor sites is important to induce rapid and specific activation of the fluorescence signal, and may be a tumor-specific receptor, an enzymatic activity, or other tumor-specific target. In the following section, we briefly review the fluorescence switching mechanisms and describe probes based on these mechanisms.

2.3 FLUORESCENCE SWITCHING MECHANISMS FOR ACTIVATABLE PROBES

2.3.1 SELF-QUENCHING

Self-quenching occurs when multiple fluorophores are located close to each other and some absorb the energy from the excited state of others, thus diminishing the fluorescence emission. Weissleder et al. have developed self-quenched but enzymatically activatable probes, designated as the ProSense series, in which multiple NIR fluorophores are conjugated on a synthetic graft copolymer through enzyme-cleavable linkers such as poly-L-lysine linker.[4–5] The fluorescence of the polymers is self-quenched, but the polymers tend to be accumulated at tumor sites via enhanced permeability retention (EPR). Then, the poly-L-lysine chain is cleaved by tumor-residing proteases such as cathepsins or matrix metalloproteinase, releasing the fluorophores from quenching (fluorescence activation). Indeed, tumors in mouse models could be visualized after intravenous injection of the probes.

In a similar manner, Kobayashi et al. have developed an activatable probe by conjugating multiple X-rhodamine (ROX) moieties to avidin, a homotetrameric glycoprotein known to bind to D-galactose receptor expressed on cancer cells. The probe is quenched but after internalization into the cell by endocytosis, the protein structure is degraded in lysosomes, liberating the fluorophore, which is thus dequenched. When the ROX-conjugated probe was applied to a mouse model with a peritoneal tumor, tumors were clearly detected with high sensitivity and specificity.

2.3.2 FÖRSTER RESONANCE ENERGY TRANSFER

Förster resonance energy transfer (FRET) is energy transfer from the excited state of the fluorophore (donor) to an adjacent molecule (acceptor). In order to design FRET-based probes, it is important to select a fluorophore or chromophore pair so that the donor emission spectrum overlaps well with the acceptor absorbance spectrum. Since the energy transfer takes place only when donor and acceptor are in close vicinity to each other, most FRET-based probes are designed so that the distance between the two fluorophores changes before and after reaction with the target molecule. Combinations of fluorophore (donor) and quencher (acceptor) are used in the design of activatable probes targeted to cancer-related enzymes.

Bogyo et al. developed FRET-based fluorescently quenched activity-based probes (qABPs) for detecting the activity of cathepsin by conjugating Cy5 (fluorophore) and QSY21 (quencher) to a cathepsin substrate peptide, (acyloxy)methyl ketone (AOMK). The probes produced a specific signal in the tumors and showed virtually no background signal. The tumor signal increased with time and reached a maximum 6–8 h after probe injection. qABPs produced a tumor-specific signal in mouse models and are also applicable to clinical samples or to fresh-frozen tissues from patients, enabling efficient diagnosis by topical spraying.[6–8]

2.3.3 Photoinduced Electron Transfer

Another mechanism available for developing small-molecule-based probes is photoinduced electron transfer (PeT). PeT is an electron transfer from the PeT donor to the excited fluorophore, thereby diminishing the fluorescence of the fluorophore. The rate of PeT can be determined from the Marcus equation, and its major determinant ΔG_{PeT} can be calculated from the Rehm–Weller equation: $\Delta G_{PeT} = E_{ox} - E_{red} - \Delta E_{00} - w_p$, where E_{ox} and E_{red} are the oxidation and reduction potentials of the electron donor and acceptor, ΔE_{00} is the singlet excited energy, and w_p is the work term for the charge separation state.

PeT quenching had been used for controlling the fluorescence emission of UV-excitable fluorophores such as anthracene, but more recently, it has been expanded to long-wavelength fluorophores, ranging from visible to NIR. We and other groups have established a new design strategy to expand the variety of PeT-based activatable fluorescence probes. In subsequent sections, we will introduce our rational design strategy of activatable fluorescence probes for biological research based on the PeT mechanism (Section 2.4.1) and also our strategies for highly sensitive fluorescence imaging of cancer (Section 2.5).

2.3.4 Intramolecular Spirocyclization

Intramolecular spirocyclization is a ground-state equilibrium between a colorless/non-fluorescent spirocyclic form and a colored/fluorescent form of a molecule. Since this equilibrium enables complete quenching of fluorescence by breaking the π-conjugation of the fluorophore scaffold, probes based on intramolecular spirocyclization can exhibit significant fluorescent activation when the equilibrium of the two forms is appropriately shifted. For example, fluorescein diacetate (FDA) for intracellular esterase[9] or rhodamine spiroamide-based probes for metal ions[10] exist in the colorless and non-fluorescent spirolactone or spiroamide form, but are converted to the colored and fluorescent xanthene form upon reaction with the targets. Recently, we have expanded the design strategy based on intramolecular spirocyclization, by changing the intramolecular nucleophile at the 2′ position of rhodamine or rhodol derivatives from carboxylate or amide to other nucleophiles, and we have developed a series of new activatable probes as tools for biological research (see Section 2.4.2). In Section 2.5, we will introduce our recently established strategies for faster and more sensitive imaging of tumors by using small-molecular fluorescence probes that show dramatic fluorescence activation upon reaction with tumor-specific enzymes.

2.4 DESIGN STRATEGIES OF ACTIVATABLE FLUORESCENCE PROBES FOR BIOIMAGING

2.4.1 PeT-based Fluorescence Imaging Probes

PeT is a widely accepted mechanism for fluorescence quenching of fluorophores. However, it had been believed that PeT quenching was applicable only to UV-excitable fluorophores such as anthracene, but not to long-wavelength fluorophores in the visible to NIR range. In this context, our research group demonstrated that the fluorescence properties of probes could be precisely controlled by means of intramolecular PeT, and we developed various kinds of fluorescence probes. We considered that the structure of fluorescein, widely used as the fluorescent core of fluorescence probes, could be divided into two parts, i.e., the benzoic acid moiety as the potential PeT donor and the xanthene ring as the fluorophore. The basis for this idea was that only small alterations in absorbance are observed among fluorescein derivatives, and the dihedral angle between the benzoic acid moiety and the xanthene ring is almost 90°, which strongly suggests that there would be little ground-state interaction between these two parts. We also found that fluorescence emission from the xanthene fluorophore can be modulated by varying the Highest Occupied Molecular Orbital (HOMO) level of the benzoic acid moiety; when the HOMO level is higher than a certain threshold, the molecule shows little fluorescence, which suggests the occurrence of PeT quenching. On the other hand, when the HOMO level is lower than the threshold, as in the original fluorescein molecule for example, strong fluorescence emission is observed. The occurrence of PeT is clearly evidenced by transient absorption spectroscopy, which showed bands of the radical cation of the electron donor moiety and the radical anion of the xanthene moiety.[11] The PeT rates and the rates of back electron transfer follow the Marcus parabolic dependence of electron transfer rate on the driving force. These observations provided a quantitative basis for modulating the fluorescence properties of fluorescein derivatives, and thus a clear rationale for designing fluorescein-based probes was established for the first time. This strategy has been proved to be applicable not only to fluorescein, but also to a wide range of long-wavelength excitable fluorophores such as BODIPYs,[12] rhodamines,[13] silicon rhodamines,[14] and cyanines.[15] Further, the opposite direction of electron transfer also works in visible light-excitable fluorophores: the fluorescence properties can be modulated via PeT from the excited fluorophore to a reducible benzene moiety (donor-excited PeT; d-PeT).[16] In other words, when the Lowest Unoccupied Molecular Orbital (LUMO) energy of the benzoic acid moiety is lower than a certain threshold, the rate of d-PeT is quite fast and hence the derivative will be non-fluorescent. This finding also provided the basis for a new and practical strategy for rational design of novel functional fluorescence probes.

Based on these PeT-based strategies, we developed many kinds of novel fluorescence probes, such as DPAXs and DMAXs for singlet oxygen,[17] DAFs and DAMBOs and DACals for nitric oxide,[18–20] HPF and APF for highly reactive oxygen species (hROS),[21] NiSPYs for peroxynitrite,[22] and DNAT-Me for the activity of glutathione S-transferase (GST).[23]

Further, in the light of our photo-physical findings, we found that the carboxylic group at the benzene moiety of traditional fluorescein derivatives could be replaced with another functional group. Our research group then developed novel fluorescein derivatives, called TokyoGreen (TG),[24] by breaking out of the traditional structure of fluorescein and introducing a methyl or methoxy group instead of the carboxylic group (Figure 2.2). TG dyes could be easily synthesized in high yields by means of the conventional C-C bond-coupling reaction. Further, precise control of the HOMO energy of the benzene moiety represents another rational design strategy for fluorescence probes. The value of this approach is exemplified by its application to develop a novel, highly sensitive, and membrane-permeable fluorescence probe for β-galactosidase, called TG-βGal. Further, based on the same strategy, TG-phos for alkaline phosphatase,[25] which is also a widely used reporter enzyme, and TG-NPE,[26] a highly efficient caged fluorophore, were successfully developed using the TG scaffold.

2.4.2 Spirocyclization-based Fluorescence Imaging Probes

Intramolecular spirocyclization has been utilized as a fluorescence switching mechanism since the 1960s, as exemplified by fluorescein-based probes such as fluorescein diacetate (FDA), fluorescein di-β-galactoside (FDG), and fluorescein diphosphate (FDP).[9] These probes, which have two reactive sites in the molecule, exist in the colorless and non-fluorescent spirolactone form in

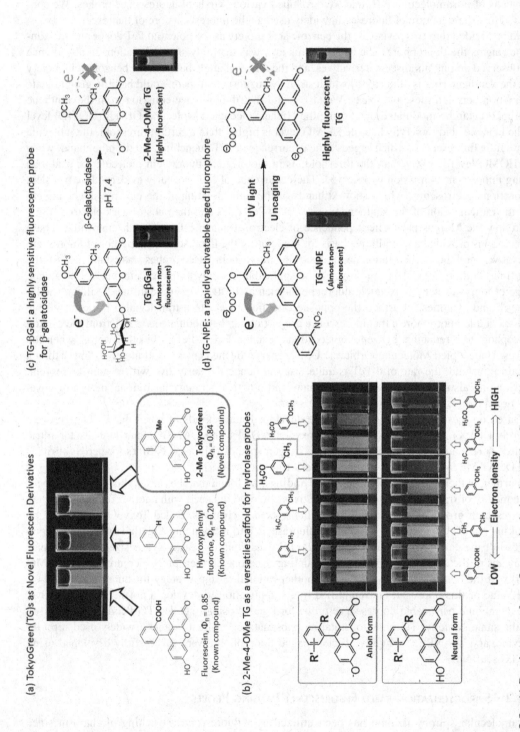

FIGURE 2.2 (a) Development of new fluorescent derivatives, TokyoGreen. (b) A series of TokyoGreen derivatives having benzene moieties with different electron density, demonstrating that 2-Me-4-OMe-TokyoGreen can be a versatile scaffold for designing hydrolase probes. (c) Design and reaction scheme of the fluorescence probe for β-galatosidase based on the TokyoGreen scaffold. (d) Design and reaction scheme of a photoactivatable caged fluorophore based on the TokyoGreen scaffold. Adapted with permission from reference 24. Copyright 2005 American Chemical Society.

which the π-conjugation of the fluorophore scaffold is broken, but are converted to the colored and fluorescent xanthene form upon reaction with esterase, β-galactosidase, or phosphatase, respectively. Rhodamine 110 has also been used to prepare activatable fluorescence probes for peptidases, such as serine proteases and caspases, by incorporating a peptide substrate at the amino groups of Rhodamine 110.[27] More recently, it was reported that silicon rhodamine derivatives with a 2′-carboxylic acid moiety tend to show a turn-on fluorescence property upon binding to tag proteins such as SNAP-tag or Halo-tag.[28] These derivatives exist in the colorless and non-fluorescent spirolactone form, but the ring opens when they bind to the protein surface, and they become fluorescent, providing a wash-free staining procedure for target structures. Rhodamine spiroamide, in which the carboxylic acid at the 2′ position of regular rhodamines is converted to amide, has also been utilized as a scaffold of activatable probes for detecting metal ions, and more recently for super-resolution imaging.

We have expanded the design strategy based on intramolecular spirocyclization, by switching the intramolecular nucleophile at the 2′ position of rhodamine or rhodol derivatives from carboxylate/amide to a more nucleophilic group such as hydroxymethyl, aminomethyl, or mercaptomethyl. Indeed, we used this approach to develop activatable probes for hypochlorous acid,[29] glycosidase,[30] aminopeptidases,[31] and super-resolution imaging.[32]

We also developed a tetramethylrhodamine analogue with a mercaptomethyl group at the 2′ position as a probe for hypochlorous acid, HySOx[29]; it exists in a completely colorless and non-fluorescent spirocyclic form in aqueous solutions over a wide range of pH values due to the strong nucleophilicity of thiol (Figure 2.3). But, when the thiol group is oxidized by hypochlorous acid (HOCl), one of the reactive oxygen species (ROS), the absorbance and fluorescence emission in the visible range recovers quickly and quantitatively due to the formation of the sulfonate derivative, $HySO_3H$, which exists completely in the colored and fluorescent xanthene form in aqueous solution.

(a) HySOx: A fluorescence probe for hypochlorous acid

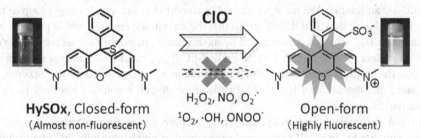

(b) Real-time monitoring of hypochlorite production with HySOx

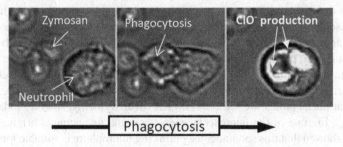

FIGURE 2.3 Development of a new fluorescence probe for hypochlorous acid employing the intramolecular spirocyclization reaction. (a) Molecular design of a fluorescence probe for hypochlorous acid, HySOx, and its reaction with hypochlorous acid. (b) Time lapse imaging of hypochlorite production during phagocytosis of zymozan by porcine neutrophils. Adapted with permission from reference 29. Copyright 2007 American Chemical Society.

Therefore, the thiol group in HySOx works not only as a cyclization enhancer, but also as a center of redox reaction. Further, HySOx is able to detect HOCl specifically among various reactive oxygen species under biological conditions and can be modified to obtain red-shifted analogues by changing the core fluorophore.

To develop probes for glycosidases, we focused on the distinctive spirocyclic nature of hydroxymethyl diethylrhodol (HMDER) derivatives (Figure 2.4).[30] The pK_a of the phenolic hydroxyl group of HMDER was determined to be 5.4, while pK_{cycl} (the pH value at which the absorbance of the compound decreases to half the maximum absorbance as a result of the spirocyclization) was calculated to be 11.3, meaning that HMDER is present as a colored and fluorescent open form at the physiological pH of 7.4. In contrast, the methyl ether derivative of HMDER (HMDER-Me) has a pK_{cycl} of 6.6, and is present mostly in the colorless, non-fluorescent, spirocyclic form. These results show that the hydroxymethyl group stabilizes the spirocyclic form of HMDER-Me but not that of HMDER at physiological pH. We considered that this remarkable difference could be utilized to design a rhodol-based probe for glycosidase by incorporating an enzyme-reactive moiety at the phenolic hydroxyl group of HMDER. As a target enzyme, we selected β-galactosidase, a widely used reporter enzyme encoded by the LacZ gene, as the activating enzyme. It has been shown that the expression of β-galactosidase in many organisms, including *Drosophila*, *Caenorhabditis elegans*, zebrafish, and mice, can be controlled to a specific region, organ, or cell line by genetically placing it under an appropriate promoter/enhancer. Thus, we incorporated a β-galactopyranoside group into HMDER to prepare HMDER-βGal, which mainly exists in its non-fluorescent spirocyclic form, but is converted to fluorescent HMDER upon reaction with β-galactosidase. This probe has good cellular permeability, enabling visualization of β-galactosidase activity in cultured cells and in tissues.

We further modified HMDER-βGal in order to achieve selective labeling of lacZ(+) cells with single-cell resolution in living tissues. For this purpose, we set out to integrate quinine methide chemistry into our design strategy. Quinone methide chemistry is widely used in the design of enzyme inhibitors, chromogenic substrates, activity-based probes, prodrugs, and for chemical selection of catalytic antibodies. We incorporated a fluoromethyl group as a leaving group at position 4 of HMDER-βGal to design SPiDER-βGal, so that the enzymatic reaction would produce a quinone methide intermediate, which is then trapped by intracellular nucleophiles, thus preventing diffusion of the activated fluorophore (Figure 2.5).[33] SPiDER-βGal exhibited simultaneous activation of fluorescence and the ability to label intracellular components upon reaction with the enzyme, enabling selective and rapid detection of live lacZ(+) cells at single-cell resolution not only in culture, but also in live tissues.

We utilized intramolecular spirocyclization to prepare a new class of spontaneously blinking fluorophore suitable for live-cell super-resolution imaging under physiological conditions (Figure 2.6).[32] Our approach makes use of the fact that the fluorescent open form and non-fluorescent closed form of the probe interconvert via an intramolecular spirocyclization reaction in the ground state, and we used this thermal equilibrium to achieve spontaneous fluorescence blinking that was suitable for super-resolution imaging. We optimized the intramolecular nucleophile and fluorophore unit to obtain an appropriate pK_{cycl} so that only a small subset of individual fluorophores should exist in the fluorescent state at physiological pH for any given frame, and a suitable lifetime for the fluorescent open form so that it lasts for at least several camera frames to enable detection of sufficient photons for accurate localization. Among various candidate rhodamine derivatives bearing different intramolecular nucleophiles, we selected HMSiR (hydroxymethyl silicon rhodamine) due to its appropriate pK_{cycl}, lifetime of the fluorescent open state, and its adequate brightness for accurate localization. We showed that this spontaneously blinking fluorophore is suitable for single-molecule localization microscopy imaging deep inside cells and also for tracking the motion of structures in living cells.[32] We also succeeded in developing a spontaneously blinking probe, HEtetTFER, that can be excited at 488 nm, and we performed dual-color super-resolution imaging of microtubules and mitochondria with using HMSiR and HEtetTFER in fixed cells.[34]

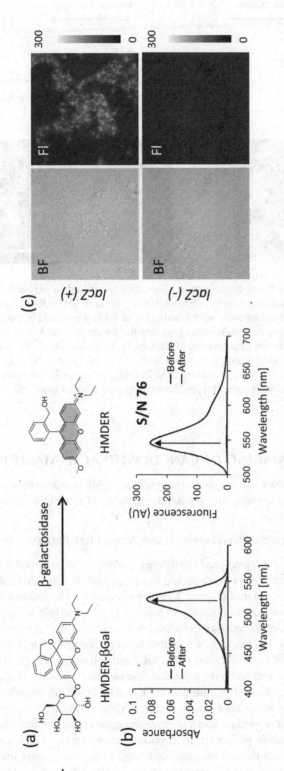

FIGURE 2.4 Development of a fluorescence probe for β-galactosidase. (a) Enzymatic activation of HMDER-βGal by β-galactosidase based on intramolecular spirocyclization reaction. (a) Enzymatic activation of HMDER-βGal by β-galactosidase. (b) Changes in absorption and fluorescence spectra of HMDER-βGal upon reaction with β-galactosidase. (c) Live imaging of HEK-lacZ(+) cells and HEK-lacZ(−) cells with HMDER-βGal. Adapted with permission from reference 30. Copyright 2011 American Chemical Society.

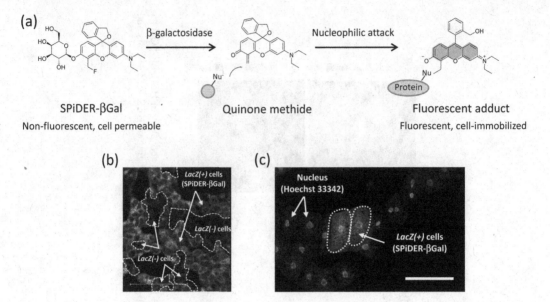

FIGURE 2.5 Development of a fluorescence probe for β-galatosidase, based on intramolecular spirocycliza-tion reaction and quinone methide chemistry. (a) Enzymatic activation of SPiDER-βGal by β-galactosidase. When the enzyme cleaves the β-galactoside moiety, activation of fluorescence and activation of the binding ability to intracellular proteins occur simultaneously, so that the fluorescent product is immobilized in liv-ing cells and cannot diffuse away. (b) Live imaging of HEK-lacZ(+) cells and HEK cells with SPiDER-βGal, enabling us to distinguish lacZ(+)cells and lacZ(-) cells with single-cell resolution. (c) Ex vivo imaging of *Drosophila melanogaster* tissue. Fluorescent staining of lacZ(+) flip-out clones in live larval fat body; nuclear co-staining with Hoechst 33342. Scale bars, 100 μm. Adapted with permission from reference 33.

2.5 FLUORESCENCE IMAGING OF CANCER WITH ACTIVATABLE PROBES

In this section, we will introduce our imaging strategies for highly sensitive fluorescence detection of cancer with PeT-based or intramolecular spirocyclization-based activatable probes.

2.5.1 COMBINATION OF ENZYME PRE-TARGETING AND ACTIVATABLE PROBES

Inspired by antibody-directed enzyme prodrug therapy (ADEPT), we adopted a two-step proce-dure in which an activating enzyme is first targeted to cancer, and then an activatable fluorescence probe for the enzyme is administered (Figure 2.7). We selected β-galactosidase as an activating enzyme, and avidin as a tumor-targeting moiety to localize β-galactosidase to cancer, since avidin is known to bind to the D-galactose receptor expressed on cancer cells.[35–37] As a fluorescence probe for β-galactosidase, we newly developed an activatable fluorescence probe for *in vivo* use, AM-TG-βGal, which is an analogue of our PeT-based probe for β-galactosidase (TG-βGal) with improved cellular retention. When we injected avidin-β-galactosidase into a mouse model of peritoneal metastasis of ovarian cancer, followed by administration of AM-TG-βGal, remarkable fluorescence activation was observed only at the tumor mass within 1 h after administration of AM-TG-βGal. Since the tumor-targeted β-galactosidase can hydrolyze numerous substrate molecules due to enzy-matic turnover, many fluorescent molecules are produced, enabling rapid and sensitive detection of the tumor *in vivo* with a high tumor-to-background ratio. Moreover, careful investigation using fluorescence microscopy revealed that cancer microfoci as small as 200 μm could be visualized by this method.[38]

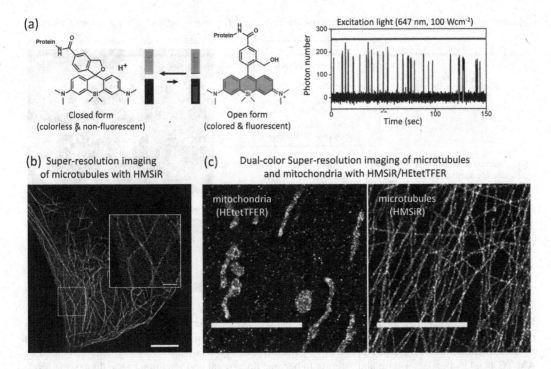

FIGURE 2.6 Development of spontaneously blinking fluorophores for super-resolution imaging based on intramolecular spirocyclization reaction. (a) Thermal equilibrium of intramolecular spirocyclization of HMSiR between the fluorescent open form and the non-fluorescent closed form. (Right) Single-molecule fluorescence time traces of antibody-bound HMSiR, which shows spontaneous blinking. (b) Super-resolution image of microtubules in living cells obtained with HMSiR. (c) Dual-color super-resolution images of microtubules and mitochondria in fixed cells with HMSiR (red-light emission) and HEtetTFER (green-light emission). Scale bars: 5 µm for main image; 1 µm for insets. Adapted with permission from references 32, 34.

2.5.2 COMBINATION OF TUMOR-TARGETING ANTIBODY AND ACIDIC-pH-ACTIVATABLE PROBES

In general, tumor-targeting antibodies show high tumor selectivity and tumor accumulation, but sometimes suffer from slow clearance from the body, which results in a considerable number of unbound probes and a high background signal. In order to overcome this problem, we developed a conjugate of cancer-targeting antibody and a small-molecular fluorophore activatable only within cancer cells in order to minimize the background signal and maximize the tumor-to-normal tissue (T/N) ratio (Figure 2.8).

As a cancer-targeting antibody, we selected a monoclonal antibody directed toward human epidermal growth factor receptor type 2 (HER2), trastuzumab, which is internalized via endocytosis after binding to HER2. As the small-molecular fluorophore, we newly designed a probe whose fluorescence is activated at acidic pH; thus, after cellular internalization, the probe is activated by the low pH (pH 5.0–6.0) of the lysosome. We developed a pH-sensitive fluorescence probe, DiEtN-BDP, by incorporating N,N-diethylaniline as the pH-responsive PeT donor into the BODIPY fluorophore, and used it to label trastuzumab, obtaining a probe-antibody conjugate, DiEtN-BDP–trastuzumab. This conjugate is almost non-fluorescent at neutral pH due to PeT quenching from the aniline moiety to the fluorophore, but becomes highly fluorescent at acidic pH, showing a substantial increase in fluorescence emission. We then evaluated the performance of this conjugate by applying it for *ex vivo* fluorescence imaging of freshly resected lungs bearing metastatic genetically engineered NIH3T3 mouse fibroblast tumors overexpressing HER2 receptors. The probe produced signals only from HER2-positive tumors, and the background fluorescence from normal lung or heart tissue was well suppressed, resulting in a high TBR ratio.[39]

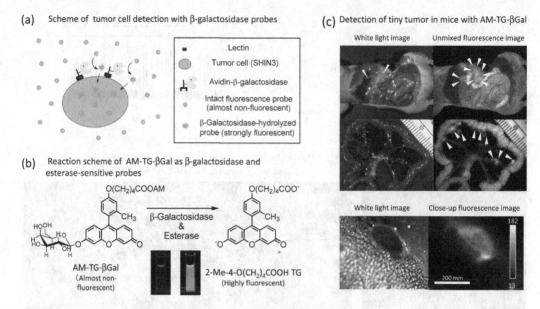

FIGURE 2.7 Fluorescence imaging of tumor by enzyme pre-targeting and applying an activatable fluorescence probe for the enzyme. (a) Tumor imaging strategy: pre-targeting of β-galactosidase, followed by administration of fluorescence probe for β-galactosidase, AM-TG-βGal. (b) Enzymatic activation of AM-TG-βGal by β-galactosidase and esterase to become fluorescent and retention of the fluorescent product in cells. (c) Ex vivo fluorescence imaging of intraperitoneally disseminated SHIN3 tumors by means of a pre-targeting strategy. Tumors (arrows) were clearly visualized with a high tumor-to-non-tumor signal ratio. (Middle) Magnified image of a peritoneal tumor. Strong fluorescence signals reveal tiny peritoneal tumor nodules (arrows). (Bottom) Fluorescent microfoci as small as 200 μm in diameter can be seen by means of fluorescence microscopy. Adapted with permission from reference 38. Copyright 2007 American Chemical Society.

2.5.3 ACTIVATABLE FLUORESCENCE PROBES TARGETED TO GLYCOSIDASES

The spectrum of glycosidase activities at tumor sites appears to differ from that of normal tissues. For example, Chatterjee et al. reported enhanced activity of β-galactosidase in primary ovarian cancers compared with normal ovaries.[40] Therefore, we focused on β-galactosidase as a target for fluorescence probes to visualize metastases originated from ovarian cancers. We first tried to use our membrane-permeable probe, TG-βGal,[24] but it failed to detect intracellular β-galactosidase activity, because the fluorescent product TG was exported from the cells by organic anion transporters, which are often overexpressed in metastatic cancers and cause multidrug resistance. Therefore, we next tried our other membrane-permeable probe, HMDER-βGal,[30] whose fluorescent product HMDER has a net charge of zero. HMDER-βGal successfully detected β-galactosidase activity in cultured cancer cells, but we found that peritoneal metastases could not be specifically visualized, owing to the high background fluorescence in the mouse tumor model. Therefore, we designed a new probe, HMRef-βGal, with carefully optimized spirocyclization behavior, and found that this probe enables highly sensitive detection of β-galactosidase activity inside living cells (Figure 2.9). To validate the potential of this newly developed probe for preclinical application, we demonstrated its suitability for *in vivo* laparotomic and endoscopic detection of small peritoneal metastases in mouse models of ovarian cancer.[41] We also reported that the N-acetyl-D-hexosamine analogue of HMRef-βGal, HMRef-βGlcNAc, which targets hexosaminidase, could visualize tiny metastatic nodules (smaller than 1 mm) in a mouse model of disseminated human peritoneal colorectal cancer, as well as cancerous tissue in colorectal cancer specimens from cancer patients, with a high TBR ratio.[42]

(a) Fluorescence Probes for Acidic pH

Key reaction: Protonation of Anilines

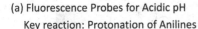

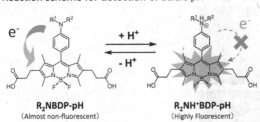

High HOMO E.   Low HOMO E.

Reaction scheme for detection of acidic pH

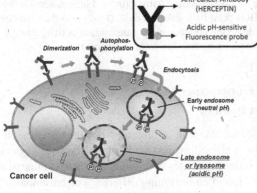

R₂NBDP-pH
(Almost non-fluorescent)

R₂NH⁺BDP-pH
(Highly Fluorescent)

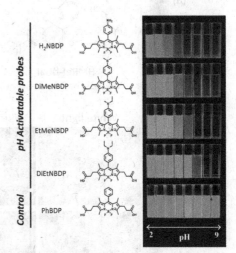

pH Activatable probes: H₂NBDP, DiMeNBDP, EtMeNBDP, DiEtNBDP

Control: PhBDP

pH 2 → 9

(b) Strategy for selective imaging of viable cancer cells by utilizing endocytotic uptake of anti-cancer antibodies bearing acidic pH-activatable probes

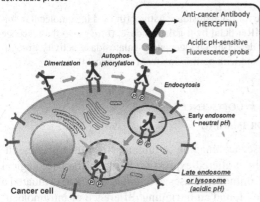

Anti-cancer Antibody (HERCEPTIN)

Acidic pH-sensitive Fluorescence probe

Dimerization — Autophosphorylation — Endocytosis

Early endosome (~neutral pH)

Late endosome or lysosome (acidic pH)

Cancer cell

(c) Application of the probes to HER2 expressing cells

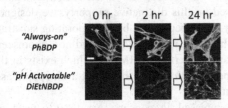

0 hr   2 hr   24 hr

"Always-on" PhBDP

"pH Activatable" DiEtNBDP

(d) In vivo tumor detection with targeted activatable fluorescence probes in HER2-positive lung metastases model mice

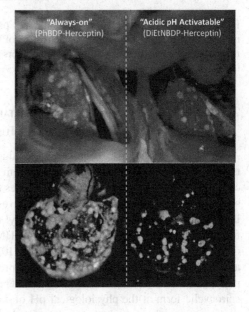

"Always-on" (PhBDP-Herceptin)   "Acidic pH Activatable" (DiEtNBDP-Herceptin)

FIGURE 2.8 Fluorescence imaging strategy for selective imaging of viable target cancer cells with an acidic pH-sensitive small-molecular probe and cancer-targeting antibody. (a) Development of an acidic pH-activatable fluorescence probe based on the BODIPY scaffold. (b) Schematic illustration of the strategy for selective imaging of viable target cancer cells with an acidic pH-sensitive small-molecular probe conjugated with cell-surface molecule-targeted monoclonal antibodies. (c) Confocal microscope images obtained just after the addition and at 2 and 24 h postaddition of always-on, pH-activatable BDP-conjugated trastuzumab directed against HER2. (d) *In vivo* tumor detection with targeted activatable fluorescent probe in a mouse model of HER2-positive lung metastasis. The pH-activatable probe produces a fluorescence signal only from tumors in the lung. However, the control always-on probe produces a fluorescence signal not only from tumors, but also from the background normal lung and heart. Adapted with permission from reference 39.

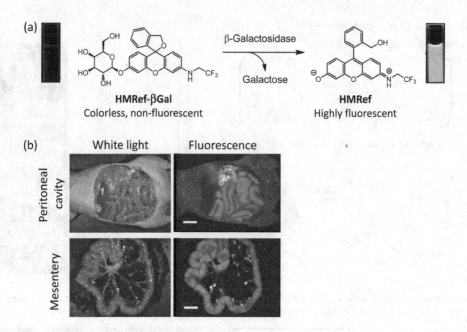

FIGURE 2.9 Development of new fluorescence probe for β-galatosidase with optimized intramolecular spirocyclization reaction. (a) Enzymatic activation of HMRef-βGal by β-galactosidase. (b) *Ex vivo* fluorescence imaging of intraperitoneally disseminated tumors by detecting endogenous β-galatosidase activity upregulated in some types of cancer with HMRef-βGal. Adapted with permission from reference 41.

2.5.4 AMINOPEPTIDASE-TARGETED ACTIVATABLE FLUORESCENCE PROBES FOR RAPID AND SENSITIVE TUMOR IMAGING

Aminopeptidases are also good targets for fluorescence imaging of tumors, since they play essential roles in many diseases and some of them exhibit altered expression levels in the pathological context. In order to develop activatable probes for aminopeptidases, we synthesized and evaluated a series of hydroxymethyl rhodamine derivatives and found an intriguing difference of intramolecular spirocyclization behavior: hydroxymethyl rhodamine green (HMRG) takes an open xanthene form, whereas the acetylated derivative, Ac-HMRG, exists as a closed spirocyclic structure in aqueous solution at physiological pH (Figure 2.10).[31] Based on this distinctive property, we designed and developed highly sensitive fluorogenic probes for aminopeptidases by incorporating substrate peptide moieties into the HMRG scaffold. These probes exist in the colorless and non-fluorescent spirocyclic form at the physiological pH of 7.4, but are converted to HMRG, which exists in the fluorescent xanthene form, by one-step hydrolysis upon reaction with the target aminopeptidase, resulting in a rapid and significant fluorescence activation.

Among various aminopeptidases that are overexpressed in cancers, we first focused on γ-glutamyltranspeptidase (GGT), which is overexpressed in several human tumors including cervical cancer, ovarian cancers, hepatocarcinoma, and lung cancer.[43–45] GGT is a cell-surface enzyme involved in glutathione biosynthesis, and is also known to promote tumor progression, invasion, and drug resistance, possibly through modulation of intracellular redox metabolism. We developed a fluorescent probe for GGT, γ-glutamyl hydroxymethyl rhodamine green (gGlu-HMRG), by incorporating a γ-glutamyl group into the HMRG scaffold.[46] When gGlu-HMRG encounters GGT on the surface of a cancer cell, it is hydrolyzed by the enzyme to afford HMRG, which is highly fluorescent. HMRG is sufficiently hydrophobic to permeate into cancer cells and is accumulated mainly in lysosomes, thus providing a clear intracellular fluorescence signal. Further, we evaluated the availability of gGlu-HMRG for *in vivo* tumor imaging of peritoneal metastasis of ovarian cancer

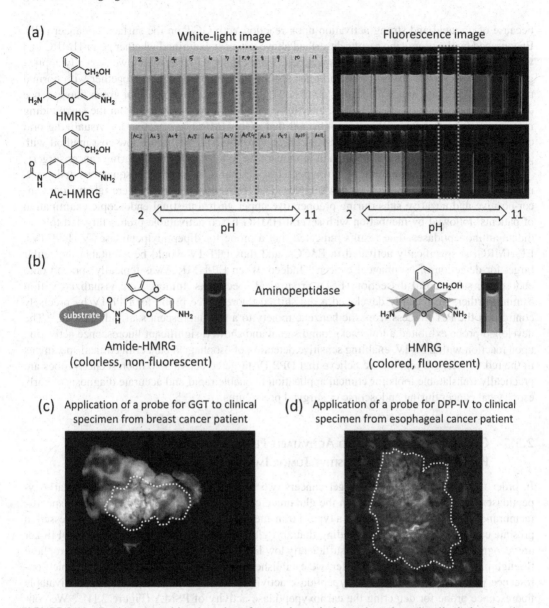

FIGURE 2.10 Rapid and sensitive detection of cancer by topical spraying of a rationally designed activatable fluorescence probe targeted for aminopeptidases. (a) pH dependency of absorbance and fluorescence intensity of HMRG and Ac-HMRG in 0.2 M sodium phosphate buffer at various pHs. (b) Design of fluorescence probe for sensitive detection of aminopeptidase activity based on the HMRG scaffold by utilizing the significant difference of fluorescence between HMRG and Ac-HMRG at the physiological pH of 7.4. (c) *Ex vivo* fluorescence imaging of cancer in clinical specimens from breast cancer patients by topical application of γ-glutamyl transpeptidase (GGT)-activatable probe, gGlu-HMRG. (d) *Ex vivo* fluorescence imaging of cancer in clinical specimens from esophageal cancer patients by topical application of dipeptidylpeptidase-IV (DPP-IV)-activatable probe, EP-HMRG. Adapted with permission from references 47, 51.

in a mouse model. The tumor site was clearly visualized with a high TBR within 10 min after probe injection, and tiny tumor sites less than 1 mm in diameter on the mesentery and peritoneal wall could also be visualized. Crucially, the obtained fluorescence intensity was high enough to be detected with the naked eye, which would be very useful during surgical resection. Therefore, gGlu-HMRG should be a practical tool for clinical application during surgical or endoscopic procedures, because of its rapid and strong activation upon reaction with GGT on the surface of cancer cells. Encouraged by the promising results described above, we next examined whether gGlu-HMRG can visualize cancerous lesions in clinical specimens from cancer patients. When we topically applied gGlu-HMRG to freshly excised human breast specimens containing lesions together with normal tissues, we found that tumorous lesions exhibited a time-dependent increase of green fluorescence derived from cleavage of gGlu-HMRG, and this clearly distinguished them from the surrounding mammary gland and fat.[47] We also confirmed that gGlu-HMRG is effective for visualizing oral cancer,[48] head and neck cancer,[49] and liver cancer.[50] We believe this rapid, low-cost method with gGlu-HMRG represents a breakthrough in intraoperative margin assessment during cancer surgery.

We next carried out screening of aminopeptidase activities of esophageal squamous cell carcinoma (ESCCs) by using patient biopsy samples. The fresh biopsy samples were taken from cancer-positive and negative sites during preoperative upper gastrointestinal endoscopic examination of patients, followed by incubation with several HMRG-based activatable probes targeted to candidate aminopeptidases. The results indicated that a probe for dipeptidylpeptidase-IV (DPP-IV), EP-HMRG, is specifically activated in ESCCs, and thus DPP-IV would be a suitable molecular target for detection of esophageal cancer.[51] Indeed, when EP-HMRG was topically sprayed onto endoscopic submucosal dissection (ESD) or surgical specimens, tumors were visualized within 5 min. Further, we recently developed a red-shifted fluorescence probe for DPP-IV by precisely controlling the PeT process from the benzene moiety to a red fluorescent scaffold, SiR600.[52] The developed probe exhibited a low background signal and showed significant fluorescence activation upon reaction with DPP-IV, enabling sensitive detection of esophageal cancer in clinical specimens in the red wavelength region. We believe that DPP-IV-targeted activatable fluorescence probes are practically translatable tools for clinical application to enable rapid and accurate diagnosis of early esophageal cancer during endoscopic or surgical procedures.

2.5.5 CARBOXYPEPTIDASE-TARGETED ACTIVATABLE FLUORESCENCE PROBES FOR RAPID AND SENSITIVE TUMOR IMAGING

In order to expand the range of target cancers, we next focused on cancer-associated carboxypeptidases. Especially, we focused on the glutamate carboxypeptidase activity of prostate-specific membrane antigen (PSMA), which is a type II transmembrane glycoprotein that is overexpressed in prostate cancer.[53–55] Based on our finding that aryl glutamate conjugates with an azoformyl linker are recognized by PSMA and have a sufficiently low LUMO energy level to quench the fluorophore through photoinduced electron transfer, we established a new design strategy for activatable fluorescence probes to visualize carboxypeptidase activity, and developed a first-in-class activatable fluorescence probe for detecting the carboxypeptidase activity of PSMA (Figure 2.11).[56] We confirmed that the developed probe allowed us to visualize the CP activity of PSMA in living cells and in clinical specimens from prostate cancer patients. This probe is expected to be useful for rapid intraoperative detection and diagnosis of prostate cancer.

2.6 CONCLUSION

The use of fluorescent probes targeted to tumor-specific molecules could provide a paradigm shift in surgical or endoscopic procedures, improving the outcome for cancer patients by providing a powerful new imaging technique for rapidly and specifically detecting cancerous lesions. A major advantage over current imaging modalities is that an intraoperative fluorescence imaging system offers a large field of view for inspection. In the past few years, intensive studies have led to the development of various types of fluorescence probes for tumor imaging. Considering the potential clinical benefits of intraoperative fluorescence imaging of tumor sites, preclinical/clinical trials of these promising probes should be conducted urgently.

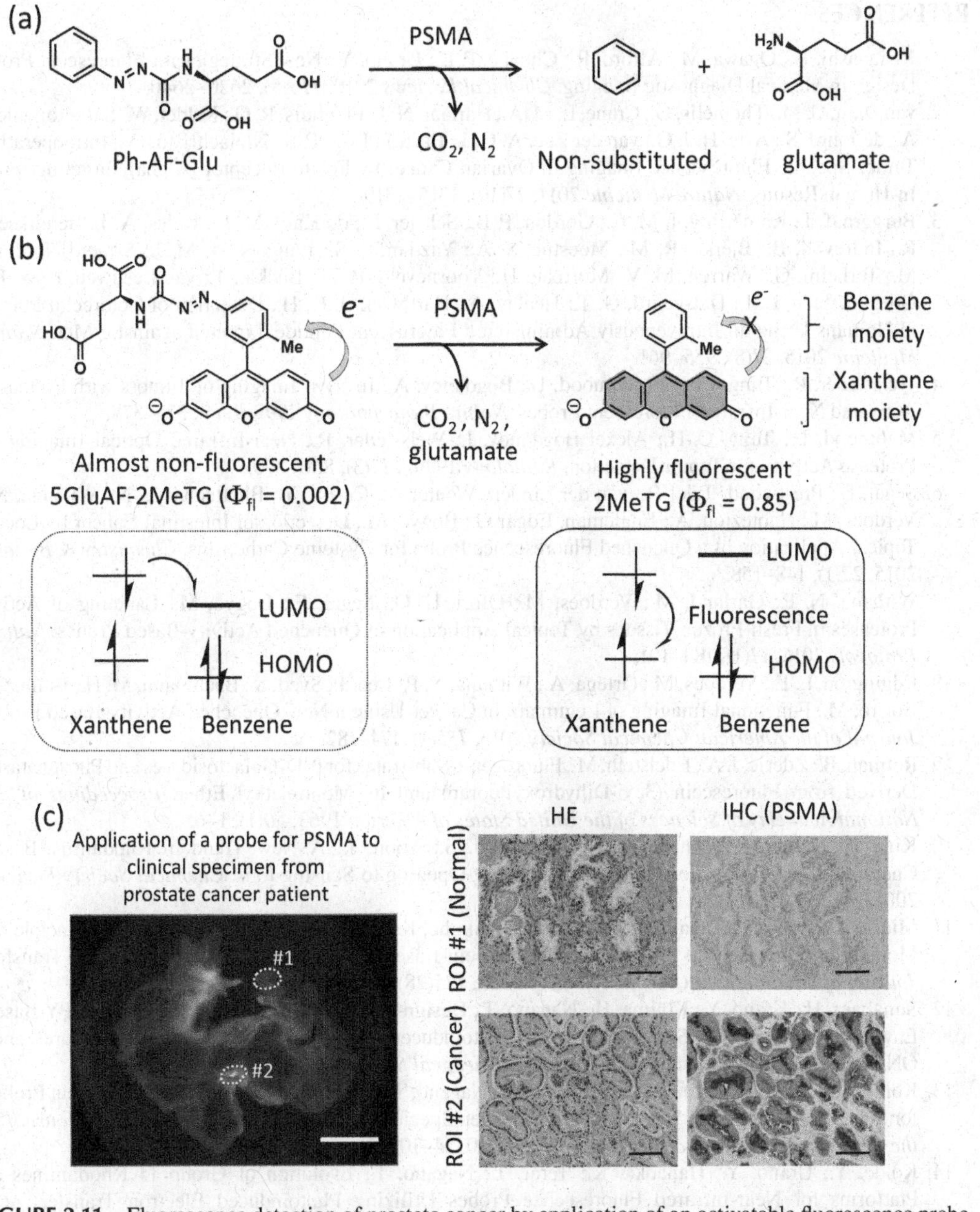

FIGURE 2.11 Fluorescence detection of prostate cancer by application of an activatable fluorescence probe to detect the carboxypeptidase activity of prostate-specific membrane antigen (PSMA). (a) Evaluation of aryl glutamate conjugate with azoformyl linker as a PSMA substrate. (b) Molecular design of activatable fluorescence probe for CP activity of PSMA based on the d-PeT process, 5GluAF-2MeTG. The probe fluorescence is quenched by d-PeT before reaction with PSMA, but PSMA catalyzes hydrolysis of the glutamate moiety, followed by the release of molecular carbon dioxide and nitrogen, to afford a highly fluorescent product. (c) *Ex vivo* fluorescence imaging of prostate cancer in clinical specimens from prostate cancer patients with 5GluAF-2MeTG. Fluorescence images were captured with a Maestro In-Vivo imaging system (PerkinElmer) using the blue-filter setting (excitation, 465/30 nm; emission, 515 nm long-pass). The fluorescence image at 540 nm was extracted and is presented here. Scale bars, 1 cm. (Bottom) Representative images of hematoxylin and eosin (HE) staining and immunohistochemistry (IHC) of PSMA. Adapted with permission from reference 56. Copyright 2019 American Chemical Society.

REFERENCES

1. Kobayashi, H.; Ogawa, M.; Alford, R.; Choyke, P. L.; Urano, Y., New Strategies for Fluorescent Probe Design in Medical Diagnostic Imaging. *Chemical Reviews* 2010, *110*(5), 2620–2640.
2. van Dam, G. M.; Themelis, G.; Crane, L. M. A.; Harlaar, N. J.; Pleijhuis, R. G.; Kelder, W.; Sarantopoulos, A.; de JongJ. S.; Arts, H. J. G.; van der Zee, A. G. J.; Bart, J.; Low, P. S.; Ntziachristos, V., Intraoperative Tumor-Specific Fluorescence Imaging in Ovarian Cancer by Folate Receptor-[Alpha] Targeting: First In-Human Results. *Nature Medicine* 2011, *17*(10), 1315–1319.
3. Burggraaf, J.; Kamerling, I. M. C.; Gordon, P. B.; Schrier, L.; de Kam, M. L.; Kales, A. J.; Bendiksen, R.; Indrevoll, B.; Bjerke, R. M.; Moestue, S. A.; Yazdanfar, S.; Langers, A. M. J.; Swaerd-Nordmo, M.; Torheim, G.; Warren, M. V.; Morreau, H.; Voorneveld, P. W.; Buckle, T.; van Leeuwen, F. W. B.; Odegardstuen, L.-I.; Dalsgaard, G. T.; Healey, A.; Hardwick, J. C. H., Detection of Colorectal Polyps in Humans Using an Intravenously Administered Fluorescent Peptide Targeted against c-Met. *Nature Medicine* 2015, *21*(8), 955–961.
4. Weissleder, R.; Tung, C.-H.; Mahmood, U.; Bogdanov, A., In Vivo Imaging of Tumors with Protease-Activated Near-Infrared Fluorescent Probes. *Nature Biotechnology* 1999, *17*(4), 375–378.
5. Mahmood, U.; Tung, C.-H.; Alexei Bogdanov, J.; Weissleder, R., Near-Infrared Optical Imaging of Protease Activity for Tumor Detection. *Radiology* 1999, *213*(3), 866–870.
6. Segal, E.; Prestwood, Tyler R.; van der Linden, Wouter A.; Carmi, Y.; Bhattacharya, N.; Withana, N.; Verdoes, M.; Habtezion, A.; Engleman, Edgar G.; Bogyo, M., Detection of Intestinal Cancer by Local, Topical Application of a Quenched Fluorescence Probe for Cysteine Cathepsins. *Chemistry & Biology* 2015, *22*(1), 148–158.
7. Withana, N. P.; Garland, M.; Verdoes, M.; Ofori, L. O.; Segal, E.; Bogyo, M., Labeling of Active Proteases in Fresh-Frozen Tissues by Topical Application of Quenched Activity-Based Probes. *Nature Protocols* 2016, *11*(1), 184–191.
8. Edgington, L. E.; Verdoes, M.; Ortega, A.; Withana, N. P.; Lee, J.; Syed, S.; Bachmann, M. H.; Blum, G.; Bogyo, M., Functional Imaging of Legumain in Cancer Using a New Quenched Activity-Based Probe. *Journal of the American Chemical Society* 2013, *135*(1), 174–182.
9. Rotman, B.; Zderic, J. A.; Edelstein, M., Fluorogenic Substrates for β-D-Galactosidases and Phosphatases Derived from Fluorescein (3, 6-Dihydroxyfluoran) and Its Monomethyl Ether. *Proceedings of the National Academy of Sciences of the United States of America* 1963, *50*(1), 1–6.
10. Kim, H. N.; Lee, M. H.; Kim, H. J.; Kim, J. S.; Yoon, J., A New Trend in Rhodamine-Based Chemosensors: Application of Spirolactam Ring-Opening to Sensing Ions. *Chemical Society Reviews* 2008, *37*(8), 1465–1472.
11. Miura, T.; Urano, Y.; Tanaka, K.; Nagano, T.; Ohkubo, K.; Fukuzumi, S., Rational Design Principle for Modulating Fluorescence Properties of Fluorescein-Based Probes by Photoinduced Electron Transfer. *Journal of the American Chemical Society* 2003, *125*(28), 8666–8671
12. Sunahara, H.; Urano, Y.; Kojima, H.; Nagano, T., Design and Synthesis of a Library of BODIPY-Based Environmental Polarity Sensors Utilizing Photoinduced Electron-Transfer-Controlled Fluorescence ON/OFF Switching. *Journal of the American Chemical Society* 2007, *129*(17), 5597–5604.
13. Koide, Y.; Urano, Y.; Kenmoku, S.; Kojima, H.; Nagano, T., Design and Synthesis of Fluorescent Probes for Selective Detection of Highly Reactive Oxygen Species in Mitochondria of Living Cells. *Journal of the American Chemical Society* 2007, *129*(34), 10324–10325.
14. Koide, Y.; Urano, Y.; Hanaoka, K.; Terai, T.; Nagano, T., Evolution of Group 14 Rhodamines as Platforms for Near-Infrared Fluorescence Probes Utilizing Photoinduced Electron Transfer. *ACS Chemical Biology* 2011, *6*(6), 600–608.
15. Sasaki, E.; Kojima, H.; Nishimatsu, H.; Urano, Y.; Kikuchi, K.; Hirata, Y.; Nagano, T., Highly Sensitive Near-Infrared Fluorescent Probes for Nitric Oxide and Their Application to Isolated Organs. *Journal of the American Chemical Society* 2005, *127*(11), 3684–3685.
16. Ueno, T.; Urano, Y.; Setsukinai, K.-I.; Takakusa, H.; Kojima, H.; Kikuchi, K.; Ohkubo, K.; Fukuzumi, S.; Nagano, T., Rational Principles for Modulating Fluorescence Properties of Fluorescein. *Journal of the American Chemical Society* 2004, *126*(43), 14079–14085.
17. Tanaka, K.; Miura, T.; Umezawa, N.; Urano, Y.; Kikuchi, K.; Higuchi, T.; Nagano, T., Rational Design of Fluorescein-Based Fluorescence Probes. Mechanism-Based Design of a Maximum Fluorescence Probe for Singlet Oxygen. *Journal of the American Chemical Society* 2001, *123*(11), 2530–2536.
18. Kojima, H.; Urano, Y.; Kikuchi, K.; Higuchi, T.; Hirata, Y.; Nagano, T., Fluorescent Indicators for Imaging Nitric Oxide Production. *Angewandte Chemie International Edition* 1999, *38*(21), 3209–3212.

19. Gabe, Y.; Urano, Y.; Kikuchi, K.; Kojima, H.; Nagano, T., Highly Sensitive Fluorescence Probes for Nitric Oxide Based on Boron Dipyrromethene Chromophore Rational Design of Potentially Useful Bioimaging Fluorescence Probe. *Journal of the American Chemical Society* 2004, *126*(10), 3357–3367.

20. Izumi, S.; Urano, Y.; Hanaoka, K.; Terai, T.; Nagano, T., A Simple and Effective Strategy to Increase the Sensitivity of Fluorescence Probes in Living Cells. *Journal of the American Chemical Society* 2009, *131*(29), 10189–10200.

21. Setsukinai, K.-I.; Urano, Y.; Kakinuma, K.; Majima, H. J.; Nagano, T., Development of Novel Fluorescence Probes That Can Reliably Detect Reactive Oxygen Species and Distinguish Specific Species. *Journal of Biological Chemistry* 2003, *278*(5), 3170–3175.

22. Ueno, T.; Urano, Y.; Kojima, H.; Nagano, T., Mechanism-Based Molecular Design of Highly Selective Fluorescence Probes for Nitrative Stress. *Journal of the American Chemical Society* 2006, *128*(33), 10640–10641.

23. Fujikawa, Y.; Urano, Y.; Komatsu, T.; Hanaoka, K.; Kojima, H.; Terai, T.; Inoue, H.; Nagano, T., Design and Synthesis of Highly Sensitive Fluorogenic Substrates for Glutathione S-Transferase and Application for Activity Imaging in Living Cells. *Journal of the American Chemical Society* 2008, *130*(44), 14533–14543.

24. Urano, Y.; Kamiya, M.; Kanda, K.; Ueno, T.; Hirose, K.; Nagano, T., Evolution of Fluorescein as a Platform for Finely Tunable Fluorescence Probes. *Journal of the American Chemical Society* 2005, *127*(13), 4888–4894.

25. Kamiya, M.; Urano, Y.; Ebata, N.; Yamamoto, M.; Kosuge, J.; Nagano, T., Extension of the Applicable Range of Fluorescein: A Fluorescein-Based Probe for Western Blot Analysis. *Angewandte Chemie International Edition* 2005, *44*(34), 5439–5441.

26. Kobayashi, T.; Urano, Y.; Kamiya, M.; Ueno, T.; Kojima, H.; Nagano, T., Highly Activatable and Rapidly Releasable Caged Fluorescein Derivatives. *Journal of the American Chemical Society* 2007, *129*(21), 6696–6697.

27. Beija, M.; Afonso, C. A. M.; Martinho, J. M. G., Synthesis and Applications of Rhodamine Derivatives as Fluorescent Probes. *Chemical Society Reviews* 2009, *38*(8), 2410–2433.

28. Lukinavičius, G.; Umezawa, K.; Olivier, N.; Honigmann, A.; Yang, G.; Plass, T.; Mueller, V.; Reymond, L.; Corrêa Jr, I. R.; Luo, Z.-G.; Schultz, C.; Lemke, E. A.; Heppenstall, P.; Eggeling, C.; Manley, S.; Johnsson, K., A Near-Infrared Fluorophore for Live-Cell Super-Resolution Microscopy of Cellular Proteins. *Nature Chemistry* 2013, *5*(2), 132–139.

29. Kenmoku, S.; Urano, Y.; Kojima, H.; Nagano, T., Development of a Highly Specific Rhodamine-Based Fluorescence Probe for Hypochlorous Acid and Its Application to Real-Time Imaging of Phagocytosis. *Journal of the American Chemical Society* 2007, *129*(23), 7313–7318.

30. Kamiya, M.; Asanuma, D.; Kuranaga, E.; Takeishi, A.; Sakabe, M.; Miura, M.; Nagano, T.; Urano, Y., β-Galactosidase Fluorescence Probe with Improved Cellular Accumulation Based on a Spirocyclized Rhodol Scaffold. *Journal of the American Chemical Society* 2011, *133*(33), 12960–12963.

31. Sakabe, M.; Asanuma, D.; Kamiya, M.; Iwatate, R. J.; Hanaoka, K.; Terai, T.; Nagano, T.; Urano, Y., Rational Design of Highly Sensitive Fluorescence Probes for Protease and Glycosidase Based on Precisely Controlled Spirocyclization. *Journal of the American Chemical Society* 2013, *135*(1), 409–414.

32. Uno, S.-N.; Kamiya, M.; Yoshihara, T.; Sugawara, K.; Okabe, K.; Tarhan, M. C.; Fujita, H.; Funatsu, T.; Okada, Y.; Tobita, S.; Urano, Y., A Spontaneously Blinking Fluorophore Based on Intramolecular Spirocyclization for Live-Cell Super-Resolution Imaging. *Nature Chemistry* 2014, *6*(8), 681–689.

33. Doura, T.; Kamiya, M.; Obata, F.; Yamaguchi, Y.; Hiyama, T. Y.; Matsuda, T.; Fukamizu, A.; Noda, M.; Miura, M.; Urano, Y., Detection of LacZ-Positive Cells in Living Tissue with Single-Cell Resolution. *Angewandte Chemie International Edition* 2016, *55*(33), 9620–9624.

34. Uno, S.-N.; Kamiya, M.; Morozumi, A.; Urano, Y., A Green-Light-Emitting, Spontaneously Blinking Fluorophore Based on Intramolecular Spirocyclization for Dual-Colour Super-Resolution Imaging. *Chemical Communications* 2018, *54*(1), 102–105.

35. Hama, Y.; Urano, Y.; Koyama, Y.; Kamiya, M.; Bernardo, M.; Paik, R. S.; Krishna, M. C.; Choyke, P. L.; Kobayashi, H., In Vivo Spectral Fluorescence Imaging of Submillimeter Peritoneal Cancer Implants Using a Lectin-Targeted Optical Agent. *Neoplasia* 2006, *8*(7), 607–612.

36. Yao, Z.; Zhang, M.; Sakahara, H.; Saga, T.; Arano, Y.; Konishi, J., Avidin Targeting of Intraperitoneal Tumor Xenografts. *JNCI: Journal of the National Cancer Institute* 1998, *90*(1), 25–29.

37. Mamede, M.; Saga, T.; Kobayashi, H.; Ishimori, T.; Higashi, T.; Sato, N.; Brechbiel, M. W.; Konishi, J., Radiolabeling of Avidin with Very High Specific Activity for Internal Radiation Therapy of Intraperitoneally Disseminated Tumors. *Clinical Cancer Research* 2003, *9*(10), 3756–3762.

38. Kamiya, M.; Kobayashi, H.; Hama, Y.; Koyama, Y.; Bernardo, M.; Nagano, T.; Choyke, P. L.; Urano, Y., An Enzymatically Activated Fluorescence Probe for Targeted Tumor Imaging. *Journal of the American Chemical Society* 2007, *129*(13), 3918–3929.

39. Urano, Y.; Asanuma, D.; Hama, Y.; Koyama, Y.; Barrett, T.; Kamiya, M.; Nagano, T.; Watanabe, T.; Hasegawa, A.; Choyke, P. L.; Kobayashi, H., Selective Molecular Imaging of Viable Cancer Cells with pH-Activatable Fluorescence Probes. *Nature Medicine* 2009, *15*(1), 104–109.

40. Chatterjee, S. K.; Bhattacharya, M.; Barlow, J. J., Glycosyltransferase and Glycosidase Activities in Ovarian Cancer Patients. *Cancer Research* 1979, *39*(6 Part 1), 1943–1951.

41. Asanuma, D.; Sakabe, M.; Kamiya, M.; Yamamoto, K.; Hiratake, J.; Ogawa, M.; Kosaka, N.; Choyke, P. L.; Nagano, T.; Kobayashi, H.; Urano, Y., Sensitive β-Galactosidase-Targeting Fluorescence Probe for Visualizing Small Peritoneal Metastatic Tumours In Vivo. *Nature Communications* 2015, *6*, 6463.

42. Matsuzaki, H.; Kamiya, M.; Iwatate, R. J.; Asanuma, D.; Watanabe, T.; Urano, Y., Novel Hexosaminidase-Targeting Fluorescence Probe for Visualizing Human Colorectal Cancer. *Bioconjugate Chemistry* 2016, *27*(4), 973–981.

43. Hanigan, M. H.; Frierson, H. F.; Brown, J. E.; Lovell, M. A.; Taylor, P. T., Human Ovarian Tumors Express γ-Glutamyl Transpeptidase. *Cancer Research* 1994, *54*(1), 286–290.

44. Yao, D.; Jiang, D.; Huang, Z.; Lu, J.; Tao, Q.; Yu, Z.; Meng, X., Abnormal Expression of Hepatoma Specific γ-Glutamyl Transferase and Alteration of γ-Glutamyl Transferase Gene Methylation Status in Patients with Hepatocellular Carcinoma. *Cancer* 2000, *88*(4), 761–769.

45. Pompella, A.; De Tata, V.; Paolicchi, A.; Zunino, F., Expression of γ-Glutamyltransferase in Cancer Cells and Its Significance in Drug Resistance. *Biochemical Pharmacology* 2006, *71*(3), 231–238.

46. Urano, Y.; Sakabe, M.; Kosaka, N.; Ogawa, M.; Mitsunaga, M.; Asanuma, D.; Kamiya, M.; Young, M. R.; Nagano, T.; Choyke, P. L.; Kobayashi, H., Rapid Cancer Detection by Topically Spraying a γ-Glutamyltranspeptidase–Activated Fluorescent Probe. *Science Translational Medicine* 2011, *3*(110), 110ra119.

47. Ueo, H.; Shinden, Y.; Tobo, T.; Gamachi, A.; Udo, M.; Komatsu, H.; Nambara, S.; Saito, T.; Ueda, M.; Hirata, H.; Sakimura, S.; Takano, Y.; Uchi, R.; Kurashige, J.; Akiyoshi, S.; Iguchi, T.; Eguchi, H.; Sugimachi, K.; Kubota, Y.; Kai, Y.; Shibuta, K.; Kijima, Y.; Yoshinaka, H.; Natsugoe, S.; Mori, M.; Maehara, Y.; Sakabe, M.; Kamiya, M.; Kakareka, J. W.; Pohida, T. J.; Choyke, P. L.; Kobayashi, H.; Ueo, H.; Urano, Y.; Mimori, K., Rapid Intraoperative Visualization of Breast Lesions with γ-Glutamyl Hydroxymethyl Rhodamine Green. *Scientific Reports* 2015, *5*, 12080.

48. Shimane, T.; Aizawa, H.; Koike, T.; Kamiya, M.; Urano, Y.; Kurita, H., Oral Cancer Intraoperative Detection by Topically Spraying a γ-Glutamyl Transpeptidase-Activated Fluorescent Probe. *Oral Oncology* 2016, *54*, e16–e18.

49. Mizushima, T.; Ohnishi, S.; Shimizu, Y.; Hatanaka, Y.; Hatanaka, K. C.; Hosono, H.; Kubota, Y.; Natsuizaka, M.; Kamiya, M.; Ono, S.; Homma, A.; Kato, M.; Sakamoto, N.; Urano, Y., Fluorescent Imaging of Superficial Head and Neck Squamous Cell Carcinoma Using a γ-Glutamyltranspeptidase-Activated Targeting Agent: A Pilot Study. *BMC Cancer* 2016, *16*(1), 411.

50. Miyata, Y.; Ishizawa, T.; Kamiya, M.; Yamashita, S.; Hasegawa, K.; Ushiku, A.; Shibahara, J.; Fukayama, M.; Urano, Y.; Kokudo, N., Intraoperative Imaging of Hepatic Cancers Using γ-Glutamyltranspeptidase-Specific Fluorophore Enabling Real-Time Identification and Estimation of Recurrence. *Scientific Reports* 2017, *7*(1), 3542.

51. Onoyama, H.; Kamiya, M.; Kuriki, Y.; Komatsu, T.; Abe, H.; Tsuji, Y.; Yagi, K.; Yamagata, Y.; Aikou, S.; Nishida, M.; Mori, K.; Yamashita, H.; Fujishiro, M.; Nomura, S.; Shimizu, N.; Fukayama, M.; Koike, K.; Urano, Y.; Seto, Y., Rapid and Sensitive Detection of Early Esophageal Squamous Cell Carcinoma with Fluorescence Probe Targeting Dipeptidylpeptidase IV. *Scientific Reports* 2016, *6*, 26399.

52. Ogasawara, A.; Kamiya, M.; Sakamoto, K.; Kuriki, Y.; Fujita, K.; Komatsu, T.; Ueno, T.; Hanaoka, K.; Onoyama, H.; Abe, H.; Tsuji, Y.; Fujishiro, M.; Koike, K.; Fukayama, M.; Seto, Y.; Urano, Y., Red Fluorescence Probe Targeted to Dipeptidylpeptidase-IV for Highly Sensitive Detection of Esophageal Cancer. *Bioconjugate Chemistry* 2019, *30*(4), 1055–1060.

53. Silver, D. A.; Pellicer, I.; Fair, W. R.; Heston, W. D.; Cordon-Cardo, C., Prostate-Specific Membrane Antigen Expression in Normal and Malignant Human Tissues. *Clinical Cancer Research* 1997, *3*(1), 81–85.

54. Bostwick, D. G.; Pacelli, A.; Blute, M.; Roche, P.; Murphy, G. P., Prostate Specific Membrane Antigen Expression in Prostatic Intraepithelial Neoplasia and Adenocarcinoma: A study of 184 cases. *Cancer* 1998, *82*(11), 2256–2261.

55. Su, S. L.; Huang, I.-P.; Fair, W. R.; Powell, C. T.; Heston, W. D. W., Alternatively Spliced Variants of Prostate-Specific Membrane Antigen RNA: Ratio of Expression as a Potential Measurement of Progression. *Cancer Research* 1995, *55*(7), 1441–1443.

56. Kawatani, M.; Yamamoto, K.; Yamada, D.; Kamiya, M.; Miyakawa, J.; Miyama, Y.; Kojima, R.; Morikawa, T.; Kume, H.; Urano, Y., Fluorescence Detection of Prostate Cancer by an Activatable Fluorescence Probe for PSMA Carboxypeptidase Activity. *Journal of the American Chemical Society* 2019, *141*(26), 10409–10416.

3 Molecular Vibrational Imaging by Coherent Raman Scattering

Yasuyuki Ozeki, Hideaki Kano, and Naoki Fukutake

CONTENTS

3.1 Introduction 38
3.2 Principles of Spontaneous Raman Scattering and Coherent Raman Scattering 38
3.3 Basic Implementation of CARS and SRS Microscopy 40
    3.3.1 CARS Microscopy 40
    3.3.2 SRS Microscopy 41
    3.3.3 Two-Color Pulse Sources for CARS and SRS Microscopy 42
3.4 Hyperspectral CARS Microscopy 43
    3.4.1 Wavelength Scanning 43
    3.4.2 Spectral Focusing 44
    3.4.3 Multiplex Method 44
    3.4.4 Fourier-Transformation 44
    3.4.5 Retrieval of Raman Spectrum from the CARS Spectrum 44
    3.4.6 Implementation of Multiplex CARS Microscopy/Microspectroscopy 45
3.5 Hyperspectral SRS Microscopy 46
    3.5.1 Wavelength Scanning 46
    3.5.2 Spectral Focusing 46
    3.5.3 Wavelength Multiplexing 46
    3.5.4 Modulation-Frequency-Division Multiplexing 48
    3.5.5 Implementation of Wavelength-Tuning Based Hyperspectral SRS Microscopy 48
3.6 Applications of CRS Microscopy 49
    3.6.1 Applications of CARS Microscopy 49
    3.6.2 Applications of SRS Microscopy 51
3.7 Summary 53
Appendix A1 Classical Picture of the CRS Process 54
    A1.1 Driving Force of Two-Color Optical Field on Molecules 54
    A1.2 Time-Domain and Frequency-Domain Responses of Dielectric Permittivity 55
    A1.3 CARS and SRS as Optical Phase Modulation 56
    A1.4 Comparison of Signal-to-Noise Ratios of CARS and SRS 57
    A1.5 Number of Photons of CARS and SRS 57
Appendix A2 Quantum-Mechanical Picture of CRS 58
Appendix A3 Image Formation in CRS Microscopy 64
    A3.1 Formulation of the Image-Forming System 64
    A3.2 Optical Resolution 66
References 68

3.1 INTRODUCTION

Raman scattering is an inelastic light scattering phenomenon involving the molecular vibration of substances that interact with light, providing the vibrational spectrum of molecules. It has been applied to biological microscopic imaging to acquire the spatial distribution of biomolecules. However, the intensity of Raman scattering is quite weak, requiring a long acquisition time. Coherent Raman scattering (CRS) microscopy using two-color and/or broadband laser pulses drastically improved the imaging speed and sensitivity of Raman imaging. CRS microscopy can be classified into several types, including coherent anti-Stokes Raman scattering (CARS) microscopy [1, 2] and stimulated Raman scattering (SRS) microscopy [3–5]. Recent advancements in the instrumentations of CRS microscopy have drastically enhanced its chemical specificity and imaging speed. Currently, the advantages and disadvantages of CARS and SRS are growing conspicuous: CARS microscopy is advantageous in acquiring broadband vibrational spectra [6, 7], whereas SRS is useful for high-speed and sensitive analysis of vibrational spectra in a relatively narrow bandwidth [8–12]. Currently, CRS microscopy is widely applied to label-free imaging of metabolites and drugs in different types of biological cells and tissues. Furthermore, with the recent advent of Raman-detectable labeling technologies, CRS microscopy has been finding novel applications, such as metabolic analysis of small biomolecules [13–15], which were previously difficult with the earlier fluorescent labeling techniques.

This chapter describes the general concepts and recent applications of CRS microscopy. Specifically, we explain the basic principles of CARS and SRS microscopy and practical instrumentations, and introduce their applications to label-free imaging, metabolic imaging, and supermultiplex/ultramultiplex imaging. In addition, we explain the classical and the quantum-mechanical principles in detail for various CRS processes, as well as the theory of image formation in CRS microscopy.

The structure of this chapter is as follows: Section 3.2 describes the principle of spontaneous Raman scattering and CRS. In Section 3.3, we introduce the basic instrumentation of CRS microscopy. Section 3.4 and Section 3.5 describe hyperspectral CARS microscopy and hyperspectral SRS microscopy, respectively. Section 3.6 reviews the applications of CARS microscopy and SRS microscopy. Section 3.7 concludes the chapter. In the Appendix, we describe the classical and the quantum-mechanical pictures of CRS, as well as the image formation in CRS microscopy for those interested in the detailed physics of CRS. Numerous review papers on CRS microscopy are available, some of which are listed in the References [11, 16–23].

3.2 PRINCIPLES OF SPONTANEOUS RAMAN SCATTERING AND COHERENT RAMAN SCATTERING

Before describing CRS, we introduce spontaneous Raman scattering, which allows us to measure the spectrum of molecular vibrations with light. As shown in Figure 3.1(a), spontaneous Raman scattering is an interaction between photons at an angular optical frequency ω_p and molecules with a vibrational resonance frequency ω_R. As a result of this interaction, photons of Raman scattering are generated at $\omega_S = \omega_p - \omega_R$. Generally, molecules possess various vibrational resonance frequencies based on their molecular structures, and spontaneous Raman scattering can occur simultaneously for all the molecular vibrational modes, whereas the strength of interaction differs for each vibrational mode. Therefore, the optical spectrum of Raman scattering provides molecular vibrational signatures, as schematically shown in Figure 3.1(b).

Recently, the interest toward Raman imaging has been increasing, where Raman scattering from a large number of locations was successively measured to obtain the vibrational spectroscopic information at every image pixel; however, the applications of Raman imaging are limited by its long acquisition time. Typically, the detection of spontaneous Raman scattering requires one second for each pixel and several tens of minutes for an image because the intensity of Raman scattering

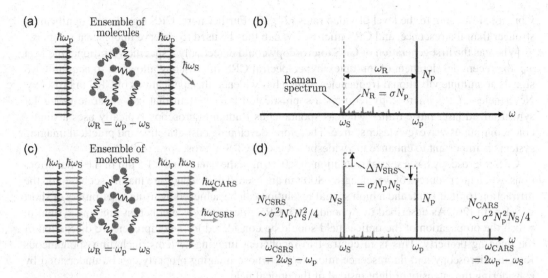

FIGURE 3.1 Schematic of spontaneous and coherent Raman scattering. (a) In spontaneous Raman scattering, a bunch of photons at an angular optical frequency ω_p and energy $\hbar\omega_p$ interact with an ensemble of molecules with a vibrational frequency ω_R. As a result, a few photons are scattered and their optical frequency shifts to $\omega_S = \omega_p - \omega_R$. Because molecules have various vibrational frequencies based on their structures, Raman scattering at various frequencies can occur simultaneously, forming a Raman spectrum as shown in (b). (c) In coherent Raman scattering, a bunch of photons at different angular optical frequencies of ω_p and ω_S interact with an ensemble of molecules with a vibrational frequency of ω_R. When $\omega_R = \omega_p - \omega_S$ is satisfied, the photons at ω_p disappear and the photons at ω_S appear as a result of stimulated Raman scattering (SRS). The probability of SRS is higher than that of spontaneous Raman scattering by a factor of number of Stokes photons. Simultaneously, photons at $2\omega_p - \omega_S$ and at $2\omega_S - \omega_p$ appear, which are known as coherent anti-Stokes Raman scattering (CARS) and coherent Stokes Raman scattering (CSRS), respectively. The number of photons of SRS is larger than that of spontaneous Raman scattering by a factor of the number of photons of Stokes pulses. The numbers of CARS and CSRS photons can be calculated according to the classical picture described in Appendix A1.5.

is relatively low. Furthermore, spontaneous Raman scattering is easily overwhelmed by the fluorescence of biological samples, as well as of optics or cover slips. Therefore, spontaneous Raman microscopy requires special attention in instrumentation, sample preparation, measurement, and data analysis.

On the contrary, as shown in Figure 3.1(c), CRS is an interaction between photons at two frequencies (i.e., ω_p and ω_S) and molecules with a vibrational resonance frequency ω_R. Furthermore, CRS occurs when $\omega_R = \omega_p - \omega_S$ is satisfied. As shown in Figure 3.1(d), CRS causes various spectral changes as follows:

- Generation of photons at a new frequency of $2\omega_p - \omega_S$: CARS
- Decrease in the number of photons at ω_p: stimulated Raman loss (SRL)
- Increase in the number of photons at ω_S: stimulated Raman gain (SRG)
- Generation of photons at a new frequency of $2\omega_S - \omega_p$: coherent Stokes Raman scattering (CSRS)

Moreover, SRL and SRG are called stimulated Raman scattering (SRS).

Currently, CARS and SRS are the two major mechanisms used in CRS microscopy. Although CRS provides the same spectroscopic information as spontaneous Raman scattering, the CRS process can be further drastically enhanced using pulsed lasers of high peak intensity. This allows the acquisition of molecular vibrational signatures in a very short duration to realize real-time molecular vibrational imaging to the level of video rates [24, 25]. Furthermore, CRS signals are significantly stronger than fluorescence, and CRS microscopy can thus be used to observe fluorescent samples.

Whereas the first generation of CRS microscopy could detect only a specific vibrational mode at ω_R, the recent development of multiplex/hyperspectral CRS microscopy allows us to acquire CRS signals at multiple vibrational frequencies, which has widened the applications of CRS microscopy. Nevertheless, CRS microscopy requires a sophisticated laser system that can generate two-color synchronized laser pulse trains, whereas spontaneous Raman microscopy typically uses a single-color continuous-wave (cw) laser source. Therefore, developing cost-effective and practical imaging systems is important to enhance the widespread use of CRS microscopy.

CRS microscopy has a spatial resolution of <0.5 μm in the lateral and <1.5 μm in the axial directions, when laser sources of wavelength ~800 nm are used. These values are mainly set based on the diffraction limit of light, and a high spatial resolution can be achieved by using laser pulses of short wavelength [26]. As described in Appendix A3, CRS microscopy is a coherent imaging system in which the propagation of the optical field should be considered for a comprehensive discussion of the imaging property. This is in contrast to incoherent imaging systems, including spontaneous Raman microscopy and fluorescence microscopy, whose imaging property can be understood by considering the intensity of light instead of the optical field.

3.3 BASIC IMPLEMENTATION OF CARS AND SRS MICROSCOPY

Figure 3.2(a) and 3.2(b) show the typical schematics of CARS (a) and SRS (b) microscopes, respectively. Both in CARS and SRS microscopy, two pulsed laser sources of different colors (referred to as pump and Stokes laser, whose optical angular frequencies are ω_p and ω_S, respectively) are used. The CARS and SRS microscopic systems are described in the following sections.

3.3.1 CARS MICROSCOPY

As shown in Figure 3.2(a), a typical CARS microscopy system employs two-color synchronized optical pulses known as pump and Stokes pulse trains, whose optical angular frequencies are ω_p and ω_S, respectively. The typical parameters of the laser pulses are as follows: pulse duration of

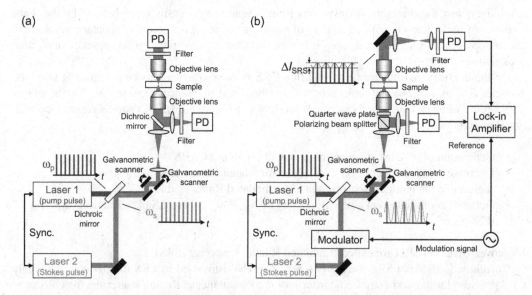

FIGURE 3.2 Schematic of typical CRS microscopy. (a) CARS microscopy. (b) SRS microscopy. Note that the synchronization of two lasers is achieved either passively or actively.

2–7 ps, repetition rate of ~80 MHz, wavelengths of ~800 nm and ~1000 nm for pump and Stokes, respectively, and average power of 10–100 mW for each pulse. Then, the pump and the Stokes pulses are combined in time and space by a dichroic mirror and focused on a sample by an objective lens. When the frequency difference between pump and Stokes pulses matches the vibrational frequency of molecules, i.e., $\omega_p = \omega_S - \omega_R$, a CARS signal at $2\omega_p - \omega_S$ is generated in the focal volume, which is detected by either another objective lens (transmission mode) or the same objective lens (epi-detection, backscattering mode, or reflection mode). After blocking the pump and the Stokes pulses using an optical filter, the CARS signal is detected using an optical detector, such as a photomultiplier tube (PMT) or an avalanche photodiode. For CARS imaging, the focal position with respect to the sample is scanned by laser scanning and/or stage scanning. In laser scanning, the propagation direction of the laser beam incident on the objective lens is controlled by a pair of galvanometric scanners in the optical path. In stage scanning, the position of the stage is controlled with mechanical actuators. The CARS signal is thus obtained as a function of 2-D or 3-D coordinates, from which a CARS image can be produced at a vibrational frequency of $\omega_p - \omega_S$.

In the CARS process, considering the phase-matching condition is important for an efficient generation of CARS signal. Phase-matching indicates that the electric fields of CARS signals emitted from all the molecules in the focal volume interfere constructively (see Appendix A2 for more information). However, this condition is not always satisfied because the phase of the CARS field depends on the phases of the pump pulse and Stokes pulse, whose wavelengths are different from those of the CARS signal, and their phases are thus affected by the different refractive indices at the respective wavelengths, especially when the interaction length (i.e., the length of the focus) is long. The first study on CARS microscopy [27] used folded BOXCARS configuration, where pump and Stokes beams are loosely focused at a certain angle to satisfy the phase matching condition. Later, collinear propagation of ω_p and ω_S pulses into microscope objective was found to satisfy the phase-matching condition automatically [1, 2] because the large numerical aperture (NA) of microscope objectives provides large angular distribution of the incident laser beam and shortens the interaction length. Another consequence of phase matching is that an intense CARS radiation can be generated in the forward direction [28]; in addition, epi-detection is possible by observing small particles or highly light-scattering samples including tissues.

The signal enhancement factor of CARS has been discussed previously. Scully's group evaluated signal amplitudes of CARS and spontaneous Raman both theoretically and experimentally, and concluded that the CARS signal is stronger by 100 times compared with the spontaneous Raman signal for microscopy [29], and by 10^5 times for bulk material because of the long coherence length [30].

It must be noted that CARS microscopy detects not only vibrationally resonant CARS signals but also nonresonant background (NRB), which is described in detail in Appendix A1 and Appendix A2. To suppress the contribution of NRB, several techniques have been proposed, such as polarization control [31], time-resolved detection [32], and back scattering configuration [33]. Further, NRB not only leads to a signal offset but also distorts the CARS spectrum, making it difficult to quantitatively measure the density of molecules only with a CARS signal at a single frequency. However, this problem has been mitigated by multiplex CARS microscopy and spectral analysis, which are discussed in Section 3.4.3.

3.3.2 SRS MICROSCOPY

Figure 3.2(b) shows the schematic of typical SRS microscopy. SRS microscopy differs from CARS microscopy in the following aspects: (1) Stokes pulse intensity is modulated in time using an optical modulator. (2) In the focal volume inside the sample, SRS occurs when the optical frequency difference between the pump and the Stokes pulses matches the molecular vibrational resonance frequency. As a result, the intensity modulation of Stokes pulses is transferred to the pump pulses through the SRS process. (3) Pump and Stokes pulses transmitted through the sample are detected by another objective lens, and only the pump pulses are filtered by an optical filter, which are then led to a photodiode (PD). (4) The output signal of the PD is measured with a lock-in amplifier, which extracts the modulation of the PD signal at the modulation frequency to obtain the SRS signal. In the epi-detection mode, the pump pulses back-scattered from the sample are detected with another PD through a polarizing beam splitter and a quarter wave plate; alternatively, an annular photodiode placed between the objective lens and the sample can be used [25]. In SRS imaging, the focal spot is scanned by laser scanning or stage scanning. It must be noted that SRS microscopy should use a sophisticated photodiode circuit with low noise and a high tolerance of signal saturation [11] to detect small intensity modulation in intense pump pulses. Furthermore, to detect small intensity modulation caused by SRS, the intensity noise of pump pulses should be as low as the shot noise limit at the lock-in frequency, which is satisfied using typical solid-state lasers and optical parametric oscillators (OPOs).

The most important advantage of SRS microscopy is that we can obtain the same spectroscopic information as that from spontaneous Raman scattering, and the concentration of molecules of interest can thus be measured even with a single-frequency SRS signal. On the contrary, the detection system is more complex than that of CARS microscopy, thus making broadband multiplex detection of SRS signals difficult. In cases of signal enhancement factor and signal-to-noise ratio, SRS and CARS are comparable, as discussed in Appendix A1.

3.3.3 Two-Color Pulse Sources for CARS and SRS Microscopy

To date, various techniques have been developed for generating two-color synchronized optical pulses, which are categorized into two types, namely, active synchronization and passive synchronization. In active synchronization, two independent lasers are used to generate two-color pulses, and the repetition rate of one laser is controlled to synchronize it with the other laser. In passive synchronization, two-color laser pulses can be generated without external active control by utilizing some interaction between two lasers. Previous studies on CARS microscopy [1, 2] used passively synchronized laser sources, including an optical parametric amplifier (OPA), which can generate synchronized wavelength-tunable pulses. However, OPAs typically have a low repetition rate of <1 MHz, and therefore are incompatible for high-speed imaging. Then, actively synchronized Ti:sapphire lasers were adopted in CARS microscopy, and significant efforts have been made to realize synchronization with a low timing jitter [34, 35]. Ti:sapphire oscillators can generate wavelength-tunable picosecond pulses with a narrow spectral width at a high repetition rate of ~80 MHz; thus, they are found to be suitable for CARS microscopy. However, active synchronization is typically very sensitive to environmental perturbations, including physical shock and temperature change, which hinders the widespread use of CRS microscopy. Then, OPOs [36, 37] were developed, which have become the most popular laser sources for CARS and SRS microscopy systems because OPO allows passive synchronization, which is typically significantly more stable than active synchronization. The OPO is excited by a pulse train from a master laser, and the pulse train can give rise to an optical parametric gain in a nonlinear optical crystal in the OPO cavity only when the pulses pass through the crystal; therefore, optical pulses that are synchronized with the master laser pulses can be generated. Further, various laser configurations have been adopted as master laser sources, including mode-locked solid-state laser oscillators using $Nd:YVO_4$ as a gain material [37]. Currently, mode-locked Yb fiber lasers, which can generate ~3 ps pulses of wavelength ~1032 nm at a repetition rate of ~80 MHz, are the most popular [36]. If an OPO pumped by frequency-doubled Yb fiber laser pulses (516 nm) is used as a laser source, the signal from the OPO (690–990 nm) and the fundamental output from the Yb fiber laser (1032 nm) can be used as optical pulses at ω_p and ω_s, respectively. Although OPO has a high wavelength tunability, wavelength tuning requires temperature control of the OPO crystal, as well as the mechanical control of birefringent filter inside the OPO cavity; thus, wavelength tuning requires several tens of seconds to several minutes.

In addition to a picosecond laser source, femtosecond and sub-nanosecond laser sources are used for CARS microscopy [38–40]. Furthermore, great efforts have been made to realize compact and practical laser sources based on fiber lasers and wavelength conversion technologies [41–43], fiber optical parametric oscillators [44], optical parametric amplifiers [45], time-lens sources [46], and semiconductor lasers [47]. For application to SRS microscopy, the intensity noise of fiber laser pulses has been importantly considered because it directly contributes to the noise of the SRS signal. Therefore, various techniques have been developed to eliminate the effect of laser intensity noise based on balanced detection [43, 48]. Furthermore, several research groups are attempting to develop fiber laser sources with high pulse energy in the laser cavity to generate low-noise optical pulses [49, 50].

3.4 HYPERSPECTRAL CARS MICROSCOPY

Hyperspectral CARS/SRS imaging refers to the acquisition of images in which each pixel contains the information of the vibrational spectrum, i.e., the CARS/SRS signal as a function of vibrational frequency. Hyperspectral imaging enables the discrimination of different molecules with a subtle difference in the vibrational spectrum and allows for a detailed analysis of biological samples. This is in contrast to the basic implementation of CARS microscopy and SRS microscopy described in Section 3.3, where images reflect only a specific vibrational mode at a single frequency that is determined by the frequency difference between pump and Stokes pulses. In this subsection, vibrational spectroscopic imaging using hyperspectral CARS microscopy is described.

Figure 3.3 illustrates a summary of various hyperspectral CARS imaging methods, namely, wavelength scanning, spectral focusing, multiplex method, and Fourier-transformation. Each technique is described here.

3.4.1 WAVELENGTH SCANNING

In the wavelength scanning method, CARS images at different vibrational frequencies are successively obtained by scanning the laser wavelength, as shown in Figure 3.3(a). To achieve high spectral resolution, two-color picosecond pulse sources with narrow linewidth are typically used. A disadvantage of this approach is that the wavelength tuning of OPOs requires a longer period, which limits the imaging speed. Currently, fast wavelength-tunable lasers have been developed [44, 51], which have made the wavelength scanning method practical. To obtain the Raman spectrum from CARS spectrum, adequate signal processing is required, as discussed in Section 3.4.5.

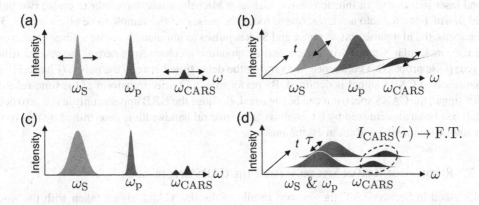

FIGURE 3.3 Hyperspectral CARS methods. (a) Wavelength scanning. (b) Spectral focusing. (c) Multiplex CARS. (d) Fourier-transform CARS.

3.4.2 Spectral Focusing

The spectral focusing method uses two-color femtosecond laser pulses as pump and Stokes pulses for hyperspectral CARS imaging. This method does not require wavelength tuning and can achieve a high spectral resolution [52, 53]; its requirements for the laser sources are thus not stern compared with those of the wavelength tuning method. In the spectral focusing method, pump and Stokes pulses are passed through dispersive elements such as glass rods to enable them to possess the same amount of frequency chirp (i.e., time dependence of instantaneous optical frequency) and a long period of time of typically several picoseconds. When the pump and the Stokes pulses are over-lapped in time, their difference frequency is kept constant for picoseconds. Furthermore, by tuning the delay between them, the frequency difference can be adjusted, as shown in Figure 3.3(b). When the pulses are focused on a sample, a particular Raman band can be excited, which is chosen by the delay between pump and Stokes pulses, leading to the generation of a CARS signal.

3.4.3 Multiplex Method

The multiplex or broadband CARS method [6, 7] enables the acquisition of the CARS spectrum in a broad spectral range at once by utilizing broadband optical pulses and narrowband pulses, as shown in Figure 3.3(c). The CARS spectrum is measured using a spectrometer equipped with a multi-channel detector, such as a charge-coupled device (CCD) or a complementary metal oxide semiconductor (CMOS) camera. A broadband laser source, such as a supercontinuum (SC) light source [54] or a femtosecond laser source [55], enables the typical spectral coverage of the multiplex CARS signal to reach approximately 3000 cm^{-1}, which is sufficiently broad to detect all the vibrational modes, including the fingerprint region and C-H and O-H stretching regions. For SC generation, a portion of the master laser output is seeded into a tapered fiber [6] or a photonic crystal fiber (PCF) [7]. In addition, a cost-effective sub-nanosecond laser source has been used as the master laser source for SC generation [56]. In the multiplex method, the image-acquisition speed depends mainly on the exposure time and the readout time of the multichannel photodetector, by which the pixel dwell time is restricted to at least ~1 ms [57, 58]. One of the advantages of the multiplex CARS is its robustness against the motion of samples and the intensity fluctuation of the incident laser, the effect of which can be suppressed by accumulating the CARS signal.

3.4.4 Fourier-Transformation

Similar to that in the Fourier-transformation infra-red (FT-IR) spectrometer, the CARS spectrum can be obtained using the FT technique. The FT-CARS method typically employs a single femto-second laser source and an interferometer, such as a Michelson interferometer to create two pulse replicas, which are fed into a microscope to focus the pulses on the sample to be observed [59, 39]. At the focus, the first pulse acts as pump and Stokes pulses to impulsively excite a vibrational coher-ence (i.e., molecular vibration in the context of quantum mechanics), whereas the second (time-delayed) pulse probes this coherence. By varying the delay time between the pulses (Figure 3.3(d)), a time-resolved CARS signal is obtained. By performing a Fourier transform of the time-resolved CARS signal, the CARS spectrum can be obtained. Because the NRB appears only at the zero delay time, it can be easily removed by FT analysis. The spectral bandwidth is determined by that of the laser source for impulsive Raman excitation.

3.4.5 Retrieval of Raman Spectrum from the CARS Spectrum

As described in Section 3.3.1, the spectral profile of the raw CARS signal taken with the wave-length scanning method, the spectral focusing method, and the multiplex method is different from that of spontaneous Raman scattering; therefore, adequate signal processing must be employed.

Specifically, the spectral profile of the raw CARS signal is similar to $\mathrm{Re}[\chi^{(3)}]$, whereas the spectral profile of spontaneous Raman scattering is proportional to $\mathrm{Im}[\chi^{(3)}]$. Here, $\chi^{(3)}$ is a complex function of vibrational frequency and is termed the third order nonlinear susceptibility, which is explained in Appendix A1 and Appendix A2. As mentioned in Section 3.3.1, the CARS signal has two contributions: NRB and vibrationally resonant CARS, whose $\chi^{(3)}$ are real and complex, respectively. Further, the spectral profile of CARS can be understood as the interference of these effects. Several techniques have been currently proposed to retrieve the $\mathrm{Im}[\chi^{(3)}]$ spectrum from the CARS spectrum (i.e., $\mathrm{Re}[\chi^{(3)}]$ spectrum), such as the maximum entropy method (MEM) [60], the time-domain Kramers–Kronig formulation (TDKK) [61], and the phase-corrected KK approach (PCKK) [53]. In FT-CARS, the NRB can be directly removed in the signal processing, as mentioned in Section 3.4.4.

3.4.6 Implementation of Multiplex CARS Microscopy/Microspectroscopy

The experimental setup of multiplex CARS microscopy developed by Kano et al. is shown in Figure 3.4(a). The wavelength, temporal duration, and repetition rate of the laser source are 1064 nm, 85 ps, and 0.82 MHz, respectively. The laser pulses are divided into two: one part is used as the pump pulse (ω_p) in the CARS process, and the other is seeded into a PCF to generate an SC, which is used as the Stokes pulse (ω_S) in the CARS process. The SC radiation is collimated by an off-axis parabolic mirror to suppress the chromatic aberration. Because the spectral density around NIR should be high for efficient CARS signal generation, the PCF was fabricated to efficiently generate the spectral components between 1100 nm and 1700 nm. The pulse energies of the ω_p and the ω_S pulses were approximately 2 µJ and 1 µJ, respectively, corresponding to peak powers of approximately 24 kW and 12 kW, respectively. After blocking the spectral components shorter than 1064 nm from the SC using a long-pass filter, the ω_p and the ω_S laser pulses were superimposed by a sharp dichroic mirror, and then introduced into a modified inverted microscope. The incident laser pulses were tightly focused onto the sample by the first objective lens. The sample was placed on a piezoelectric stage for microscopic imaging, and the CARS signal was collected using the second objective lens and was dispersed by the spectrometer. Finally, the CARS signal was detected by the CCD camera.

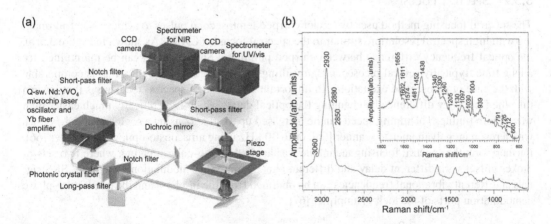

FIGURE 3.4 (a) Schematic of multiplex CARS microscope. (b) Intracellular-averaged Im[$\chi^{(3)}$] spectrum of HeLa cell in G1 phase. (Reprinted from H. Yoneyama, K. Sudo, P. Leproux, V. Couderc, A. Inoko, and H. Kano, "CARS molecular fingerprinting using sub-100-ps microchip laser source with fiber amplifier," *APL Photon.* **3**, 092408 (2018); licensed under a Creative Commons Attribution (CC BY) license.)

Figure 3.4(b) shows the intra-cellular averaged Raman-equivalent $\text{Im}[\chi^{(3)}]$ spectrum of a HeLa cell in the G1 phase. The $\text{Im}[\chi^{(3)}]$ spectrum at each spatial position was calculated from the raw CARS spectra using the maximum entropy method (MEM) [54, 60]. The MEM does not require any *a priori* knowledge of the vibrational bands, but can still retrieve the phase information of $\chi^{(3)}$. The CARS signal was clearly observed between 600 and 3200 cm^{-1}. The assignments of the corresponding vibrational bands are listed in Table 3.1 [62–64].

3.5 HYPERSPECTRAL SRS MICROSCOPY

Figure 3.5 schematically illustrates the methods of hyperspectral SRS microscopy, where SRS signals at various vibrational frequencies are obtained to perform SRS imaging of multiple constituents. Specifically, the methods are categorized into (a) wavelength scanning, (b) spectral focusing, (c) wavelength-division multiplexing, and (d) modulation-frequency multiplexing. Each method is described below.

3.5.1 WAVELENGTH SCANNING

In the wavelength scanning method, a series of SRS images are obtained by changing the laser wavelength [8–11], similar to the wavelength scanning CARS. In this method, the use of a high-speed wavelength-tunable laser is important because the time required for wavelength tuning can limit the imaging speed. An advantage of the wavelength scanning method in SRS is the flexibility of data acquisition. For example, by continuously scanning the laser wavelength, the SRS spectra can be obtained, using which small difference in the spectral shape for observation. If *a priori* knowledge about the sample is available, we can discretely change the laser wavelength to reduce the number of spectral points and the acquisition time. Because the SRS signal directly produces $\text{Im}[\chi^{(3)}]$, we can directly analyze the SRS data even when SRS is measured at discrete spectral points. On the contrary, the wavelength scanning approach has the problem of motion artifacts when fast-moving samples are imaged. To mitigate this problem, several groups are developing fast wavelength-tunable laser pulse sources [8, 9, 11, 12]. In particular, Suzuki et al. have reported a four-color, pulse-pair-resolved, and wavelength-switchable laser, which can be applied to SRS imaging flow cytometry at a flow speed as high as 20 mm/s [12].

3.5.2 SPECTRAL FOCUSING

The spectral focusing method uses two-color chirped femtosecond pulses to achieve spectral imaging with high spectral resolution, similar to the spectral focusing CARS. As shown in Figure 3.5(b), the optical frequency difference between chirped pump and Stokes pulses can be maintained for a long time (typically several picoseconds), resulting in high spectral resolution that is comparable with the case of picosecond excitation. An important advantage of spectral focusing is that we can tune the frequency difference by changing the optical delay, which is technically much easier than wavelength tuning. The tuning speed can be increased up to 24 kHz using a sophisticated 4-f optics with a resonant galvanometric scanner [65], or to 30 kHz using an acousto-optic tunable filter [66]. Furthermore, the spectral focusing technique was applied in two-color imaging, where two sets of Stokes pulses with different delay and different phase in intensity modulation were used. SRS signals at different vibrational frequencies can be obtained from the in-phase and the quadrature-phase demodulation ports of the lock-in amplifier [67].

3.5.3 WAVELENGTH MULTIPLEXING

As shown in Figure 3.5(c), the wavelength multiplexing method employs picosecond Stokes pulses and broadband pump pulses. Although the latter need not necessarily be transform-limited, its

TABLE 3.1
Vibrational Bands and Their Assignments for Intracellular Im[$\chi^{(3)}$] Spectrum

Typical Raman shift (cm^{-1})	Assignment	Main molecular component
3427	O-H a-stretch.	Water
3200	O-H s-stretch.	Water
3066	C-H stretch. (aromatic)	Proteins
3010	=C-H stretch.	Lipids
2953	CH$_3$ a-stretch.	DNAs and RNAs
2930	CH$_3$ s-stretch.	Proteins/lipids
2902	CH$_2$ a-stretch.	Lipids/proteins
2872	Overtone of CH$_3$ deform. in Fermi resonance with CH$_3$ s-stretch.	Proteins/Lipids
2854	CH$_2$ s-stretch.	Lipids
2580	S-H stretch.	Proteins
1741	C=O stretch. (ester)	Lipids
1655	cis C=C stretch./Amide I	Lipids/proteins
1611	Tyr, Trp	Proteins
1602	Phe	Proteins
1574	Purine ring (A and G)	DNAs and RNAs
1481	Purine ring (A and G)	DNAs and RNAs
1452	CH$_3$ deg. deform.	Proteins
1438	CH$_2$ scis.	Lipids
1376	T, A, and C	DNAs
1340	CH deform.	Proteins
1335	Purine ring (A and G)	DNAs and RNAs
1300	CH$_2$ twist.	Lipids
1260	=C-H bend.	Lipids
1240	Amide III	Proteins
1205	C-C$_6$H$_5$ (phenyl ring) stretch.	Proteins
1130	Skeletal C-C (trans) stretch.	Lipids
1097	PO$_2^-$ stretch.	DNAs and RNAs
1084	Skeletal C-C (gauche) stretch.	Lipids
1063	Skeletal C-C (trans) stretch.	Lipids
1030	In-plane phenyl ring deform.	Proteins
1004	Phenyl ring breath.	Proteins
939	C-C	Proteins
892	C$_2$-C$_1$ (trans) stretch./CH$_3$ rock.	Lipids
876	C$_2$-C$_1$ (gauche) stretch./CH$_3$ rock.	Lipids
790	Pyrimidine ring (C, T, and U)/bk. (OPO)	DNAs and RNAs
726	CH$_2$ rock.	Lipids
719	Head group (choline (H$_3$C)N$^+$ sym. stretch.) of phosphatidylcholine	Lipids
709	cholesterol	Lipids
660	C-S stretch.	Proteins

Abbreviations: stretch.: stretching; deform.: deformation; rock.: rocking; twist.: twisting; scis.: scissors; bend.: bending; s-: symmetric; a-: antisymmetric; deg.: degenerate; breath.: breathing; bk.: DNA backbone; A: adenine; C: cytosine; G: guanine; T: thymine; U: uracil; Phe: phenylalanine; Trp: tryptophan; Tyr: tyrosine.

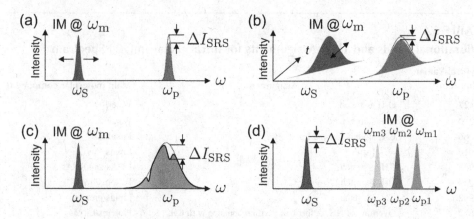

FIGURE 3.5 Methods of hyperspectral SRS microscopy. (a) Wavelength scanning. (b) Spectral focusing. (c) Wavelength-division multiplexing. (d) Modulation-frequency-division multiplexing.

duration must be significantly shorter than the Stokes pulses, so that pump and Stokes pulses have sufficient temporal overlap for the SRS effect. Moreover, SRS at various vibrational frequencies can occur simultaneously if such pulses are used. The resultant spectral change of the pump pulses is detected by a spectrometer equipped with a PD array, and the photocurrent of each PD in the array is detected with a lock-in amplifier or a combination of an RF filter and an RF detector to obtain the SRS signal [68]. This method has been applied not only to microscopy but also to flow cytometry, where the SRS spectrum in each sample flowing in a microfluidic device is measured at a high throughput of 10,000 particles/s [69].

   This method is advantageous in that the SRS signal at various vibrational frequencies can be obtained simultaneously. Therefore, this method is more tolerant to motion artifact compared with the wavelength-scanning approach. However, a disadvantage is that the photodetector for SRS requires a high dynamic range, thus necessitating a bulk PD array and a lock-in amplifier array.

3.5.4 MODULATION-FREQUENCY-DIVISION MULTIPLEXING

In the modulation-frequency-division multiplexing technique, different spectral components are modulated at different frequencies. Through SRS, modulation transfer occurs simultaneously at various frequencies, as shown in Figure 3.5(d). Such sophisticated intensity modulation is realized using a rotating polygon mirror, 4-f optics, and a specially designed mask pattern [70]. This technique is advantageous in that the SRS spectrum can be obtained without using a spectrometer. Consequently, the imaging speed is decreased by the number of modulation frequencies because the imaging bandwidth is limited by the spacing between neighboring modulation frequencies.

3.5.5 IMPLEMENTATION OF WAVELENGTH-TUNING BASED HYPERSPECTRAL SRS MICROSCOPY

To realize high-speed hyperspectral microscopy, Ozeki et al. developed a wavelength-tunable laser based on the spectral filtering of Yb fiber laser pulses [8, 11]. This laser can generate picosecond pulses with a spectral width of ~3 cm^{-1} in the wavelength range 1014–1046 nm, which corresponds to a tuning bandwidth of 300 cm^{-1}. The Yb fiber laser is actively synchronized with a Ti:sapphire laser that produces pump pulses at ~800 nm. We realized video-rate spectral imaging using this system, where SRS images can be obtained at a rate of 30 fps, whereas the wavelength is tuned in a frame-by-frame manner, as schematically shown in Figure 3.6(a) [11]. The resultant hyperspectral image data contains the SRS spectra at every pixel. If the SRS spectra of all the constituents are known, the hyperspectral image data can be decomposed into various components using the Moore–Penrose pseudo-inverse matrix calculation. Figures 3.6(b)

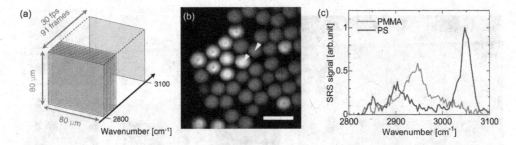

FIGURE 3.6 Hyperspectral SRS imaging of a mixture of polymer beads composed of poly (methyl methacrylate) (PMMA) and polystyrene (PS). (a) Schematic of wavelength-tuning based hyperspectral SRS imaging. SRS images at different wavenumbers were obtained at a rate of 30 fps. (b) Merged SRS image of PMMA (dark gray) and PS (light gray) beads. Scale bar: 20 μm. (c) SRS spectra of PMMA and PS obtained at the locations indicated by arrows in Fig. 6(b).

and 3.6(c) illustrate hyperspectral SRS imaging. We obtained the hyperspectral images of two types of polymer beads, one composed of poly(methyl methacrylate) (PMMA) and the other of polystyrene (PS), and the hyperspectral images were decomposed into the images of PMMA and PS to obtain a merged SRS image, as shown in Figure 3.6(b). Figure 3.6(c) shows the SRS spectra at the locations indicated by the arrows in Figure 3.6(b), which were used for decomposing the hyperspectral data.

3.6 APPLICATIONS OF CRS MICROSCOPY

3.6.1 APPLICATIONS OF CARS MICROSCOPY

Since the early stage of the application to living cells and tissues, CARS microscopy has been proven to provide excellent image contrast with metabolites such as lipids. Because of the long aliphatic chains of lipids, CH_2 symmetric stretching vibrational mode at ~2850 cm^{-1} can be used to visualize intracellular lipid droplets and tissue-specific lipids such as myelin sheath. Using the intense band caused by the CH_2 symmetric stretching vibrational mode, Hellerer et al. reported 3-D visualization and quantitation of lipid droplets in *Caenorhabditis elegans* [71]. For the CARS study using the fingerprint region, Petrov et al. reported CARS spectra of single bacterial endospores (*Bacillus subtilis*) in the fingerprint region and intra-cellular dipicolinic acid (DPA) [29].

As a typical application to tissue imaging [72], Evans et al. reported the development of video-rate CARS microscopy and visualized a penetration process of mineral oil into mouse skin through intercellular lipid [24]. In addition to such *in vivo* CARS imaging, CARS imaging has been applied to pathological measurement and diagnosis of living tissues. Because of the advantage of high image contrast of CH_2 stretching vibrational mode at lipids, CARS imaging has been applied to cirrhosis [73], cancer [74], and arteriosclerosis [75].

Figure 3.7 summarizes the recent advancements in CARS imaging. Evans et al. demonstrated imaging of brain structure and pathology *ex vivo* [76] (Figure 3.7A). Because the brain myelin located in the oligodendrocyte sheaths coating the axonal bundles is lipid-rich, the CARS image at 2846 cm^{-1} exhibits excellent image contrast. Using fresh unfixed *ex vivo* mouse brain, they identified normal gray and white matter structures on several scales to distinguish glioma from normal brain tissue and to clearly identify the tumor margin.

Camp et al. reported multiplex CARS images of healthy murine liver and pancreas tissues, as well as interfaces between xenograft brain tumors, and the surrounding healthy brain matter [55]. Owing to the molecular fingerprint, clear vibrational contrast is achieved for the nucleus, collagen, arterial wall, and lipid body (Figure 3.7B).

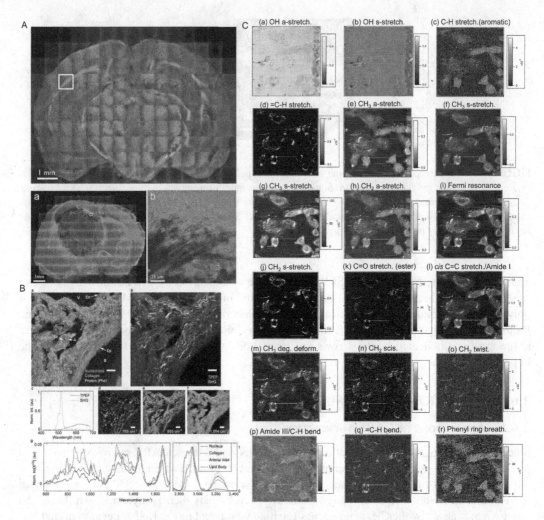

FIGURE 3.7 (A) **a**: Mosaic CARS image of a healthy mouse brain at 2846 cm⁻¹. Each image has a size of 700×700 μm. The brain slice was sectioned at 2.8 mm posterior to the bregma. **b**: Low-resolution and large field-of-view mosaic CARS image of astrocytoma at 2846 cm⁻¹. **c**: Higher resolution CARS image corresponding to the area enclosed by rectangle in **b**. The image visualizes microscopic infiltration at the boundary between the tumor and the normal tissue. (Reprinted after modification with permission from C. L. Evans, X. Xu, S. Kesari, X. S. Xie, S. T.C. Wong, and G. S. Young, "Chemically-selective imaging of brain structures with CARS microscopy," *Optics Express* 15(19), 12076–12087 (2007), OSA Publishing.) (B) Ultra-multiplex CARS imaging of A549 cells at (a) 3427 cm⁻¹, (b) 3200 cm⁻¹, (c) 3066 cm⁻¹, (d) 3017 cm⁻¹, (e) 2953 cm⁻¹, (f) 2939 cm⁻¹, (g) 2921 cm⁻¹, (h) 2902 cm⁻¹, (i) 2872 cm⁻¹, (j) 2854 cm⁻¹, (k) 1744 cm⁻¹, (l) 1657 cm⁻¹, (m) 1451 cm⁻¹, (n) 1438 cm⁻¹, (o) 1303 cm⁻¹, (p) 1284 cm⁻¹, (q) 1265 cm⁻¹, and (r) 1009 cm⁻¹, respectively. The exposure time per pixel was ~1.8 ms. (C) **a**: Multiplex CARS images of a portal triad within murine liver tissue with the nuclei represented in dark gray, collagen in medium gray, and protein content in light gray; A: port artery; B: bile duct; V: portal vein; Ep: epithelial cell; En: endothelial cell. **b**: SHG and two-photon excitation fluorescence (TPEF) image. C: Spectral profile of TPEF and SHG signals. **d–f**: Multiplex CARS images of individual vibrational modes represented by the color channels at (**d**) 785 cm⁻¹; (**e**) 855 cm⁻¹; (**f**) 1004 cm⁻¹. **g**: Single-pixel Im[χ⁽³⁾] spectra from the nucleus (DNA), collagen fiber, arterial wall and a lipid droplet. Scale bars: 20 μm.

Kano et al. reported ultramultiplex CARS spectroscopic imaging using a CCD camera with a fast readout time (<1 ms) [58]. Further, CARS images of 161×161 pixels of polymer beads were obtained with a total data-acquisition time of approximately 28 s. Label-free and ultramultiplex (18 colors) imaging of living cells was additionally performed at an effective exposure time of 1.8 ms, based on the molecular fingerprint and the C-H and the O-H stretching vibrational modes using a master oscillator fiber amplifier (MOFA) laser source (Figure 3.4(a)). By analyzing the spectral profile at each cell position, the CARS images are reconstructed for various Raman bands. Figure 3.7C summarizes the results of ultramultiplex CARS imaging. It is apparent that the microscopic intracellular structures were successfully visualized based on their molecular fingerprint. Particle-like structures with a diameter of a few micrometers were observed in the cytoplasm, as shown in Figure 3.7C (d), (j), (k), (n), (o), and (q). These bands are assigned as the C-H stretching vibrational mode of =C-H bonds (d), CH_2 symmetric stretching vibrational mode (j), C=O stretch mode of the ester (k), CH_2 scissoring mode (n), CH_2 twisting mode (o), and =C-H bending mode (q), respectively, all of which are typically observed in lipids. These organelles are, therefore, safely assigned as lipid droplets. As clearly shown in the CARS images (d) and (q), the lipid droplets contain rich unsaturated lipids. These results are consistent with the characteristics of the A549 cell, which is clearly known to produce unsaturated fats. On the contrary, the CARS image at 2939 cm^{-1} and 2921 cm^{-1} exhibit a dark round structure in the cell; this structure corresponds to the cell nucleus. The intra-nuclear bright spots correspond to the nucleolus, and introduce high vibrational contrast in the CARS images (c), (e), (f), (g), (i), (m), and (r). These vibrational modes correspond to the aromatic CH stretching (c), CH_3 asymmetric stretching (e), CH_3 symmetric stretching ((f) and (g)), overtone of CH_3 deformation in Fermi resonance with CH_3 symmetric stretching (i), CH_3 degenerate deformation (m), and phenyl ring breathing (r) vibrational modes. On the other hand, the CARS images at 3427 cm^{-1} and 3200 cm^{-1}, which correspond to OH stretching vibrational modes, exhibit low contrast in the field of view.

3.6.2 APPLICATIONS OF SRS MICROSCOPY

Various applications of SRS microscopy have been demonstrated. Here, we classify the applications based on the following three wavenumber regions: (i) carbon-hydrogen (CH)-stretching region (2800–3100 cm^{-1}); (ii) fingerprint region (500–1800 cm^{-1}); and (iii) silent region (2000–2300 cm^{-1}).

The CH-stretching region is the most often employed region in SRS microscopy. Because almost all the biomolecules contain carbon and hydrogen, they can have SRS signals in this region. Moreover, previously, the molecular specificity in this region was considered to be low. In actuality, lipids, proteins, and nucleic acids have different spectral shapes, and can be discriminated by careful analysis of their spectral difference. Furthermore, the signal level is higher than that in other regions because the density of CH bonds is high. For these reasons, SRS imaging in the CH stretching region has various applications, including quantification of lipid accumulation [77–79], stain-free imaging of tissue [3, 8, 80–84], skin [3, 25, 85–87], nucleic acids [88], food products [89], and tablets [90]. In particular, SRS can be used to monitor lipid desaturation or carbon–carbon double bonds [3], and the accumulation of saturated lipid is observed in cancer stem cells [91] and cancer tissue [92]. Further, it has also been reported that SRS can measure the membrane potential of nerve cells [93].

In the fingerprint region, we observe various vibrational modes to discriminate different types of biomolecules, such as cholesterol [94, 95], retinoids [96], amyloid plaques [97], drug penetrating into skin [3, 98, 99], anti-cancer drug in leukemia cells [100], DNA [101], neurotransmitters [102], fingerprints [103], cell walls in plants [104–107], and teeth [108, 109]. Recently, SRS in the fingerprint region was applied to analyze ion dynamics in electrolytes [110], microfossils [111], and microplastics [112].

The silent region, where no endogenous biomolecules have the Raman signal, has recently been gaining considerable attention. By using engineered molecules that have Raman response in this region, we can realize the following two interesting applications. The first application is the monitoring of metabolic activity, where cells uptake small biomolecules, including glucose, fatty acids, amino acids and nucleotides, synthesize functional biomolecules including lipids, proteins, and nucleic acids by biochemical reactions, and digest them. By replacing hydrogen atoms in the small biomolecules with deuterium atoms (i.e., deuterium labeling), we can detect the biomolecules with SRS microscopy because they exhibit carbon-deuterium (CD) stretching vibrations at around 2200 cm^{-1} [13, 15, 113–116]. Furthermore, by adding deuterated glucose [13, 15] or deuterated water [116] to cells, deuterium atoms are incorporated to the inside of cells, and converted to various biomolecules, allowing us to monitor metabolic processes. In addition to deuterium labeling, isotopes such as ^{13}C can be used for metabolic analysis [117, 118]. It must be noted that it is difficult to label small biomolecules with fluorescent molecules, which are big in size and can alter the biochemical properties of biomolecules. Labeling of small biomolecules with alkyne (carbon–carbon triple bond) [119] is another interesting approach because alkynes have much sharper spectral response at ~2100 cm^{-1}, thereby making detection using SRS rather than deuterated molecules easier [14, 120–124]. The vibrational frequency of alkynes can be modulated by the existence of neighboring functional groups; this phenomenon was used to develop a Raman-based ratiometric probe to detect the concentration of ions [125].

Another application in the silent region is supermultiplex imaging, where various Raman-active molecules with distinct spectral peaks are used to label biomolecules, enabling 20-color imaging with SRS microscopy [126, 127]. The vibrational frequencies of the molecules were designed by chromophores attached with isotope-edited alkyne, nitrile, or azide groups, or by incorporating various lengths of polyynes with carbon isotopes. This is in contrast to fluorescence microscopy, where the spectral overlap of fluorescence limits the number of available colors to ~5. This technology will be advantageous in exploring the complex interactions between heterogeneous cells and biomolecules.

Figures 3.8(a)–(c) show the results of SRS imaging of a microalgal cell [79] and liver tissue [82] in the CH-stretching region. Figure 3.8(a) shows the SRS spectra of lipids, polysaccharides, proteins, and chlorophyll obtained from the hyperspectral SRS data of fixed *Euglena gracilis*. These spectra were used to decompose the hyperspectral SRS images into the SRS images of different constituents and the merged image shown in Figure 3.8(b). Figure 3.8(c) shows an SRS image of acetaminophen (APAP)-overdosed mouse liver tissue. A hyperspectral data cube was processed by principal component analysis, and some principal components were manually selected to produce a merged image. Cellular morphologies of cytoplasm (medium gray), nucleus (dark gray), and lipids (light gray) are clearly seen, indicating the loss of cellular morphologies and accumulation of lipids in the central area. Figure 3.8(d) and 3.8(e) illustrate two-color metabolic analysis of glucose uptake and glucose incorporation [128]. Figure 3.8(d) shows the Raman spectra of various glucose analogues, including deuterated glucose (dark gray) and 3-O-propargyl-D-glucose (3-OPG)-^{13}C$_3$ (black); the former is metabolically active, and the latter is metabolically inactive. Importantly, 3-OPG-$_{13}$C$_3$ has isotopically modified alkyne, whose vibrational frequency of 2053 cm^{-1} (black) is significantly shifted from 2129 cm^{-1} in normal 3-OPG (medium gray) to avoid spectral overlap of glucose and 3-OPG-^{13}C$_3$. Figure 3.8(e) shows SRS images of co-culture of human prostate cancer cells (PC-3) and normal human prostate cells (RWPE-1) treated with D7-glucose and 3-OPG-^{13}C$_3$, which indicate the lower density of D7-glucose in cancer cells and a comparable density of 3-OPG-^{13}C$_3$. This result suggests that both types of cells actively incorporate glucose, but the incorporated glucose is more actively digested in cancer cells. Raman-labeled molecules thus allow us to monitor metabolic activities with single-cell resolution in this way. Figure 3.8(f) shows the Raman spectra of Raman-labeling molecules based on polyynes (i.e., long chains of alkynes) [127], whose vibrational frequencies can be controlled by changing the chain length and/or by introducing stable carbon isotopes. These molecules can be attached to various types of organelles to realize supermultiplex imaging. Combined with fluorescent labeling, Hu et al. demonstrated ten-color imaging, as shown in Figure 3.8(g). This technology will be useful in analyzing complex intracellular and extracellular interactions.

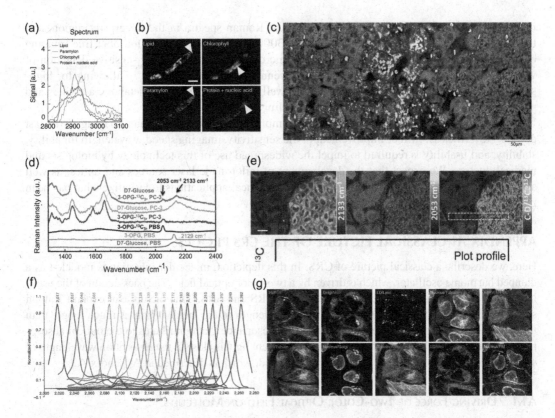

FIGURE 3.8 Selected applications of SRS microscopy. (a) SRS spectra of lipids, polysaccharide called paramylon, chlorophyll, and protein obtained from a hyperspectral SRS image of a microalgal cell of *Euglena gracilis*. (b) Decomposed SRS images of lipids, paramylon, chlorophyll, and protein of *Euglena gracilis*. Light gray arrows indicate the locations where SRS spectra in (a) were extracted. (c) Multicolor SRS image of APAP-overdosed mouse liver tissue. (d) and (e) metabolic analysis of cancer cells with alkyne-labeling and deuterium labeling. (c) (d) Raman spectra of various glucose analogues. (e) SRS images of the co-culture of PC-3 and RWPE-1. SRS images of glucose incorporation (C-D, light gray hot) and glucose uptake (13C-13C, dark gray hot), and ratiometric image (C≡D/13CR13C). (Reproduced from R. Long, L. Zhang, L. Shi, Y. Shen, F. Hu, C. Zeng, and W. Min, "Two-color vibrational imaging of glucose metabolism using stimulated Raman scattering," *Chem Commun (Camb)* **54**(2), 152–155 (2018).) (f) Raman spectra of engineered polyynes. (g) Supermultiplex imaging with polyyne labeling and fluorescent labeling. (Reproduced from F. Hu, C. Zeng, R. Long, Y. Miao, L. Wei, Q. Xu, and W. Min, "Supermultiplexed optical imaging and barcoding with engineered polyynes," *Nat Methods* **15**(3), 194–200 (2018).)

Most recently, several studies have been conducted that combine SRS imaging with other microscopy methods such as tissue clearing [129], endoscopy [130, 131], handy probe [132], and super-resolution [133]. Furthermore, it has been shown that molecular-selective fluorescence excitation is possible by stimulated Raman excited fluorescence (SREF) [134]. In SREF, fluorescent molecules are excited to a vibrationally excited state by two-color pulses, and then further excited to an electronically excited state. The resultant fluorescence provides an extreme sensitivity down to the single-molecule level. This technology allows for supermultiplexed imaging with unprecedented sensitivity.

3.7 SUMMARY

Coherent Raman scattering microscopy including CARS microscopy and SRS microscopy allows for sensitive detection of molecular-vibrational signature of cells and tissues, and widens its range of applications to label-free imaging, metabolic imaging, and ultramultiplex or supermultiplex imaging. CARS microscopy can obtain broadband Raman spectra in the fingerprint region, and the CH-stretching and OH-stretching regions (>3000 cm^{-1}), thus allowing for label-free imaging of cells and tissues with high specificity. SRS microscopy can be used for sensitive detection of vibrational signatures in a relatively narrow wavenumber region (~300 cm^{-1}), allowing for high-speed label-free imaging of cells and tissues, as well as the monitoring of metabolic activities and multiplexed imaging with Raman-detectable labeling molecules.

In this way, CRS microscopy has drastically improved the speed of Raman imaging in the last decade. Nevertheless, further improvement of the sensitivity, imaging speed, wavelength tunability, stability, and usability is required to impel the widespread use of this technology by biologists and medical doctors. We expect that the development of tailor-made laser sources and optics, as well as Raman-labeling molecules, will create novel biomedical applications to visualize what has been impossible to be visualized with previous technologies.

APPENDIX A1 CLASSICAL PICTURE OF THE CRS PROCESS

Here, we describe a classical picture of CRS. In this depiction, molecular vibration is modeled as a damped harmonic oscillator, which is driven by a two-color optical field. The back action of the oscillator leads to coherent Raman effects, including CARS and SRS. This picture can explain Raman scattering from transparent samples, where the wavelength of the incident light is much longer than the absorption wavelength; therefore, the electronic resonance that contributes to the optical absorption is not involved. When the wavelength of the incident light is close to the absorption wavelength of samples, the quantum-optical description must be used, which is described in Appendix A2.

A1.1   DRIVING FORCE OF TWO-COLOR OPTICAL FIELD ON MOLECULES

The driving force provided by the two-color optical field in the coherent Raman effect is explained below. Assume that the light interacts with a dielectric material of dielectric permittivity ε, and that the molecules that constitute the material have a vibrational resonance at a frequency ω_R. We additionally assume that ε depends on the vibrational coordinate q, which is the displacement of the internuclear distance of molecular bonds, i.e., $d\varepsilon/dq \neq 0$. In an isotropic material, the electric displacement D is related to the electric field E of light as follows:

$$D = \varepsilon E. \tag{A1.1}$$

The energy of the electric field of light U per unit volume is given by,

$$U = \frac{DE}{2} = \frac{D^2}{2\varepsilon}. \tag{A1.2}$$

When ε changes, the polarization density P of the material changes too, leading to a change in E, whereas D remains constant. According to the principle of virtual displacement [135], the force on the displacement is described as,

$$F = -\frac{dU}{dq} = -\frac{d}{dq}\frac{D^2}{2\varepsilon} = \frac{d\varepsilon}{dq}\frac{D^2}{2\varepsilon^2} = \frac{d\varepsilon}{dq}\frac{E^2}{2}. \tag{A1.3}$$

Therefore, the light can apply a force on the vibrational mode such that ε becomes larger. Furthermore, the force is proportional to the square of the electric field, i.e., the intensity of light. We can understand that the increase in ε originates from the increase in the polarizability α of the molecules that constitute the material, according to the Clausius–Mossotti relation given below [135],

$$\frac{\varepsilon - 1}{\varepsilon + 2} = \frac{N\alpha}{3}, \tag{A1.4}$$

where N is the density of molecules.

A1.2   Time-Domain and Frequency-Domain Responses of Dielectric Permittivity

The intensity of the two-color optical field oscillates in time at its difference frequency at $\omega_p - \omega_S$ (i.e., beating). When $d\varepsilon/dq \neq 0$, the two-color optical field can drive the vibrational mode so that q and ε oscillates in time. However, the amplitude and the phase of the oscillation of ε depend on the difference frequency $\omega_p - \omega_S$.

Because the frequency response of a linear system is given by the Fourier transform of its impulse response, let us discuss the impulse response of ε to the optical intensity. To this extent, let us assume that, instead of a two-color optical field, an ultrashort optical pulse interacts with the material at a time $t=0$. The time-domain response of the change in ε can be expressed by,

$$\Delta\varepsilon(t) = \Delta\varepsilon_R e^{-\Gamma t} \sin\omega_R t + \Delta\varepsilon_{NR}\delta(t) \quad (t > 0), \tag{A1.5}$$

where $\Delta\varepsilon_R$ and $\Delta\varepsilon_{NR}$ are the real values that denote the amplitudes of vibrational response and non-resonant electronic response, respectively, and Γ is the decay constant of the molecular vibration. Figure 3.9(a) shows the waveform of Eq. (A1.5). This equation considers the following effects: the pulse can apply an impulsive force on molecules to increase ε. Because of the inertia of molecules,

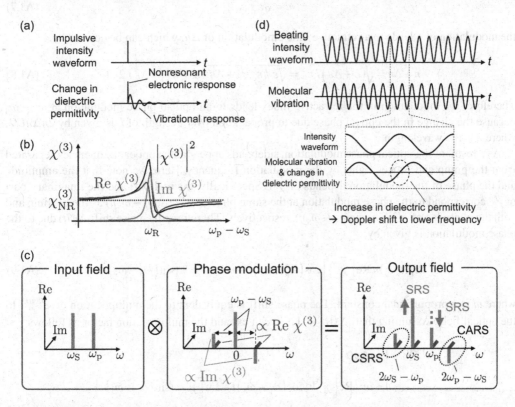

FIGURE 3.9   Classical picture of coherent Raman scattering. (a) Change in dielectric permittivity induced by impulsive optical pulses. (b) Fourier transform of (a). (c) Beating of two-color light drives molecular vibrations at the same frequency as the beat, whereas its phase lags because of vibrational resonance. (d) Change in optical spectrum as a result of the phase modulation caused by the change in dielectric permittivity. The phase modulation spectrum is given by the Fourier transform of Eq. (A1.10).

ε does not increase instantaneously, but starts oscillating at ω_R, and the oscillation decays at a rate of Γ. Furthermore, the Kerr effect, which originates from the nonlinear response of electrons, leads to an impulsive change in ε. Performing a Fourier transform of Eq. (A1.5), we obtain,

$$\Delta\varepsilon(\omega) = \int_{-\infty}^{\infty} \Delta\varepsilon(t)e^{i\omega t}dt = \frac{\Delta\varepsilon_R/2}{\omega_R - \omega - i\Gamma} + \frac{\Delta\varepsilon_R/2}{\omega_R + \omega + i\Gamma} + \Delta\varepsilon_{NR}. \tag{A1.6}$$

In the frequency domain, we consider the intensity waveform in time oscillating at a specific frequency at $\omega_p - \omega_S$. Because the third-order nonlinear susceptibility $\chi^{(3)}$ is the proportional constant that relates the oscillation in the intensity and the change in the dielectric permittivity (i.e., $\chi^{(3)}(\omega_p - \omega_S) \propto \Delta\varepsilon(\omega_p - \omega_S)$), we can plot $\chi^{(3)}$ as a function of $\omega_p - \omega_S$, as shown in Figure 3.9(b). We can see various characteristic features of $\chi^{(3)}$ in Eq. (A1.6) and Figure 3.9(b): $\text{Im}[\chi^{(3)}]$ has a single peak only in the vicinity of ω_R because of the first term in Eq. (A1.6). In the vicinity of ω_R, $\text{Re}[\chi^{(3)}]$ has a dispersive shape because of the first term in Eq. (A1.6) and an offset caused by the third term in Eq. (A1.6).

A1.3 CARS AND SRS AS OPTICAL PHASE MODULATION

In the previous sections, we described that ε can be modulated by a two-color optical field. Because ε is related to the refractive index n by,

$$n = \sqrt{\varepsilon/\varepsilon_0}, \tag{A1.7}$$

the modulation of ε leads to a refractive index modulation of Δn, which can be derived by,

$$n + \Delta n = \sqrt{(\varepsilon + \Delta\varepsilon)/\varepsilon_0} = \sqrt{\varepsilon/\varepsilon_0}\sqrt{1 + \Delta\varepsilon/\varepsilon} = n(1 + \Delta\varepsilon/2\varepsilon). \tag{A1.8}$$

Therefore, the modulation in the refractive index leads to an optical phase modulation at $\omega_p - \omega_S$ because the change in the optical phase due to propagation over a length of L is given by $2\pi\Delta nL/\lambda$, where λ is the wavelength.

As a result of the optical phase modulation, sidebands appear at the optical frequencies separated from the pump and Stokes lights by the modulation frequency. Here, we note that the amplitude and the phase of the modulation depends on $\chi^{(3)}$. Specifically, the real part and the imaginary part of $\chi^{(3)}$ correspond to the phase modulation at the same phase (i.e., $\cos(\omega_p - \omega_S)t$) as the beating, and with the phase lag of $\pi/2$ (i.e., $\sin(\omega_p - \omega_S)t$), respectively. Therefore, the phase shift $\Delta\phi(t)$ due to the phase modulation is given by,

$$\Delta\phi(t) = m\,\text{Re}\left[\chi^{(3)}\right]\cos\left(\omega_p - \omega_S\right)t + m\,\text{Im}\left[\chi^{(3)}\right]\sin\left(\omega_p - \omega_S\right)t, \tag{A1.9}$$

where m is a proportional constant. The phase shift is equivalent to the multiplication of $e^{i\Delta\phi(t)}$ to the optical field. Assuming that $|\Delta\phi(t)| \ll 1$, we can expand the multiplication factor as follows:

$$\begin{aligned}
e^{i\Delta\phi(t)} &\sim 1 + i\Delta\phi(t) \\
&= 1 + im\left\{\text{Re}\left[\chi^{(3)}\right]\cos\left(\omega_p - \omega_S\right)t + \text{Im}\left[\chi^{(3)}\right]\sin\left(\omega_p - \omega_S\right)t\right\} \\
&= 1 + \frac{m}{2}\left\{-\text{Im}\left[\chi^{(3)}\right] + i\text{Re}\left[\chi^{(3)}\right]\right\}e^{-i(\omega_p - \omega_S)t} \\
&\quad + \frac{m}{2}\left\{\text{Im}\left[\chi^{(3)}\right] + i\text{Re}\left[\chi^{(3)}\right]\right\}e^{i(\omega_p - \omega_S)t}
\end{aligned} \tag{A1.10}$$

The last line of Eq. (A1.10) comprises a zero frequency component, positive frequency component, and negative frequency component. The last two components are responsible for the generation of the upper sideband and the lower sideband, respectively, generated by the phase modulation. Specifically, the upper sideband is proportional to $-\text{Im}\left[\chi^{(3)}\right] + i\text{Re}\left[\chi^{(3)}\right] = i\chi^{(3)}$, and the lower sideband is proportional to $\text{Im}\left[\chi^{(3)}\right] + i\text{Re}\left[\chi^{(3)}\right] = i\chi^{(3)*}$.

Figure 3.9(c) shows the spectral change after the phase modulation. The lower sideband of the pump light appears at ω_S, which interferes with the Stokes light. The upper sideband of the Stokes light appears at ω_p, which interferes with the pump light. These result in SRS, i.e., amplification of the Stokes light and attenuation of the pump light, which are proportional to $\text{Im}[\chi^{(3)}]$. The upper sideband of pump light appears at $2\omega_p - \omega_S$, which is known as coherent anti-Stokes Raman scattering (CARS). The lower sideband of Stokes light appears at $2\omega_S - \omega_p$, which is known as coherent Stokes Raman scattering (CSRS).

For SRS to occur, $\chi^{(3)}$ must essentially have an imaginary component, i.e., the molecular vibration is affected by vibrational resonance. According to the theory of the harmonic oscillator, when an oscillator is driven at its resonant frequency, the phase of the oscillator lags by $\pi/2$ with respect to the force. Similarly, when $\omega_p = \omega_S - \omega_R$, the molecular vibration driven by the two-color intensity beat lags by $\pi/2$. Therefore, in addition, the oscillation of ε and hence the modulation phase of optical phase modulation lag, which affects the phases of the sidebands. We can apparently observe the consequence of vibrational resonance in the time domain, as shown in Figure 3.9(b). Due to the phase lag in the molecular vibration, the refractive index increases with time, when the beat has maximum intensity. This reduces the average optical frequency through the Doppler effect, resulting in SRS effects.

A1.4 Comparison of Signal-to-Noise Ratios of CARS and SRS

Here, we explain that the signal-to-noise ratios of SRS and CARS are comparable. Many studies have demonstrated high-speed imaging with CARS and SRS at rates up to the video-rate. The equivalence in the SNR is intuitively understood as follows. When the same pump and Stokes pulses are used, CARS and SRS appear as sidebands created by optical phase modulation induced by molecular vibration; therefore, their amplitudes are almost the same. Quantum optics indicates that the quantum-limited SNR is given by the ratio of the electric field and vacuum fluctuation (with an energy of $\hbar\omega/2$). Therefore, the quantum-limited SNRs of SRS and CARS are almost the same, whereas various technical factors affect SNR.

A1.5 Number of Photons of CARS and SRS

Let us estimate the number of photons appearing in the CRS process, as shown in Figure 3.1. We denote the number of photons of the pump pulse as N_p, and that of spontaneous Raman scattering generated in the same bandwidth as the Stokes pulse as N_R. Considering that N_R is proportional to N_p, we can relate them as $N_R = \sigma N_p$, where σ is the probability of Raman scattering. Quantum mechanics indicates that when the number of photons of Stokes pulses is N_S, the probability of Raman scattering is accelerated by N_S. Therefore, the number of SRS photons ΔN_{SRS} is given by $\Delta N_{SRS} = \sigma N_p N_S$.

By considering the phase modulation sidebands that give rise to SRS, we can further estimate the number of photons of CARS and CSRS. Hereafter, let us express the amplitude as the square root of the number of photons; the pump and the Stokes fields are $\sqrt{N_p}$ and $\sqrt{N_S}$, respectively. Due to SRL, the Stokes field becomes $\sqrt{N_p - \Delta N_{SRS}} = \sqrt{N_p}\sqrt{1 - \sigma N_S} \sim \sqrt{N_p} - \sigma N_S\sqrt{N_p}/2$, where the last term is the upper sideband of Stokes field. Thus, the phase modulation amplitude is derived as

$\sigma N_S \sqrt{N_p} / 2\sqrt{N_S} = \sigma\sqrt{N_p N_S} / 2$. Because CARS is the upper sideband of the pump, its amplitude is given by $\sqrt{N_p}\sigma\sqrt{N_p N_S} / 2$, which corresponds to the number of photons of $\sigma^2 N_p^2 N_S / 4$. Similarly, the amplitude of CSRS is given by $\sqrt{N_S}\sigma\sqrt{N_p N_S} / 2$, which corresponds to the number of photons of $\sigma^2 N_p N_S^2 / 4$.

APPENDIX A2 QUANTUM-MECHANICAL PICTURE OF CRS

To derive a quantum-mechanical picture of the CRS processes, we introduce the density matrix $\rho(t)$, and describe the interaction between light and molecules [136]. The density matrix can address mixed states, i.e., an ensemble of molecules or statistics of quantum states. Specifically, the diagonal elements of the density matrix provide populations (probability) of the corresponding eigenstate and the off-diagonal elements provide the coherence (or degree of superposition) of different eigenstates. The density matrix obeys the Liouville variant (quantum Liouville equation) of the Schrödinger equation as follows:

$$\frac{\partial \rho}{\partial t} = \frac{1}{i\hbar}\left[H(t), \rho\right], \tag{A2.1}$$

where $H(t)$ is the total Hamiltonian of the system. It is convenient to separate the total Hamiltonian of the system into the following two contributions:

$$H(t) = H_0 + \lambda H_1(t), \tag{A2.2}$$

Here, the Hamiltonian H_0 describes the isolated molecular systems; it specifies the zero-th order energy eigenstates. The interactions between light and molecules are described by the interaction Hamiltonian $H_1(t)$. The dimensionless parameter λ is used to keep track of the order of the perturbation.

We then introduce the interaction picture of the density matrix $\rho'(t)$ and the interaction Hamiltonian $H_1(t)$, given by,

$$\rho' = e^{+iH_0 t/\hbar}\rho e^{-iH_0 t/\hbar} \tag{A2.3}$$

$$H_1'(t) = e^{+iH_0 t/\hbar}H_1(t)e^{-iH_0 t/\hbar}. \tag{A2.4}$$

Substituting Eq. (A2.2), (A2.3), and (A2.4) into Eq. (A2.1), the quantum Liouville equation in the interaction picture is obtained as follows:

$$\frac{\partial \rho'}{\partial t} = \frac{1}{i\hbar}\left[\lambda H_1'(t), \rho'\right]. \tag{A2.5}$$

Here ρ' is expanded around $t = t_0$ as follows:

$$\rho'(t) = \rho'^{(0)}(t_0) + \lambda\rho'^{(1)}(t) + \lambda^2\rho'^{(2)}(t) + \lambda^3\rho'^{(3)}(t) + \cdots. \tag{A2.6}$$

Substituting (A2.6) into Eq. (A2.5), and comparing the n-th order of λ on both sides, the following equation is obtained:

$$\frac{\partial \rho'^{(n)}}{\partial t} = \frac{1}{i\hbar}\left[H_1'(t), \rho'^{(n-1)} \right]. \tag{A2.7}$$

Equation (A2.7) can be formally rewritten in the integral form, indicated as follows for up to $n=3$.

$$\rho'^{(1)}(t) = \frac{1}{i\hbar}\int_{t_0}^{t} dt_1 \left[H_1'(t_1), \rho'^{(0)}(t_0) \right] \tag{A2.8}$$

$$\rho'^{(2)}(t) = \left(\frac{1}{i\hbar}\right)^2 \int_{t_0}^{t} dt_2 \int_{t_0}^{t} dt_1 \left[H_1'(t_2), \left[H_1'(t_1), \rho'^{(0)}(t_0) \right] \right] \tag{A2.9}$$

$$\rho'^{(3)}(t) = \left(\frac{1}{i\hbar}\right)^3 \int_{t_0}^{t} dt_3 \int_{t_0}^{t_3} dt_2 \int_{t_0}^{t_2} dt_1 \left[H_1'(t_3), \left[H_1'(t_2), \left[H_1'(t_1), \rho'^{(0)}(t_0) \right] \right] \right] \tag{A2.10}$$

Although Eq. (A2.10) provides eight terms in total, only two terms contribute to the coherent Raman processes as follows:

$H_1'(t_3)H_1'(t_2)H_1'(t_1)\rho'^{(0)}(t_0)$: corresponds to CARS and SRL

$H_1'(t_3)\rho'^{(0)}(t_0)H_1'(t_1)H_1'(t_2)$: corresponds to CSRS and SRG

This point can be understood as follows: $H_I'(t_2)H_I'(t_1)\rho'^{(0)}$ and $\rho'^{(0)}H_I'(t_2)H_I'(t_1)$ indicate that the molecules interacting with two-color light have a vibrational coherence (i.e., superposition of the vibrational ground state and the vibrational excited state). Moreover, the former and the latter correspond to positive and negative vibrational frequencies, respectively, which changes the optical frequency positively and negatively with respect to the third light. Figure 3.10 summarizes the energy-level diagrams of four CRS processes, namely, CARS, SRL, SRG, and CSRS.

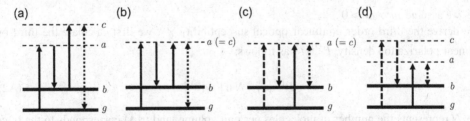

FIGURE 3.10 Energy diagrams of four coherent Raman processes. (a) CARS, (b) SRL, (c) SRG, and (d) CSRS. The solid and the broken arrows represent the interaction of the radiation field with molecules. In detail, the solid (upward facing) and the broken (downward facing) arrows correspond to the radiation at positive frequencies with the phase factor of $e^{-i\omega t}$. Similarly, the solid (downward facing) and broken (upward facing) arrows correspond to the radiation at negative frequencies with the phase factor of $e^{i\omega t}$. The arrows facing up and down represent the resulting third-order nonlinear polarization.

Here, we focus on CARS and SRL, both of which are frequently detected in CRS microscopy. The third-order perturbation term, $\rho^{(3)}(t)$ can be expanded as follows:

$$\rho(t) = \rho^{(0)}(t_0) + \lambda\rho^{(1)}(t) + \lambda^2\rho^{(2)}(t) + \lambda^3\rho^{(3)}(t) + \cdots. \tag{A2.11}$$

Using the interaction picture of the density matrix, $\rho'^{(3)}(t)$ in Eq. (A2.10), $\rho^{(3)}(t)$ can be derived as,

$$\rho^{(3)}(t) = \left(\frac{1}{i\hbar}\right)^3 \int_{t_0}^{t} dt_3 \int_{t_0}^{t_3} dt_2 \int_{t_0}^{t_2} dt_2\, e^{-iH_0(t-t_3)/\hbar}\, H_1(t_3)e^{-iH_0(t_3-t_2)/\hbar} \tag{A2.12}$$

$$\times H_1(t_2)e^{-iH_0(t_2-t_1)/\hbar}\, H_1(t_1)e^{-iH_0(t_1-t_0)/\hbar}\, \rho^{(0)}(t_0)e^{+iH_0(t-t_0)/\hbar}.$$

Here, we assume that the system is in thermal equilibrium before the perturbation. Substituting $\rho^{(0)}(t_0) = \rho^{(0)}(-\infty)$ into Eq. (A2.12), Eq. (A2.12) can be rewritten as,

$$\rho^{(3)}(t) = \left(\frac{1}{i\hbar}\right)^3 \int_{-\infty}^{t} dt_3 \int_{-\infty}^{t_3} dt_2 \int_{-\infty}^{t_2} dt_1\, e^{-iH_0(t-t_3)/\hbar}\, H_1(t_3)e^{-iH_0(t_3-t_2)/\hbar} \tag{A2.13}$$

$$\times H_1(t_2)e^{-iH_0(t_2-t_1)/\hbar}\, H_1(t_1)e^{-iH_0t_1/\hbar}\, \rho^{(0)}(-\infty)e^{+iH_0t/\hbar}.$$

The interaction Hamiltonian can be described as,

$$H_I(t) = -\mu E(t, \mathbf{r}). \tag{A2.14}$$

Here, μ and $E(t, \mathbf{r})$ represent dipole moment of a molecule and optical electric field, respectively. In the coherent Raman processes, two optical electric fields with different frequencies $E(t, \mathbf{r})$ can be written as,

$$E(t,\mathbf{r}) \equiv E_1(t,\mathbf{r}) + E_2(t,\mathbf{r}) = E_1\left(e^{-i(\omega_1 t - \mathbf{k}_1\mathbf{r})} + e^{+i(\omega_1 t - \mathbf{k}_1\mathbf{r})}\right) \tag{A2.15}$$

$$+ E_2\left(e^{-i(\omega_2 t - \mathbf{k}_2\mathbf{r})} + e^{+i(\omega_2 t - \mathbf{k}_2\mathbf{r})}\right).$$

Here, we assume $\omega_1 > \omega_2 > 0$.

To derive the third-order nonlinear optical susceptibility $\chi^{(3)}$, we first calculate the third-order nonlinear polarization density $P^{(3)}(t)$ as follows:

$$P^{(3)}(t) = N\,\mathrm{tr}\left[\mu\rho^{(3)}\right]. \tag{A2.16}$$

Here, N represents the number of molecules per unit volume, and tr[$\mathbf{A}$] corresponds to the trace of matrix $\mathbf{A}$. Substituting Eq. (A2.13), (A2.14), and (A2.15) into Eq. (A2.16), we obtain,

$$P^{(3)}(t) = N\left(\frac{i}{\hbar}\right)^3 \int_{-\infty}^{t} dt_3 \int_{-\infty}^{t_3} dt_2 \int_{-\infty}^{t_2} dt_1\,\mathrm{tr}\left[\mu e^{-iH_0(t-t_3)/\hbar}\, \mu e^{-iH_0(t_3-t_2)/\hbar}\right. \tag{A2.17}$$

$$\left.\times \mu e^{-iH_0(t_2-t_1)/\hbar}\, \mu e^{-iH_0t_1/\hbar}\, \rho^{(0)}(-\infty)e^{+iH_0t/\hbar}\right]E(t_3)E(t_2)E(t_1).$$

For simplicity, we assume that all molecules before the perturbation ($\rho^{(0)}(-\infty)$) stay at electronic and vibrational ground states. We denote the intermediate states as a, b, and c, and Eq. (A2.17) can be rewritten as follows:

$$P^{(3)}(t) = N\left(\frac{i}{\hbar}\right)^3 \int_{-\infty}^{t} dt_3 \int_{-\infty}^{t_3} dt_2 \int_{-\infty}^{t_2} dt_1 \mu_{gc}\, e^{-i\omega_c(t-t_3)}\, \mu_{cb}\, e^{-i\omega_b(t_3-t_2)}\, \mu_{ba}\, e^{-i\omega_a(t_2-t_1)}$$

$$\times \mu_{ag}\, e^{-i\omega_g t_1}\, e^{+i\omega_g t}\, E(t_3)E(t_2)E(t_1)$$

$$= N\left(\frac{i}{\hbar}\right)^3 \mu_{gc}\mu_{cb}\mu_{ba}\mu_{ag}\, e^{-i\omega_{cg}t} \int_{-\infty}^{t} dt_3\, e^{+i(\omega_{cg}-\omega_{bg})t_3}\, E(t_3) \tag{A2.18}$$

$$\times \int_{-\infty}^{t_3} dt_2\, e^{+i(\omega_{bg}-\omega_{ag})t_2}\, E(t_2) \int_{-\infty}^{t_2} dt_1\, e^{+i\omega_{ag}t_1}\, E(t_1)$$

where $\omega_x\,(x=g,a,b,c)$ denotes the eigen angular frequency of the corresponding eigenstate of molecules, which is related to the eigenenergy ε_x by $\omega_x = \varepsilon_x/\hbar$, and $\mu_{pq}\,(p,q=g,a,b,c)$ corresponds to eigen transition dipole moments between eigenstates. Here, we define $\omega_{pq} = \omega_p - \omega_q\,(p,q=g,a,b,c)$.

To consider the relaxation of eigenstates, we replace the eigen angular frequency ω_{pq} with complex eigen angular frequency, $\tilde{\omega}_{pq} \equiv \omega_{pq} - i\Gamma_{pq}(\equiv \omega_p - \omega_q - i\Gamma_{pq})$, giving rise to,

$$P^{(3)}(t) = N\left(\frac{i}{\hbar}\right)^3 \mu_{gc}\mu_{cb}\mu_{ba}\mu_{ag}\, e^{-i(\omega_c-\omega_g-i\Gamma_{cg})t} \int_{-\infty}^{t} dt_3\, e^{+i(\omega_c-\omega_b-i(\Gamma_{cg}-\Gamma_{bg}))t_3}\, E(t_3)$$

$$\times \int_{-\infty}^{t_3} dt_2\, e^{+i(\omega_b-\omega_a-i(\Gamma_{bg}-\Gamma_{ag}))t_2}\, E(t_2) \int_{-\infty}^{t_2} dt_1\, e^{+i(\omega_a-\omega_g-i\Gamma_{ag})t_1}\, E(t_1). \tag{A2.19}$$

Hereafter, we discuss (I) CARS and (II) SRL.

(I) CARS

In the case of CARS, $E(t_3)E(t_2)E(t_1)$ becomes $E_1 e^{-i(\omega_1 t_3 - \mathbf{k}_1 \mathbf{r})} E_2 e^{+i(\omega_2 t_2 - \mathbf{k}_2 \mathbf{r})} E_1 e^{-i(\omega_1 t_1 - \mathbf{k}_1 \mathbf{r})}$ (see Figure 3.10(a)), resulting in,

$$P_{\text{CARS}}^{(3)}(t) = N\left(\frac{i}{\hbar}\right)^3 \mu_{gc}\mu_{cb}\mu_{ba}\mu_{ag}\, e^{-i(\omega_c-\omega_g-i\Gamma_{cg})t} \int_{-\infty}^{t} dt_3\, e^{+i(\omega_c-\omega_b-\omega_1-i(\Gamma_{cg}-\Gamma_{bg}))t_3}$$

$$\times \int_{-\infty}^{t_3} dt_2\, e^{+i(\omega_b-\omega_a+\omega_2-i(\Gamma_{bg}-\Gamma_{ag}))t_2} \int_{-\infty}^{t_2} dt_1\, e^{+i(\omega_a-\omega_g-\omega_1-i\Gamma_{ag})t_1}\, e^{i(2\mathbf{k}_1-\mathbf{k}_2)\mathbf{r}}\, E_1 E_2 E_1. \tag{A2.20}$$

Eq. (A2.20) can be calculated directly, and we can finally obtain the following equation:

$$P_{\text{CARS}}^{(3)}(t) = \frac{N}{\hbar^3} \mu_{gc}\mu_{cb}\mu_{ba}\mu_{ag}\, \frac{1}{\omega_c-\omega_g-(2\omega_1-\omega_2)-i\Gamma_{cg}}\, \frac{1}{\omega_b-\omega_g-(\omega_1-\omega_2)-i\Gamma_{bg}}$$

$$\times \frac{1}{\omega_a-\omega_g-\omega_1-i\Gamma_{ag}}\, e^{-i(2\omega_1-\omega_2)t+i(2\mathbf{k}_1-\mathbf{k}_2)\mathbf{r}}\, E_1 E_2 E_1 \tag{A2.21}$$

Eq. (A2.21) contains three complex Lorentzian functions. The first, second, and third factors correspond to $\omega_{\text{CARS}}\left(\equiv 2\omega_1 - \omega_2\right)$ resonance, vibrational ($\omega_1 - \omega_2$) resonance, and ω_1 resonance, respectively. The first and the third factors contribute to electronic resonance enhancement of the CARS signal (so-called resonance CARS).

According to the exponential factor $e^{-i(2\omega_1-\omega_2)t+i(2\mathbf{k}_1-\mathbf{k}_2)\mathbf{r}}$ in Eq. (A2.21), the third-order nonlinear polarization $P_{\text{CARS}}{}^{(3)}(t)$ has the angular frequency of $\omega_{\text{CARS}} \equiv 2\omega_1 - \omega_2$ and the wave vector of $2\mathbf{k}_1 - \mathbf{k}_2$. In dispersive materials, the relationship such as $\omega = c|\mathbf{k}|/n$ holds, where n represents the refractive index. Therefore, if the absolute value of the wave vector $\mathbf{k}_{\text{CARS}}$, $|\mathbf{k}_{\text{CARS}}|\left(\equiv n\omega_{\text{CARS}}/c\right)$, coincides with that of $P_{\text{CARS}}{}^{(3)}(t)$ ($2\mathbf{k}_1 - \mathbf{k}_2$), i.e., if $|\mathbf{k}_{\text{CARS}}| = |2\mathbf{k}_1 - \mathbf{k}_2|$ is satisfied, $P_{\text{CARS}}{}^{(3)}(t)$ at each point interferes constructively, giving rise to intense and uni-directional radiation, which is known as the phase-matching condition. As a result, CARS radiation with the angular frequency of $\omega_{\text{CARS}} \equiv 2\omega_1 - \omega_2$ and the wavevector of $\mathbf{k}_{\text{CARS}} = 2\mathbf{k}_1 - \mathbf{k}_2$ is generated. The intensity of the CARS signal I_{CARS} is proportional to the absolute square of Eq. (A2.21): $I_{\text{CARS}} \propto \left|P_{\text{CARS}}{}^{(3)}(t)\right|^2$.

Using Eq. (A2.21) and the definition of the third-order nonlinear optical susceptibility, $\chi^{(3)}$ ($P_{\text{CARS}}{}^{(3)}(t) = \varepsilon_0\chi_{\text{CARS}}^{(3)}E_1E_2E_1$), $\chi_{\text{CARS}}^{(3)}$ is finally obtained as follows:

$$\chi_{\text{CARS}}^{(3)} = \frac{N}{\varepsilon_0\hbar^3}\mu_{\text{gc}}\mu_{\text{cb}}\mu_{\text{ba}}\mu_{\text{ag}}\frac{1}{\omega_c - \omega_g - \left(2\omega_1 - \omega_2\right) - i\Gamma_{\text{cg}}}$$

$$\times\frac{1}{\omega_b - \omega_g - \left(\omega_1 - \omega_2\right) - i\Gamma_{\text{bg}}}\frac{1}{\omega_a - \omega_g - \omega_1 - i\Gamma_{\text{ag}}} \tag{A2.22}$$

$$\propto\frac{1}{\omega_c - \omega_g - \left(2\omega_1 - \omega_2\right) - i\Gamma_{\text{cg}}}.$$

Here, we assumed that all the electronic states are far-off resonance, by which only the vibrational ($\omega_1 - \omega_2$) resonance contributes to the third-order nonlinear optical susceptibility, $\chi^{(3)}$.

As discussed, the third-order nonlinear optical susceptibility due to the CARS process with vibrational resonance has been derived. Among the third-order nonlinear optical processes, however, other processes with no vibrational resonance can occur. This is termed as nonresonant background (NRB), which means that no vibrational resonance contributes to this process. One of the third-order nonlinear optical susceptibilities due to NRB, $\chi_{\text{NRB}}^{(3)}$, is described as follows:

$$\chi_{\text{NRB}}^{(3)} = \frac{N}{\varepsilon_0\hbar^3}\mu_{\text{gc}}\mu_{\text{cb}}\mu_{\text{ba}}\mu_{\text{ag}}\frac{1}{\omega_c - \omega_g - \left(2\omega_1 - \omega_2\right) - i\Gamma_{\text{cg}}}$$

$$\times\frac{1}{\omega_b - \omega_g - 2\omega_1 - i\Gamma_{\text{bg}}}\frac{1}{\omega_a - \omega_g - \omega_1 - i\Gamma_{\text{ag}}} \tag{A2.23}$$

$$\propto C$$

Here, we assumed that all the electronic states are far-off resonance, by which $\chi_{\text{NRB}}^{(3)}$ contains no resonance term. As a result, the third-order nonlinear optical susceptibilities due to NRB, $\chi_{\text{NRB}}^{(3)}$ can be described simply by the real number C.

Because the angular frequency and the wave vector of NRB is identical with those of CARS, these radiations interfere with each other. As a result, the optical electric field to be observed can be written as,

$$E_S \propto i\left(\chi_{\text{CARS}}^{(3)} + \chi_{\text{NRB}}^{(3)}\right)E_1E_2E_1. \tag{A2.24}$$

The intensity of the total signal radiation, $I_S \equiv |E_S|^2$, is thus described as follows:

$$I_S \propto \left| \chi_{CARS}^{(3)} + \chi_{NRB}^{(3)} \right|^2 I_1^2 I_2. \tag{A2.25}$$

Here, $I_1 \equiv |E_1|^2$, $I_2 \equiv |E_2|^2$.

(II) SRL

In the case of SRL, $E(t_3)E(t_2)E(t_1)$ becomes $E(t_3)E(t_2)E(t_1) = E_2 e^{-i(\omega_2 t_3 - \mathbf{k}_2 \mathbf{r})} E_2 e^{+i(\omega_2 t_2 - \mathbf{k}_2 \mathbf{r})} E_1 e^{-i(\omega_1 t_1 - \mathbf{k}_1 \mathbf{r})}$ (see Figure 3.10(b)); substituting it into Eq. (A2.19) results in,

$$P_{SRL}^{(3)}(t) = N \left(\frac{i}{\hbar} \right)^3 \mu_{gc} \mu_{ab} \mu_{ba} \mu_{ag} \, e^{-i(\omega_c - \omega_g - i\Gamma_{cg})t} \int_{-\infty}^{t} dt_3 \, e^{+i(\omega_c - \omega_b - \omega_1 - i(\Gamma_{cg} - \Gamma_{bg}))t_3}$$

$$\tag{A2.26}$$

$$\times \int_{-\infty}^{t_3} dt_2 \, e^{+i(\omega_b - \omega_a + \omega_2 - i(\Gamma_{bg} - \Gamma_{ag}))t_2} \int_{-\infty}^{t_2} dt_1 \, e^{+i(\omega_a - \omega_g - \omega_1 - i\Gamma_{ag})t_1} \, e^{i\mathbf{k}_1 \mathbf{r}} \, E_2 E_2 E_1.$$

Similar to that for CARS, it can be calculated as follows:

$$P_{SRL}^{(3)}(t) = \frac{N}{\hbar^3} \mu_{gc} \mu_{cb} \mu_{ba} \mu_{ag} \frac{1}{\omega_c - \omega_g - \omega_1 - i\Gamma_{cg}} \frac{1}{\omega_b - \omega_g - (\omega_1 - \omega_2) - i\Gamma_{bg}}$$

$$\tag{A2.27}$$

$$\times \frac{1}{\omega_a - \omega_g - \omega_1 - i\Gamma_{ag}} e^{-i\omega_1 t + i\mathbf{k}_1 \mathbf{r}} \, E_2 E_2 E_1.$$

The nonlinear polarization due to the SRL process, $P_{SRL}^{(3)}$, has an identical angular frequency and wave vector with one of the incident optical electric field, E_1. The nonlinear polarization, $P_{SRL}^{(3)}$ emits the signal radiation, E_S, which is expressed as follows [135, 137]:

$$E_S \propto i P_{SRL}^{(3)}. \tag{A2.28}$$

Eq. (A2.28) indicates that the electric field originating from the oscillating polarization aligned in a sheet lags by $\pi/2$ with respect to the polarization. Because the SRL automatically satisfies the phase-matching condition ($|\mathbf{k}_{SRL}| = |\mathbf{k}_1|$), the signal radiation E_S propagates in the same direction as the incident radiation E_1. Assuming $|E_1|^2 \equiv I_1 \gg |E_S|^2$, the intensity I to be observed can be expressed as follows:

$$I = |E_1 + E_S|^2$$

$$\simeq I_1 + 2\,\mathrm{Re}\left[E_1^* E_S \right] \tag{A2.29}$$

$$= I_1 - 2\,\mathrm{Im}\left[E_1^* P_{SRL}^{(3)} \right].$$

Here, E_1^* represents the complex conjugate of E_1. Using the intensity difference ΔI due to SRL, the change in transmittance $\Delta I / I_1$ can be expressed as follows:

$$\Delta I/I_1 = \left(I - I_1\right)/I_1$$

$$= -2\,\mathrm{Im}\left[E_1^* P^{(3)}\right]/I_1 \qquad (A2.30)$$

$$= -2\,\mathrm{Im}\left[\frac{P^{(3)}}{E_1}\right] \propto -\mathrm{Im}\left[\chi_{\mathrm{SRL}}^{(3)}\right] I_2.$$

Here, the nonlinear optical susceptibility due to SRL, $\chi_{\mathrm{SRL}}{}^{(3)}$, is defined as,

$$\chi_{\mathrm{SRL}}{}^{(3)} = \frac{N}{\varepsilon_0 \hbar^3}\,\mu_{\mathrm{gc}}\mu_{\mathrm{cb}}\mu_{\mathrm{ba}}\mu_{\mathrm{ag}}\,\frac{1}{\omega_c - \omega_g - \omega_1 - i\Gamma_{\mathrm{cg}}}$$

$$\times \frac{1}{\omega_b - \omega_g - \left(\omega_1 - \omega_2\right) - i\Gamma_{\mathrm{bg}}}\,\frac{1}{\omega_a - \omega_g - \omega_1 - i\Gamma_{\mathrm{ag}}} \qquad (A2.31)$$

$$\propto \frac{1}{\omega_b - \omega_g - \left(\omega_1 - \omega_2\right) - i\Gamma_{\mathrm{bg}}}.$$

Here, we used the same assumption as in Eq. (A.2.22). According to Eq. (A2.30), $\Delta I/I_1$ is proportional to the imaginary part of $\chi_{\mathrm{SRL}}{}^{(3)}$ ($\mathrm{Im}\left[\chi_{\mathrm{SRL}}^{(3)}\right]$) and the intensity of E_2 (I_2). Moreover, the phase of the signal radiation E_{S} is π-phase shifted with that of the incident radiation E_1, giving rise to the intensity loss, i.e., stimulated Raman loss (SRL).

APPENDIX A3 IMAGE FORMATION IN CRS MICROSCOPY

A3.1 FORMULATION OF THE IMAGE-FORMING SYSTEM

In Appendix A1 and A2, we discussed the generation of CRS signals by implicitly assuming that the excitation beams are plane waves and the light-matter interaction occurs in a homogeneous medium. As discussed in Appendix A2, the nonlinear susceptibility $\chi_{\mathrm{CRS}}^{(3)}$ is a complex number that has different values depending on the kind of CRS; specifically, the sign of $\mathrm{Im}\,\chi_{\mathrm{CRS}}^{(3)}$ is positive in CARS and SRL (see Appendix A2), and negative in CSRS and SRG. The CRS signal field E_{sig} is proportional to the polarization density $P^{(3)} = \varepsilon_0 \chi_{\mathrm{CRS}}^{(3)}\left[E_{\mathrm{ex}}^3\right]$, where the product of three excitation fields is denoted by $\left[E_{\mathrm{ex}}^3\right]$. In particular, $\left[E_{\mathrm{ex}}^3\right]$ reduces to $E_{\mathrm{p}}E_{\mathrm{S}}^*E_{\mathrm{p}}$ in CARS, $E_{\mathrm{p}}E_{\mathrm{S}}^*E_{\mathrm{S}}$ in SRL, $E_{\mathrm{p}}^*E_{\mathrm{S}}E_{\mathrm{p}}$ in SRG, and $E_{\mathrm{p}}^*E_{\mathrm{S}}E_{\mathrm{S}}$ in CSRS. Under the far off-resonant condition, $E_{\mathrm{sig}} \propto i\mathrm{Re}\,\chi_{\mathrm{CRS}}^{(3)}\left[E_{\mathrm{ex}}^3\right]$, whereas under the just resonant condition, $E_{\mathrm{sig}} \propto -\mathrm{Im}\,\chi_{\mathrm{CRS}}^{(3)}\left[E_{\mathrm{ex}}^3\right]$. The signal field in the resonant and off-resonant cases can be united into one expression: $E_{\mathrm{sig}} \propto i\chi_{\mathrm{CRS}}^{(3)}\left[E_{\mathrm{ex}}^3\right]$. This holds true in simple systems where the excitation plane waves are incident onto a sufficiently large homogeneous material, in which case the signal field is additionally emitted as a plane wave. However, in 3-D systems such as microscopy, the focused excitation beams, nonlinear susceptibility distribution $\chi_{\mathrm{CRS}}^{(3)}(x)$, signal field emitted from $\chi_{\mathrm{CRS}}^{(3)}(x)$, and Gouy phase shift [138] must be considered. The Gouy phase shift is the phase difference between the excitation field and the signal field emitted from a molecule in the focal point, which arises when traveling in the vicinity of the focal point.

To discuss 3-D systems, we first consider the signal field emitted from a single point in the sample. After excitation, a molecule at a single point emits a signal field as a spherical wave. The discussion below uses the scalar approximation, which does not consider the electric field vector, where the direction depends on the propagation direction of light; however, the physical principle/phenomenon can be captured. Irrespective of the resonant or nonresonant conditions, the initial phase of the signal field becomes equal to that of $P^{(3)}(x)$, i.e., $E_{\text{sig}}(x) \propto \chi_{\text{CRS}}^{(3)}(x)\left[E_{\text{ex}}^3(x)\right]$. To connect the simple system discussed in the preceding paragraph to the 3-D system, we imagine that each point in the sample emits a signal spherical wave propagating toward a certain detecting point $x_{\text{d}} = (x_{\text{d}}, y_{\text{d}}, z_{\text{d}})$. If the plane waves excite the sample that is homogeneous and sufficiently large, the integral of all signal waves over x results in a phase shift of $\pi/2$: $E_{\text{sig}}(x_d) \propto i\chi_{\text{CRS}}^{(3)}\left[E_{\text{ex}}^3\right]$. This result is consistent with that in Appendix A1 and Appendix A2.

To formulate the image-forming theory for microscopy, we utilize the formula $E_{\text{sig}}(x) \propto \chi_{\text{CRS}}^{(3)}(x)\left[E_{\text{ex}}^3(x)\right]$. A schematic of CRS is shown in Figure 3.11. We define the coordinate system as follows: x for position in the sample, x_{d} for detecting position, and x' for the displacement of the sample stage or the position of the focal spot of a scanned laser beam. For simplicity, the first Born approximation is applied to understand the nature of the optical resolution. In this approximation, multiple scattering and depletion of the beam are neglected, which usually hold true for the nearly transparent sample such as a biological specimen. If the multiple scattering and the depletion are nonnegligible, the image will be deformed to some extent. We assume that both the excitation and the signal-collection systems have a magnification of 1× with no aberration, which does not change the characteristics of the image-forming properties. Moreover, our model employs the scalar diffraction theory. The nonlinear susceptibility distribution $\chi^{(3)}(x - x')$ at the sample-stage displacement x' acts as the object in the imaging system. The polarization density $P^{(3)}(x) = \varepsilon_0 \chi_{\text{CRS}}^{(3)}(x - x')\left[E_{\text{ex}}^3(x)\right]$ is induced by the excitation electric field $\left[E_{\text{ex}}^3(x)\right]$ formed by the excitation system, and the induced polarization $P^{(3)}(x)$ emits the signal electric field that

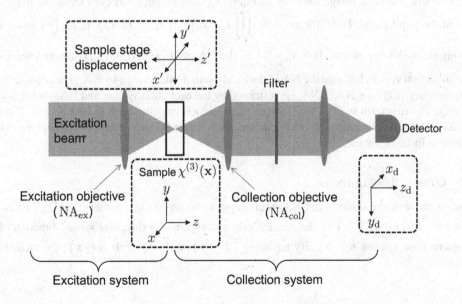

FIGURE 3.11 Schematic of CRS microscopy. NA of excitation and collection systems are NA_{ex} and NA_{col}, respectively.

propagates toward the detecting point x_d through the signal-collection system, forming an amplitude spread function $h_{col}(x_d - x)$. We assume that the NAs of the excitation and signal-collection systems are NA_{ex} and NA_{col}, respectively.

The image intensity is written as,

$$I(x') = \iiint \left| cE_{lo}(x_d) + \iiint \chi_{CRS}^{(3)}(x - x')[E_{ex}^3(x)]h_{col}(x_d - x)d^3x \right|^2 D(x_d, y_d)\delta(z_d)d^3x_d, \qquad (A3.1)$$

where $D(x_a, y_a)$ is a 2-D function representing the detector size, and $\delta(z_d)$ denotes Dirac delta function. The electric field in the detecting space $E_{lo}(x_d)$, which is the excitation beam itself in SRL and SRG or the nonresonant background in CARS and CSRS, acts as the local oscillator. In SRL and SRG, the constant c before $E_{lo}(x_d)$ becomes $-i$ that stems from the Gouy phase shift in the vicinity of the focus of the excitation beam in the sample. On the other hand, in CARS and CSRS, the constant c becomes unity because the nonresonant background is generated from the sample with an initial phase of zero. In Eq. (A3.1), we presumed that the electric permittivity ε_0 of free space is unity: $\varepsilon_0 = 1$. For simplicity, in CARS and CSRS, the nonresonant background is assumed to be spectrally flat and have homogeneous intensity over the sample.

The image intensity obtained by a detector placed at the image plane conjugate to the sample plane without confocal pinhole is proportional to

$$I(x') \propto \iiint \left| cE_{lo}(x_d) + \iiint \chi_{CRS}^{(3)}(x - x')[E_{ex}^3(x)]h_{col}(x_d - x)d^3x \right|^2 d^3x_d$$

$$\approx \iiint \left| E_{lo}(x_d) \right|^2 d^3x_d + c^*E_{lo}^* \iiint \chi_{CRS}^{(3)}(x - x')[E_{ex}^3(x)]h_{sig}(-x)d^3x + c.c., \qquad (A3.2)$$

where a term corresponding to the square of modulus of the signal was neglected, and E_{lo} is a constant. Note that the same image intensity as Eq. (A3.2) can be obtained even when the detector is placed at the pupil plane. Here, the relation $\iiint E_{lo}^*(x_d)h_{col}(x_d - x)d^3x_d \approx E_{lo}^*h_{sig}(-x)$ was used assuming the following points: If $NA_{ex} > NA_{col}$, the NA of $h_{sig}(-x)$, NA_{sig}, is determined by the signal-collection system because the NA of the local oscillator is limited by NA_{col}, i.e., $NA_{ex} > NA_{sig}$. On the contrary, if $NA_{ex} \leq NA_{col}$, NA_{sig} is restricted by the excitation system and NA_{sig} is not smaller than NA_{ex}. The first term in Eq. (A3.2) is a constant, which vanishes with lock-in detection in SRL and SRG and can be eliminated in the computer in CARS and CSRS. The cross terms (second and third terms) form an image.

A3.2 Optical Resolution

To discuss the optical resolution in CRS microscopy, we focus on the cross terms in Eq. (A3.2). The expression $[E_{ex}^3(x)]h_{sig}(-x)$ in the second term corresponds to the point spread function of the total microscope system $h_t(-x)$. By replacing $[E_{ex}^3(x)]h_{sig}(-x)$ with $h_t(-x)$, the cross terms become

$$I_{cro}(x') = c^* \iiint \chi_{CRS}^{(3)}(x - x')h_t(-x)d^3x + c.c., \qquad (A3.3)$$

where the positive real number E_{lo}^* was omitted. To address the optical transfer function (OTF), we perform a Fourier transform of Eq. (A3.3), resulting in:

$$\mathcal{F}\left[I_{cro}\right](f) = c^*\mathcal{F}\left[\chi_{CRS}^{(3)}\right](f)A(f) + c\left\{\mathcal{F}\left[\chi_{CRS}^{(3)}\right](-f)\right\}^* A^*(-f)$$

$$= \mathcal{F}\left[\mathrm{Re}\left\{\chi_{CRS}^{(3)}\right\}\right](f)\cdot\left\{c^*A(f) + cA^*(-f)\right\} \qquad (A3.4)$$

$$+ \mathcal{F}\left[\mathrm{Im}\left\{\chi_{CRS}^{(3)}\right\}\right](f)\cdot\left\{ic^*A(f) - icA^*(-f)\right\},$$

where $\mathcal{F}[\cdot]$ denotes Fourier transform and the function $A(f)$, which we term the 3-D aperture, is a Fourier transform of $h_t(-x)$. We define the OTF of the real and the imaginary parts of $\chi^{(3)}$ as $\mathrm{OTF_r}(f) = c^*A(f) + cA^*(-f)$ and $\mathrm{OTF_i}(f) = ic^*A(f) - icA^*(-f)$, respectively. In SRL and SRG, substituting $c = -i$ leads to $\mathrm{OTF_r}(f) = i\{A(f) - A^*(-f)\}$ and $\mathrm{OTF_i}(f) = -\{A(f) + A^*(-f)\}$, and in CARS and CSRS, substituting $c = 1$ causes $\mathrm{OTF_r}(f) = A(f) + A^*(-f)$ and $\mathrm{OTF_i}(f) = i\{A(f) - A^*(-f)\}$. This indicates that if $A(f) = A^*(-f)$, $\mathrm{OTF_r}(f)$ in SRL and SRG and $\mathrm{OTF_i}(f)$ in CARS and CSRS vanish, but if $A(f) \neq A^*(-f)$, they remain. Consequently, in SRL and SRG, under the condition $A(f) \neq A^*(-f)$, the real part of $\chi^{(3)}$, which additionally contains cross phase modulation (XPM, an optical process causing intensity-dependent phase shift), appears in the image. Similarly, in CARS and CSRS, under the condition $A(f) \neq A^*(-f)$, the imaginary part of $\chi^{(3)}$ is observed. These phenomena occur under the condition $NA_{ex} > NA_{col}$ as discussed below.

We consider the influence of NA on the OTF in CARS and SRL microscopy. By considering the excitation fields $\left[E_{ex}^3(x)\right]$, i.e., $E_p(x)E_s^*(x)E_p(x)$ in CARS and $E_p(x)E_s^*(x)E_s(x)$ in SRL, we analyze the key factor $h_t(-x) = \left[E_{ex}^3(x)\right]h_{sig}(-x)$ that determines the optical resolution. It must be noted that $\left[E_{ex}^3(x)\right]$ is formed by the excitation system, whereas $h_{sig}(-x)$ is restricted by NA_{sig}. In other words, Fourier transforms of $E_p(x)$ and $E_S(x)$, i.e., $P_p(f)$ and $P_S(f)$, are the spherical-shell shaped pupil functions of the excitation system, which are determined by the excitation wavelengths and NA_{ex}, and $h_{sig}(-x)$ is the Fourier transform of the pupil function corresponding to the local oscillator $P_{sig}(f)$, which is determined by the signal wavelength and NA_{sig}. Moreover, the radius of $P_{sig}(f)$ of SRL differs from that of CARS, which can be attributed to the difference in wavelength. The 3-D aperture $A(f)$, i.e., the Fourier transform of $h_t(-x)$, is calculated by convolving the four pupil functions mentioned above: $P_p(-f) \otimes P_S^*(f) \otimes P_S(-f) \otimes P_{sig}(f)$ for SRL and $P_p(-f) \otimes P_S^*(f) \otimes P_p(-f) \otimes P_{sig}(f)$ for CARS. In both cases, the magnitude relation between the NAs of the excitation and signal-collection systems affects the shape of the 3-D aperture $A(f)$. If $NA_{ex} > NA_{col}$, the 3-D aperture becomes asymmetric with respect to the f_z axis, i.e., $A(f) \neq A^*(-f)$, resulting in the appearance of XPM in SRL microscopy. In this case, $h_t(-x)$ becomes a complex function with a phase, which causes a phase shifting of the signal. On the contrary, if $NA_{ex} = NA_{col}$ (or even $NA_{ex} \leq NA_{col}$, i.e., $NA_{ex} \approx NA_{sig}$), the XPM disappears from the image, which results from the absence of the signal phase shifting. It must be noted that if the

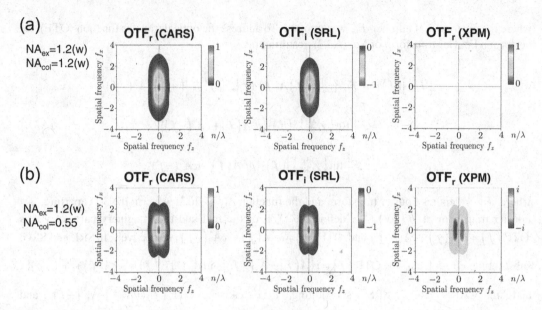

FIGURE 3.12   OTF for CARS, SRL, and XPM. n is the average refractive index in the sample. The wavelengths of pump and Stokes are assumed to be 800 nm and 1036 nm, respectively, which correspond to the vibrational frequency of 2850 cm^{-1} (CH$_2$-stretching mode).

three focal points (excitation system for the pump beam, excitation system for the Stokes beam, and signal-collection system) do not coincide with each other in the sample, XPM emerges even if $NA_{ex} = NA_{col}$.

Figure 3.12 shows the NA dependence of the OTF in CARS and SRL microscopy. We assume that, in SRL microscopy, the wavelengths of the pump and Stokes beams are tuned to a full-resonant vibrational level: in CARS microscopy, they are slightly detuned to observe Re $\chi_{CARS}^{(3)}$. As shown in Figure 3.12(a), under the same excitation conditions, CARS microscopy exhibits a slightly higher optical resolution than SRL microscopy because of the difference in the signal wavelength. As shown in Figure 3.12(b), in SRL microscopy, if $NA_{ex} > NA_{col}$, the OTF of the XPM appears, which is usually undesired. Because an XPM point spread function, defined as the Fourier transform of the OTF of XPM, becomes an odd function in the z direction, the XPM image appears to be a differential image with respect to z.

REFERENCES

1. A. Zumbusch, G. R. Holtom, and X. S. Xie, "Three-dimensional vibrational imaging by coherent anti-Stokes Raman scattering," *Phys Rev Lett* **82**(20), 4142–4145 (1999).
2. M. Hashimoto, T. Araki, and S. Kawata, "Molecular vibration imaging in the fingerprint region by use of coherent anti-Stokes Raman scattering microscopy with a collinear configuration," *Opt Lett* **25**(24), 1768–1770 (2000).
3. C. W. Freudiger, W. Min, B. G. Saar, S. Lu, G. R. Holtom, C. He, J. C. Tsai, J. X. Kang, and X. S. Xie, "Label-free biomedical imaging with high sensitivity by stimulated Raman scattering microscopy," *Science* **322**(5909), 1857–1861 (2008).
4. P. Nandakumar, A. Kovalev, and A. Volkmer, "Vibrational imaging based on stimulated Raman scattering microscopy," *New J Phys* **11**(3), 033026 (2009).
5. Y. Ozeki, F. Dake, S. Kajiyama, K. Fukui, and K. Itoh, "Analysis and experimental assessment of the sensitivity of stimulated Raman scattering microscopy," *Opt Express* **17**(5), 3651–3658 (2009).
6. T. W. Kee, and M. T. Cicerone, "Simple approach to one-laser, broadband coherent anti-Stokes Raman scattering microscopy," *Opt Lett* **29**(23), 2701–2703 (2004).

7. H. Kano, and H. Hamaguchi, "Ultrabroadband (> 2500 cm⁻¹) multiplex coherent anti-Stokes Raman scattering microspectroscopy using a supercontinuum generated from a photonic crystal fiber," *Appl Phys Lett* **86**(12), 121113 (2005).

8. Y. Ozeki, W. Umemura, Y. Otsuka, S. Satoh, H. Hashimoto, K. Sumimura, N. Nishizawa, K. Fukui, and K. Itoh, "High-speed molecular spectral imaging of tissue with stimulated Raman scattering," *Nat Photonics* **6**(12), 844–850 (2012).

9. L. J. Kong, M. B. Ji, G. R. Holtom, D. Fu, C. W. Freudiger, and X. S. Xie, "Multicolor stimulated Raman scattering microscopy with a rapidly tunable optical parametric oscillator," *Opt Lett* **38**(2), 145–147 (2013).

10. D. L. Zhang, P. Wang, M. N. Slipchenko, D. Ben-Amotz, A. M. Weiner, and J. X. Cheng, "Quantitative vibrational imaging by hyperspectral stimulated Raman scattering microscopy and multivariate curve resolution analysis," *Anal Chem* **85**(1), 98–106 (2013).

11. Y. Ozeki, T. Asai, J. Shou, and H. Yoshimi, "Multicolor stimulated Raman scattering microscopy with fast wavelength-tunable Yb fiber," *IEEE J Sel Top Quant* **25**, 7100211 (2019).

12. Y. Suzuki, K. Kobayashi, Y. Wakisaka, D. Deng, S. Tanaka, C. J. Huang, C. Lei, C. W. Sun, H. Liu, Y. Fujiwaki, S. Lee, A. Isozaki, Y. Kasai, T. Hayakawa, S. Sakuma, F. Arai, K. Koizumi, H. Tezuka, M. Inaba, K. Hiraki, T. Ito, M. Hase, S. Matsusaka, K. Shiba, K. Suga, M. Nishikawa, M. Jona, Y. Yatomi, Y. Yalikun, Y. Tanaka, T. Sugimura, N. Nitta, K. Goda, and Y. Ozeki, "Label-free chemical imaging flow cytometry by high-speed multicolor stimulated Raman scattering," *Proc Natl Acad Sci U S A* **116**(32), 15842–15848 (2019).

13. J. Li, and J. X. Cheng, "Direct visualization of de novo lipogenesis in single living cells," *Sci Rep* **4**, 6807 (2014).

14. L. Wei, F. Hu, Y. Shen, Z. Chen, Y. Yu, C. C. Lin, M. C. Wang, and W. Min, "Live-cell imaging of alkyne-tagged small biomolecules by stimulated Raman scattering," *Nat Methods* **11**(4), 410–412 (2014).

15. L. Zhang, L. Shi, Y. Shen, Y. Miao, M. Wei, N. Qian, Y. Liu, and W. Min, "Spectral tracing of deuterium for imaging glucose metabolism," *Nat Biomed Eng* **3**, 402–413 (2019).

16. W. Min, C. W. Freudiger, S. Lu, and X. S. Xie, "Coherent nonlinear optical imaging: Beyond fluorescence microscopy," *Annu Rev Phys Chem* **62**, 507–530 (2011).

17. H. Kano, H. Segawa, P. Leproux, and V. Couderc, "Linear and nonlinear Raman microspectroscopy: History, instrumentation, and applications," *Opt Rev* **21**(6), 752–761 (2014).

18. C. H. Camp Jr., and M. T. Cicerone, "Chemically sensitive bioimaging with coherent Raman scattering," *Nat Photonics* **9**(5), 295 (2015).

19. J. X. Cheng, and X. S. Xie, "Vibrational spectroscopic imaging of living systems: An emerging platform for biology and medicine," *Science* **350**(6264), aaa8870 (2015).

20. H. Kano, H. Segawa, M. Okuno, P. Leproux, and V. Couderc, "Hyperspectral coherent Raman imaging - Principle, theory, instrumentation, and applications to life sciences," *J Raman Spectrosc* **47**(1), 116–123 (2016).

21. H. Rigneault, and P. Berto, "Tutorial: Coherent Raman light matter interaction processes," *Apl Photonics* **3**(9), 091101 (2018).

22. C. Zhang, and J.-X. Cheng, "Perspective: Coherent Raman scattering microscopy, the future is bright," *Apl Photonics* **3**(9), 090901 (2018).

23. F. Hu, L. Shi, and W. Min, "Biological imaging of chemical bonds by stimulated Raman scattering microscopy," *Nat Methods* **16**(9), 830–842 (2019).

24. C. L. Evans, E. O. Potma, M. Puoris'haag, D. Cote, C. P. Lin, and X. S. Xie, "Chemical imaging of tissue in vivo with video-rate coherent anti-Stokes Raman scattering microscopy," *Proc Natl Acad Sci U S A* **102**(46), 16807–16812 (2005).

25. B. G. Saar, C. W. Freudiger, J. Reichman, C. M. Stanley, G. R. Holtom, and X. S. Xie, "Video-rate molecular imaging in vivo with stimulated Raman scattering," *Science* **330**(6009), 1368–1370 (2010).

26. Y. Bi, C. Yang, Y. Chen, S. Yan, G. Yang, Y. Wu, G. Zhang, and P. Wang, "Near-resonance enhanced label-free stimulated Raman scattering microscopy with spatial resolution near 130 nm," *Light Sci Appl* **7**, 81 (2018).

27. M. D. Duncan, J. Reintjes, and T. J. Manuccia, "Scanning coherent anti-Stokes Raman microscope," *Opt Lett* **7**(8), 350–352 (1982).

28. J. X. Cheng, A. Volkmer, and X. S. Xie, "Theoretical and experimental characterization of coherent anti-Stokes Raman scattering microscopy," *J Opt Soc Am B* **19**(6), 1363–1375 (2002).

29. G. I. Petrov, R. Arora, V. V. Yakovlev, X. Wang, A. V. Sokolov, and M. O. Scully, "Comparison of coherent and spontaneous Raman microspectroscopies for noninvasive detection of single bacterial endospores," *P Natl Acad Sci USA* **104**(19), 7776–7779 (2007).

30. D. Pestov, G. O. Ariunbold, X. Wang, R. K. Murawski, V. A. Sautenkov, A. V. Sokolov, and M. O. Scully, "Coherent versus incoherent Raman scattering: Molecular coherence excitation and measurement," *Opt Lett* **32**(12), 1725–1727 (2007).

31. J. X. Cheng, L. D. Book, and X. S. Xie, "Polarization coherent anti-Stokes Raman scattering microscopy," *Opt Lett* **26**(17), 1341–1343 (2001).

32. A. Volkmer, L. D. Book, and X. S. Xie, "Time-resolved coherent anti-Stokes Raman scattering microscopy: Imaging based on Raman free induction decay," *Appl Phys Lett* **80**(9), 1505–1507 (2002).

33. J. X. Cheng, A. Volkmer, L. D. Book, and X. S. Xie, "An epi-detected coherent anti-stokes Raman scattering (E-CARS) microscope with high spectral resolution and high sensitivity," *J Phys Chem B* **105**(7), 1277–1280 (2001).

34. E. O. Potma, D. J. Jones, J. X. Cheng, X. S. Xie, and J. Ye, "High-sensitivity coherent anti-Stokes Raman scattering microscopy with two tightly synchronized picosecond lasers," *Opt Lett* **27**(13), 1168–1170 (2002).

35. T. Minamikawa, N. Tanimoto, M. Hashimoto, T. Araki, M. Kobayashi, K. Fujita, and S. Kawata, "Jitter reduction of two synchronized picosecond mode-locked lasers using balanced cross-correlator with two-photon detectors," *Appl Phys Lett* **89**(19), 191101 (2006).

36. K. Kieu, B. G. Saar, G. R. Holtom, X. S. Xie, and F. W. Wise, "High-power picosecond fiber source for coherent Raman microscopy," *Opt Lett* **34**(13), 2051–2053 (2009).

37. M. Jurna, J. P. Korterik, H. L. Offerhaus, and C. Otto, "Noncritical phase-matched lithium triborate optical parametric oscillator for high resolution coherent anti-Stokes Raman scattering spectroscopy and microscopy," *Appl Phys Lett* **89**(25), 251116 (2006).

38. B. von Vacano, T. Buckup, and M. Motzkus, "Highly sensitive single-beam heterodyne coherent anti-Stokes Raman scattering," *Opt Lett* **31**, 2495–2497 (2006).

39. K. Isobe, A. Suda, M. Tanaka, H. Hashimoto, F. Kannari, H. Kawano, H. Mizuno, A. Miyawaki, and K. Midorikawa, "Single-pulse coherent anti-Stokes Raman scattering microscopy employing an octave spanning pulse," *Opt Express* **17**(14), 11259–11266 (2009).

40. I. Pope, W. Langbein, P. Watson, and P. Borri, "Simultaneous hyperspectral differential-CARS, TPF and SHG microscopy with a single 5 fs Ti:Sa laser," *Opt Express* **21**(6), 7096–7106 (2013).

41. G. Krauss, T. Hanke, A. Sell, D. Trautlein, A. Leitenstorfer, R. Selm, M. Winterhalder, and A. Zumbusch, "Compact coherent anti-Stokes Raman scattering microscope based on a picosecond two-color Er:fiber laser system," *Opt Lett* **34**(18), 2847–2849 (2009).

42. A. Gambetta, V. Kumar, G. Grancini, D. Polli, R. Ramponi, G. Cerullo, and M. Marangoni, "Fiber-format stimulated-Raman-scattering microscopy from a single laser oscillator," *Opt Lett* **35**(2), 226–228 (2010).

43. C. W. Freudiger, W. Yang, G. R. Holtom, N. Peyghambarian, X. S. Xie, and K. Q. Kieu, "Stimulated Raman scattering microscopy with a robust fibre laser source," *Nat Photonics* **8**(2), 153–159 (2014).

44. M. Brinkmann, A. Fast, T. Hellwig, I. Pence, C. L. Evans, and C. Fallnich, "Portable all-fiber dual-output widely tunable light source for coherent Raman imaging," *Biomed Opt Express* **10**(9), 4437–4449 (2019).

45. H. Linnenbank, T. Steinle, F. Mörz, M. Flöss, H. Cui, A. Glidle, and H. Giessen, "Robust and rapidly tunable light source for SRS/CARS microscopy with low-intensity noise," *Adv Photonics* **1**(5), 055001 (2019).

46. K. Wang, J. Q. Wang, and P. Qiu, "Synchronization maintenance of synchronized time-lens source in the presence of repetition rate drift of the mode-locked laser for coherent Raman scattering microscopy," *IEEE J Quantum Electron* **53**(1), 8600105 (2017).

47. K. Tokunaga, Y. C. Fang, H. Yokoyama, and Y. Ozeki, "Generation of synchronized picosecond pulses by a 1.06-microm gain-switched laser diode for stimulated Raman scattering microscopy," *Opt Express* **24**(9), 9617–9628 (2016).

48. K. Nose, Y. Ozeki, T. Kishi, K. Sumimura, N. Nishizawa, K. Fukui, Y. Kanematsu, and K. Itoh, "Sensitivity enhancement of fiber-laser-based stimulated Raman scattering microscopy by collinear balanced detection technique," *Opt Express* **20**(13), 13958–13965 (2012).

49. E. S. Lamb, L. G. Wright, and F. W. Wise, "Divided-pulse lasers," *Opt Lett* **39**(9), 2775–2777 (2014).

50. H. Yoshimi, K. Sumimura, and Y. Ozeki, "An Er fiber laser generating multi-milliwatt picosecond pulses with ultralow intensity noise," *Jpn J Appl Phys* **57**(10), 108001 (2018).

51. S. Begin, B. Burgoyne, V. Mercier, A. Villeneuve, R. Vallee, and D. Cote, "Coherent anti-Stokes Raman scattering hyperspectral tissue imaging with a wavelength-swept system," *Biomed Opt Express* **2**(5), 1296–1306 (2011).

52. T. Hellerer, A. M. K. Enejder, and A. Zumbusch, "Spectral focusing: High spectral resolution spectroscopy with broad-bandwidth laser pulses," *Appl Phys Lett* **85**(1), 25–27 (2004).

53. A. Karuna, F. Masia, M. Wiltshire, R. Errington, P. Borri, and W. Langbein, "Label-free volumetric quantitative imaging of the human somatic cell division by hyperspectral coherent anti-stokes Raman scattering," *Anal Chem* **91**(4), 2813–2821 (2019).

54. M. Okuno, H. Kano, P. Leproux, V. Couderc, J. P. R. Day, M. Bonn, and H. Hamaguchi, "Quantitative CARS molecular fingerprinting of single living cells with the use of the maximum entropy method," *Angew Chem Int Ed Engl* **49**(38), 6773–6777 (2010).

55. C. H. Camp Jr., Y. J. Lee, J. M. Heddleston, C. M. Hartshorn, A. R. Hight Walker, J. N. Rich, J. D. Lathia, and M. T. Cicerone, "High-speed coherent Raman fingerprint imaging of biological tissues," *Nat Photonics* **8**, 627–634 (2014).

56. M. Okuno, H. Kano, P. Leproux, V. Couderc, and H. Hamaguchi, "Ultrabroadband (> 2000 cm(−1)) multiplex coherent anti-Stokes Raman scattering spectroscopy using a subnanosecond supercontinuum light source," *Opt Lett* **32**(20), 3050–3052 (2007).

57. H. Yoneyama, K. Sudo, P. Leproux, V. Couderc, A. Inoko, and H. Kano, "Invited Article: CARS molecular fingerprinting using sub-100-ps microchip laser source with fiber amplifier," *Apl Photonics* **3**(9), 092408 (2018).

58. H. Kano, T. Maruyama, J. Kano, Y. Oka, D. Kaneta, T. Guerenne, P. Leproux, V. Couderc, and M. Noguchi, "Ultra-multiplex CARS spectroscopic imaging with 1-millisecond pixel dwell time," *OSA Continuum* **2**(5), 1693–1705 (2019).

59. J. P. Ogilvie, E. Beaurepaire, A. Alexandrou, and M. Joffre, "Fourier-transform coherent anti-Stokes Raman scattering microscopy," *Opt Lett* **31**(4), 480–482 (2006).

60. E. M. Vartiainen, H. A. Rinia, M. Müller, and M. Bonn, "Direct extraction of Raman line-shapes from congested CARS spectra," *Opt Express* **14**(8), 3622–3630 (2006).

61. Y. X. Liu, Y. J. Lee, and M. T. Cicerone, "Broadband CARS spectral phase retrieval using a time-domain Kramers-Kronig transform," *Opt Lett* **34**(9), 1363–1365 (2009).

62. T. Shimanouchi, "Tables of molecular vibrational frequencies. Consolidated volume II," *J Phys Chem Ref Data* **6**(3), 993–1102 (1977).

63. Q. Matthews, A. Brolo, J. Lum, X. Duan, and A. Jirasek, "Raman spectroscopy of single human tumour cells exposed to ionizing radiation in vitro," *Phys Med Biol* **56**(1), 19–38 (2011).

64. M. T. Cicerone, and C. H. Camp, "Histological coherent Raman imaging: A prognostic review," *Analyst* **143**(1), 33–59 (2017).

65. C.-S. Liao, K.-C. Huang, W. Hong, A. J. Chen, C. Karanja, P. Wang, G. Eakins, and J.-X. Cheng, "Stimulated Raman spectroscopic imaging by microsecond delay-line tuning," *Optica* **3**(12), 1377–1380 (2016).

66. M. S. Alshaykh, C. S. Liao, O. E. Sandoval, G. Gitzinger, N. Forget, D. E. Leaird, J. X. Cheng, and A. M. Weiner, "High-speed stimulated hyperspectral Raman imaging using rapid acousto-optic delay lines," *Opt Lett* **42**(8), 1548–1551 (2017).

67. R. Y. He, Y. K. Xu, L. L. Zhang, S. H. Ma, X. Wang, D. Ye, and M. B. Ji, "Dual-phase stimulated Raman scattering microscopy for real-time two-color imaging," *Optica* **4**(1), 44–47 (2017).

68. C. S. Liao, M. N. Slipchenko, P. Wang, J. Li, S. Y. Lee, R. A. Oglesbee, and J. X. Cheng, "Microsecond scale vibrational spectroscopic imaging by multiplex stimulated Raman scattering microscopy," *Light Sci Appl* **4**, e265 (2015).

69. C. Zhang, K.-C. Huang, B. Rajwa, J. Li, S. Yang, H. Lin, C.-S. Liao, G. Eakins, S. Kuang, V. Patsekin, J. P. Robinson, and J.-X. Cheng, "Stimulated Raman scattering flow cytometry for label-free single-particle analysis," *Optica* **4**(1), 103–109 (2017).

70. C. S. Liao, P. Wang, P. Wang, J. Li, H. J. Lee, G. Eakins, and J. X. Cheng, "Spectrometer-free vibrational imaging by retrieving stimulated Raman signal from highly scattered photons," *Sci Adv* **1**(9), e1500738 (2015).

71. T. Hellerer, C. Axäng, C. Brackmann, P. Hillertz, M. Pilon, and A. Enejder, "Monitoring of lipid storage in Caenorhabditis elegans using coherent anti-Stokes Raman scattering (CARS) microscopy," *P Natl Acad Sci USA* **104**(37), 14658–14663 (2007).

72. J. P. Pezacki, J. A. Blake, D. C. Danielson, D. C. Kennedy, R. K. Lyn, and R. Singaravelu, "Chemical contrast for imaging living systems: Molecular vibrations drive CARS microscopy," *Nat Chem Biol* **7**(3), 137–145 (2011).

73. Y.-M. Wu, H.-C. Chen, W.-T. Chang, J.-W. Jhan, H.-L. Lin, and I. Liau, "Quantitative assessment of hepatic fat of intact liver tissues with coherent anti-stokes Raman scattering microscopy," *Anal Chem* **81**(4), 1496–1504 (2009).

74. T. T. Le, T. B. Huff, and J.-X. Cheng, "Coherent anti-Stokes Raman scattering imaging of lipids in cancer metastasis," *BMC Cancer* **9**, 42–42 (2009).

75. S.-H. Kim, E.-S. Lee, J. Y. Lee, E. S. Lee, B.-S. Lee, J. E. Park, and D. W. Moon, "Multiplex coherent anti-stokes Raman spectroscopy images intact atheromatous lesions and concomitantly identifies distinct chemical profiles of atherosclerotic lipids," *Circ Res* **106**(8), 1332–1341 (2010).

76. C. L. Evans, X. Xu, S. Kesari, X. S. Xie, S. T. C. Wong, and G. S. Young, "Chemically-selective imaging of brain structures with CARS microscopy," *Opt Express* **15**(19), 12076–12076 (2007).

77. M. C. Wang, W. Min, C. W. Freudiger, G. Ruvkun, and X. S. Xie, "RNAi screening for fat regulatory genes with SRS microscopy," *Nat Methods* **8**(2), 135–138 (2011).

78. D. Fu, Y. Yu, A. Folick, E. Currie, R. V. Farese Jr., T. H. Tsai, X. S. Xie, and M. C. Wang, "In vivo metabolic fingerprinting of neutral lipids with hyperspectral stimulated Raman scattering microscopy," *J Am Chem Soc* **136**(24), 8820–8828 (2014).

79. Y. Wakisaka, Y. Suzuki, O. Iwata, A. Nakashima, T. Ito, M. Hirose, R. Domon, M. Sugawara, N. Tsumura, H. Watarai, T. Shimobaba, K. Suzuki, K. Goda, and Y. Ozeki, "Probing the metabolic heterogeneity of live Euglena gracilis with stimulated Raman scattering microscopy," *Nat Microbiol* **1**(10), 16124 (2016).

80. C. W. Freudiger, R. Pfannl, D. A. Orringer, B. G. Saar, M. Ji, Q. Zeng, L. Ottoboni, Y. Wei, C. Waeber, J. R. Sims, P. L. De Jager, O. Sagher, M. A. Philbert, X. Xu, S. Kesari, X. S. Xie, and G. S. Young, "Multicolored stain-free histopathology with coherent Raman imaging," *Lab Investig* **92**(10), 1492–1502 (2012).

81. M. B. Ji, D. A. Orringer, C. W. Freudiger, S. Ramkissoon, X. H. Liu, D. Lau, A. J. Golby, I. Norton, M. Hayashi, N. Y. R. Agar, G. S. Young, C. Spino, S. Santagata, S. Camelo-Piragua, K. L. Ligon, O. Sagher, and X. S. Xie, "Rapid, label-free detection of brain tumors with stimulated Raman scattering microscopy," *Sci Transl Med* **5**(201), 201ra119 (2013).

82. S. Satoh, Y. Otsuka, Y. Ozeki, K. Itoh, A. Hashiguchi, K. Yamazaki, H. Hashimoto, and M. Sakamoto, "Label-free visualization of acetaminophen-induced liver injury by high-speed stimulated Raman scattering spectral microscopy and multivariate image analysis," *Pathol Int* **64**(10), 518–526 (2014).

83. F. Tian, W. Yang, D. A. Mordes, J. Y. Wang, J. S. Salameh, J. Mok, J. Chew, A. Sharma, E. Leno-Duran, S. Suzuki-Uematsu, N. Suzuki, S. S. Han, F. K. Lu, M. Ji, R. Zhang, Y. Liu, J. Strominger, N. A. Shneider, L. Petrucelli, X. S. Xie, and K. Eggan, "Monitoring peripheral nerve degeneration in ALS by label-free stimulated Raman scattering imaging," *Nat Commun* **7**, 13283 (2016).

84. D. A. Orringer, B. Pandian, Y. S. Niknafs, T. C. Hollon, J. Boyle, S. Lewis, M. Garrard, S. L. Hervey-Jumper, H. J. L. Garton, C. O. Maher, J. A. Heth, O. Sagher, D. A. Wilkinson, M. Snuderl, S. Venneti, S. H. Ramkissoon, K. A. McFadden, A. Fisher-Hubbard, A. P. Lieberman, T. D. Johnson, X. S. Xie, J. K. Trautman, C. W. Freudiger, and S. Camelo-Piragua, "Rapid intraoperative histology of unprocessed surgical specimens via fibre-laser-based stimulated Raman scattering microscopy," *Nat Biomed Eng* **1**, 0027 (2017).

85. D. M. Drutis, T. M. Hancewicz, E. Pashkovski, L. Feng, D. Mihalov, G. Holtom, K. P. Ananthapadmanabhan, X. S. Xie, and M. Misra, "Three-dimensional chemical imaging of skin using stimulated Raman scattering microscopy," *J Biomed Opt* **19**(11), 111604 (2014).

86. M. Egawa, K. Tokunaga, J. Hosoi, S. Iwanaga, and Y. Ozeki, "In situ visualization of intracellular morphology of epidermal cells using stimulated Raman scattering microscopy," *J Biomed Opt* **21**(8), 86017 (2016).

87. M. Egawa, S. Iwanaga, J. Hosoi, M. Goto, H. Yamanishi, M. Miyai, C. Katagiri, K. Tokunaga, T. Asai, and Y. Ozeki, "Label-free stimulated Raman scattering microscopy visualizes changes in intracellular morphology during human epidermal keratinocyte differentiation," *Sci Rep* **9**(1), 12601 (2019).

88. F. K. Lu, S. Basu, V. Igras, M. P. Hoang, M. Ji, D. Fu, G. R. Holtom, V. A. Neel, C. W. Freudiger, D. E. Fisher, and X. S. Xie, "Label-free DNA imaging in vivo with stimulated Raman scattering microscopy," *Proc Natl Acad Sci U S A* **112**(37), 11624–11629 (2015).

89. M. B. Roeffaers, X. Zhang, C. W. Freudiger, B. G. Saar, M. van Ruijven, G. van Dalen, C. Xiao, and X. S. Xie, "Label-free imaging of biomolecules in food products using stimulated Raman microscopy," *J Biomed Opt* **15**(6), 066016 (2010).

90. M. N. Slipchenko, H. Chen, D. R. Ely, Y. Jung, M. T. Carvajal, and J. X. Cheng, "Vibrational imaging of tablets by epi-detected stimulated Raman scattering microscopy," *Analyst* **135**(10), 2613–2619 (2010).

91. J. Li, S. Condello, J. Thomes-Pepin, X. Ma, Y. Xia, T. D. Hurley, D. Matei, and J. X. Cheng, "Lipid desaturation is a metabolic marker and therapeutic target of ovarian cancer stem cells," *Cell Stem Cell* **20**(3), 303–314 e305 (2017).

92. S. Yan, S. Cui, K. Ke, B. Zhao, X. Liu, S. Yue, and P. Wang, "Hyperspectral stimulated Raman scattering microscopy unravels aberrant accumulation of saturated fat in human liver cancer," *Anal Chem* **90**(11), 6362–6366 (2018).

93. H. J. Lee, D. Zhang, Y. Jiang, X. Wu, P. Y. Shih, C. S. Liao, B. Bungart, X. M. Xu, R. Drenan, E. Bartlett, and J. X. Cheng, "Label-free vibrational spectroscopic imaging of neuronal membrane potential," *J Phys Chem Lett* **8**(9), 1932–1936 (2017).

94. J. L. Suhalim, C. Y. Chung, M. B. Lilledahl, R. S. Lim, M. Levi, B. J. Tromberg, and E. O. Potma, "Characterization of cholesterol crystals in atherosclerotic plaques using stimulated Raman scattering and second-harmonic generation microscopy," *Biophys J* **102**(8), 1988–1995 (2012).

95. P. Wang, J. Li, P. Wang, C. R. Hu, D. Zhang, M. Sturek, and J. X. Cheng, "Label-free quantitative imaging of cholesterol in intact tissues by hyperspectral stimulated Raman scattering microscopy," *Angew Chem Int Ed Engl* **52**(49), 13042–13046 (2013).

96. A. J. Chen, J. Li, A. Jannasch, A. S. Mutlu, M. C. Wang, and J. X. Cheng, "Fingerprint stimulated Raman scattering imaging reveals retinoid coupling lipid metabolism and survival," *ChemPhysChem* **19**(19), 2500–2506 (2018).

97. M. Ji, M. Arbel, L. Zhang, C. W. Freudiger, S. S. Hou, D. Lin, X. Yang, B. J. Bacskai, and X. S. Xie, "Label-free imaging of amyloid plaques in Alzheimer's disease with stimulated Raman scattering microscopy," *Sci Adv* **4**(11), eaat7715 (2018).

98. B. G. Saar, L. R. Contreras-Rojas, X. S. Xie, and R. H. Guy, "Imaging drug delivery to skin with stimulated Raman scattering microscopy," *Mol Pharm* **8**(3), 969–975 (2011).

99. N. A. Belsey, N. L. Garrett, L. R. Contreras-Rojas, A. J. Pickup-Gerlaugh, G. J. Price, J. Moger, and R. H. Guy, "Evaluation of drug delivery to intact and porated skin by coherent Raman scattering and fluorescence microscopies," *J Control Release* **174**, 37–42 (2014).

100. D. Fu, J. Zhou, W. S. Zhu, P. W. Manley, Y. K. Wang, T. Hood, A. Wylie, and X. S. Xie, "Imaging the intracellular distribution of tyrosine kinase inhibitors in living cells with quantitative hyperspectral stimulated Raman scattering," *Nat Chem* **6**(7), 614–622 (2014).

101. X. Zhang, M. B. Roeffaers, S. Basu, J. R. Daniele, D. Fu, C. W. Freudiger, G. R. Holtom, and X. S. Xie, "Label-free live-cell imaging of nucleic acids using stimulated Raman scattering microscopy," *ChemPhysChem* **13**(4), 1054–1059 (2012).

102. D. Fu, W. Yang, and X. S. Xie, "Label-free imaging of neurotransmitter acetylcholine at neuromuscular junctions with stimulated Raman scattering," *J Am Chem Soc* **139**(2), 583–586 (2017).

103. B. Figueroa, Y. Chen, K. Berry, A. Francis, and D. Fu, "Label-free chemical imaging of latent fingerprints with stimulated Raman scattering microscopy," *Anal Chem* **89**(8), 4468–4473 (2017).

104. B. G. Saar, Y. Zeng, C. W. Freudiger, Y. S. Liu, M. E. Himmel, X. S. Xie, and S. Y. Ding, "Label-free, real-time monitoring of biomass processing with stimulated Raman scattering microscopy," *Angew Chem Int Ed Engl* **49**(32), 5476–5479 (2010).

105. J. C. Mansfield, G. R. Littlejohn, M. P. Seymour, R. J. Lind, S. Perfect, and J. Moger, "Label-free chemically specific imaging in planta with stimulated Raman scattering microscopy," *Anal Chem* **85**(10), 5055–5063 (2013).

106. G. R. Littlejohn, J. C. Mansfield, D. Parker, R. Lind, S. Perfect, M. Seymour, N. Smirnoff, J. Love, and J. Moger, "In vivo chemical and structural analysis of plant cuticular waxes using stimulated Raman scattering microscopy," *Plant Physiol* **168**(1), 18–28 (2015).

107. B. Liu, P. Wang, J. I. Kim, D. Zhang, Y. Xia, C. Chapple, and J. X. Cheng, "Vibrational fingerprint mapping reveals spatial distribution of functional groups of lignin in plant cell wall," *Anal Chem* **87**(18), 9436–9442 (2015).

108. Z. Wang, W. Zheng, S. C. Hsu, and Z. Huang, "Optical diagnosis and characterization of dental caries with polarization-resolved hyperspectral stimulated Raman scattering microscopy," *Biomed Opt Express* **7**(4), 1284–1293 (2016).

109. M. Ando, C. S. Liao, G. J. Eckert, and J. X. Cheng, "Imaging of demineralized enamel in intact tooth by epidetected stimulated Raman scattering microscopy," *J Biomed Opt* **23**(10), 1–9 (2018).

110. Q. Cheng, L. Wei, Z. Liu, N. Ni, Z. Sang, B. Zhu, W. Xu, M. Chen, Y. Miao, L. Q. Chen, W. Min, and Y. Yang, "Operando and three-dimensional visualization of anion depletion and lithium growth by stimulated Raman scattering microscopy," *Nat Commun* **9**(1), 2942 (2018).

111. A. Golreihan, C. Steuwe, L. Woelders, A. Deprez, Y. Fujita, J. Vellekoop, R. Swennen, and M. B. J. Roeffaers, "Improving preservation state assessment of carbonate microfossils in paleontological research using label-free stimulated Raman imaging," *PLoS One* **13**(7), e0199695 (2018).

112. L. Zada, H. A. Leslie, A. D. Vethaak, G. H. Tinnevelt, J. J. Jansen, J. F. de Boer, and F. Ariese, "Fast microplastics identification with stimulated Raman scattering microscopy," *J Raman Spectrosc* **49**(7), 1136–1144 (2018).

113. A. Alfonso-Garcia, S. G. Pfisterer, H. Riezman, E. Ikonen, and E. O. Potma, "D38-cholesterol as a Raman active probe for imaging intracellular cholesterol storage," *J Biomed Opt* **21**(6), 61003 (2016).

114. F. Hu, L. Wei, C. Zheng, Y. Shen, and W. Min, "Live-cell vibrational imaging of choline metabolites by stimulated Raman scattering coupled with isotope-based metabolic labeling," *Analyst* **139**(10), 2312–2317 (2014).

115. Y. Shen, Z. Zhao, L. Zhang, L. Shi, S. Shahriar, R. B. Chan, G. Di Paolo, and W. Min, "Metabolic activity induces membrane phase separation in endoplasmic reticulum," *Proc Natl Acad Sci U S A* **114**(51), 13394–13399 (2017).

116. L. Shi, C. Zheng, Y. Shen, Z. Chen, E. S. Silveira, L. Zhang, M. Wei, C. Liu, C. de Sena-Tomas, K. Targoff, and W. Min, "Optical imaging of metabolic dynamics in animals," *Nat Commun* **9**(1), 2995 (2018).

117. Y. Shen, F. Xu, L. Wei, F. Hu, and W. Min, "Live-cell quantitative imaging of proteome degradation by stimulated Raman scattering," *Angew Chem Int Ed Engl* **53**(22), 5596–5599 (2014).

118. N. Ota, Y. Yonamine, T. Asai, Y. Yalikun, T. Ito, Y. Ozeki, Y. Hoshino, and Y. Tanaka, "Isolating Single Euglena gracilis Cells by Glass microfluidics for Raman Analysis of paramylon Biogenesis," *Anal Chem* **91**(15), 9631–9639 (2019).

119. H. Yamakoshi, K. Dodo, A. Palonpon, J. Ando, K. Fujita, S. Kawata, and M. Sodeoka, "Alkyne-tag Raman imaging for visualization of mobile small molecules in live cells," *J Am Chem Soc* **134**(51), 20681–20689 (2012).

120. S. Yamaguchi, T. Matsushita, S. Izuta, S. Katada, M. Ura, T. Ikeda, G. Hayashi, Y. Suzuki, K. Kobayashi, K. Tokunaga, Y. Ozeki, and A. Okamoto, "Chemically-activatable alkyne-tagged probe for imaging microdomains in lipid bilayer membranes," *Sci Rep* **7**, 41007 (2017).

121. H. J. Lee, W. D. Zhang, D. L. Zhang, Y. Yang, B. Liu, E. L. Barker, K. K. Buhman, L. V. Slipchenko, M. J. Dai, and J. X. Cheng, "Assessing cholesterol storage in live cells and C-elegans by stimulated Raman scattering imaging of phenyl-diyne cholesterol," *Sci Rep-UK* **5**, 7930 (2015).

122. X. Li, M. Jiang, J. W. Y. Lam, B. Z. Tang, and J. Y. Qu, "Mitochondrial imaging with combined fluorescence and stimulated Raman scattering microscopy using a probe of the aggregation-induced emission characteristic," *J Am Chem Soc* **139**(47), 17022–17030 (2017).

123. W. J. Tipping, M. Lee, A. Serrels, V. G. Brunton, and A. N. Hulme, "Imaging drug uptake by bioorthogonal stimulated Raman scattering microscopy," *Chem Sci* **8**(8), 5606–5615 (2017).

124. M. M. Gaschler, F. Hu, H. Feng, A. Linkermann, W. Min, and B. R. Stockwell, "Determination of the subcellular localization and mechanism of action of ferrostatins in suppressing ferroptosis," *ACS Chem Biol* **13**(4), 1013–1020 (2018).

125. C. Zeng, F. Hu, R. Long, and W. Min, "A ratiometric Raman probe for live-cell imaging of hydrogen sulfide in mitochondria by stimulated Raman scattering," *Analyst* **143**(20), 4844–4848 (2018).

126. L. Wei, Z. Chen, L. Shi, R. Long, A. V. Anzalone, L. Zhang, F. Hu, R. Yuste, V. W. Cornish, and W. Min, "Super-multiplex vibrational imaging," *Nature* **544**(7651), 465–470 (2017).

127. F. Hu, C. Zeng, R. Long, Y. Miao, L. Wei, Q. Xu, and W. Min, "Supermultiplexed optical imaging and barcoding with engineered polyynes," *Nat Methods* **15**(3), 194–200 (2018).

128. R. Long, L. Zhang, L. Shi, Y. Shen, F. Hu, C. Zeng, and W. Min, "Two-color vibrational imaging of glucose metabolism using stimulated Raman scattering," *Chem Commun (Camb)* **54**(2), 152–155 (2018).

129. M. Wei, L. Shi, Y. Shen, Z. Zhao, A. Guzman, L. J. Kaufman, L. Wei, and W. Min, "Volumetric chemical imaging by clearing-enhanced stimulated Raman scattering microscopy," *Proc Natl Acad Sci U S A* **116**(14), 6608–6617 (2019).

130. S. Brustlein, P. Berto, R. Hostein, P. Ferrand, C. Billaudeau, D. Marguet, A. Muir, J. Knight, and H. Rigneault, "Double-clad hollow core photonic crystal fiber for coherent Raman endoscope," *Opt Express* **19**(13), 12562–12568 (2011).

131. B. G. Saar, R. S. Johnston, C. W. Freudiger, X. S. Xie, and E. J. Seibel, "Coherent Raman scanning fiber endoscopy," *Opt Lett* **36**(13), 2396–2398 (2011).

132. C.-S. Liao, P. Wang, C. Y. Huang, P. Lin, G. Eakins, R. T. Bentley, R. Liang, and J.-X. Cheng, "In vivo and in situ spectroscopic imaging by a handheld stimulated Raman scattering microscope," *ACS Photonics* **5**(3), 947–954 (2018).

133. L. Gong, W. Zheng, Y. Ma, and Z. Huang, "Saturated stimulated-Raman-scattering microscopy for far-field superresolution vibrational imaging," *Phys Rev Appl* **11**(3), 034041 (2019).

134. H. Xiong, L. Shi, L. Wei, Y. Shen, R. Long, Z. Zhao, and W. Min, "Stimulated Raman excited fluorescence spectroscopy and imaging," *Nat Photonics* **13**(6), 412–417 (2019).

135. R. P. Feynman, R. B. Leighton, and M. Sands, *The Feynman Lectures on Physics* (Addison Wesley, 1971).

136. S. Mukamel, *Principles of Nonlinear Optical Spectroscopy* (Oxford University Press, 1995).

137. M. D. Levenson, and S. Kano, *Introduction to Nonlinear Laser Spectroscopy* (Academic Press, 1989).

138. R. W. Boyd, "Intuitive explanation of the phase anomaly of focused light beams," *J Opt Soc Am* **70**(7), 877–880 (1980).

4 Magnetic Resonance Imaging
Principles and Applications

Masaki Sekino and Shoogo Ueno

CONTENTS

4.1 Introduction 75
4.2 Basic Principles of MRI 75
    4.2.1 Hardware 75
    4.2.2 Nuclear Magnetic Moment and Macroscopic Magnetization 77
    4.2.3 Relaxation of Magnetization...80
    4.2.4 Bloch Equation 81
    4.2.5 Excitation of Magnetization Using an RF Pulse 86
    4.2.6 Pulse Sequence and Image Reconstruction..87
    4.2.7 Contrast..90
4.3 Numerical Analyses of the Bloch Equation 91
4.4 Functional MRI 95
    4.4.1 Neurovascular Coupling 95
    4.4.2 Applications of Functional MRI..96
4.5 Recent Advances in Imaging Techniques and Applications 97
    4.5.1 Diffusion..97
    4.5.2 Flow and Pressure..97
    4.5.3 Electric Properties of Tissue...98
    4.5.4 MRI-Guided Surgery and Treatment 98
    4.5.5 Image Processing...98
References...99

4.1 INTRODUCTION

Magnetic resonance imaging (MRI) is a technique to obtain images based on nuclear magnetic resonance (NMR) signals generated by nuclei under a strong magnetic field. The MRI technique was demonstrated by Paul Lauterbur in 1973 [1]. The method later progressed through the development of magnets that stably generate strong magnetic fields, developments of diverse imaging techniques and increases in scan speed, the application of diagnosing various diseases, and the development of image processing methods, until it became an essential technology for imaging diagnosis. Features of MRI that distinguish it from other diagnostic imaging methods include the high contrast between soft tissues and the acquisition of functional information such as metabolic processes and brain activity in addition to morphological information. Furthermore, it involves no radiation exposure and is noninvasive. This chapter introduces the basic principles of MRI and the mechanism of imaging methods such as functional MRI and diffusion MRI.

4.2 BASIC PRINCIPLES OF MRI

4.2.1 HARDWARE

In principle, MRI requires a magnet to apply a magnetic field to the measured object such as the human body. Currently, two types are used: superconducting magnets and permanent magnets.

Superconducting magnets are advantageous in that they can generate a strong magnetic field approaching a maximum of 10 T and the generated magnetic field is highly stable over time. Because permanent magnets do not require cooling, they are easy to maintain and are used in low magnetic field devices.

Figures 4.1(a) and (b) show a 7 T MRI system for small animals and a schematic diagram of an MRI apparatus using the most common horizontal field type superconducting magnet, respectively. It has a superconducting magnet on the outside and a shim coil, gradient coil, and radio frequency (RF) coil on the inside. In the superconducting magnet, a stationary current flows through a coil housed in a cryogenic vessel, and a strong magnetic field (main field) is generated. The stronger the main field, the larger the MRI signal and the higher the signal-to-noise ratio [2]. Inside the super-conducting magnet, there is a shim coil that decreases disturbances in the uniformity of the main field. In addition to the shim coil, magnetic field correction may also be performed by attaching iron pieces inside the magnet. Three gradient coils are arranged within the magnet to handle each of the three axes x, y, and z. By applying a current in the opposite direction to a pair of coils, a magnetic field with an intensity that changes in proportion to the position coordinates x, y, and z, that is, a gradient magnetic field, is generated. Figures 4.2(a) and (b) show the basic forms of the gradient coils for the z- and y-axis, respectively. If multiple gradient coils are operated at the same time, a gradient inclined in an arbitrary direction can be obtained. The RF coil placed inside the gradient coil irradiates the object to be measured with a Larmor frequency electromagnetic wave, which is described later, and then detects the electromagnetic wave generated from the object with the same frequency. In some cases, transmission and reception are performed by separate RF coils. Because the MRI apparatus uses a strong magnetic field, it is necessary to shield the outside of the laboratory from magnetic field leakage to ensure safety. Effective methods include adding a shielding coil to

(a)

(b)

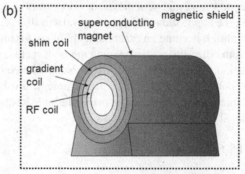

FIGURE 4.1   (a) A 7 T MRI system for small animals. (b) Hardware of MRI consisting of a superconducting magnet, a set of shim and gradient coils, an RF coil, and a magnetically shielded room.

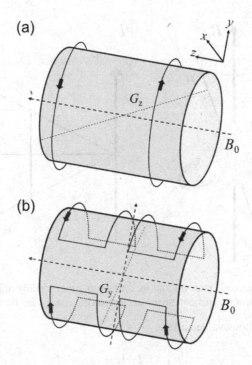

FIGURE 4.2 Design of gradient coils. The coils in (a) and (b) produce magnetic fields with the z components proportional to z and y coordinates, respectively.

the outermost periphery of the superconducting magnet or establishing a magnetic layer in the wall of the examination room.

4.2.2 Nuclear Magnetic Moment and Macroscopic Magnetization

A nucleus has spin and generates a magnetic moment derived from this. Intuitively, spin corresponds to the rotation of a nucleus, and magnetic moment means that the nucleus behaves as a small magnet with N and S poles. NMR is caused by the interaction between the nuclear magnetic moment and an external magnetic field.

Figure 4.3(a) shows a nucleus placed in a static magnetic field. The following relation holds true between the spin angular momentum vector $\boldsymbol{J}$ and the magnetic moment vector $\boldsymbol{\mu}$ of the nucleus.

$$\mu = \gamma J \tag{4.1}$$

Here, γ is called the gyromagnetic ratio and its value is specific to the nuclide. Furthermore, each nuclide has a spin quantum number I, which has the value of an integer or half-integer. The magnitude of the spin angular momentum vector $\boldsymbol{J}$ is given by $\sqrt{I(I+1)}\hbar$ using the Planck constant $\hbar = h/2\pi$. In MRI, hydrogen ^{1}H (a proton) is measured in most cases because of the following reasons: it gives high sensitivity, a large amount of water exists in the human body, and signal attenuation due to relaxation is relatively slow, which is described later. For hydrogen, $\gamma = 42.58$ MHz/T and $I = 1/2$. In addition, because of the quantum-theoretical effect, the direction of spin angular momentum cannot be specified, but when observed in a certain direction, such as the z direction, its component is given by

$$J_z = m_I \hbar \tag{4.2}$$

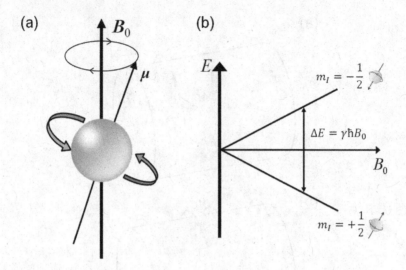

FIGURE 4.3 (a) Nuclear magnetic moment and its rotation in an externally applied static magnetic field. (b) Zeeman effect on the nuclear magnetic moment and a resulting energy gap between −1/2 and + 1/2 spins.

where m_I takes a discontinuous value such as

$$m_I = I, I-1, \cdots -I. \tag{4.3}$$

When the main field oriented in the z direction

$$B_0 = B_0 e_z \tag{4.4}$$

is applied to such a nucleus, an interaction energy E between the static magnetic field and the magnetic moment is generated.

$$E = -\mu \cdot B_0 = -\gamma J_z B_0$$
$$= -\gamma m_I \hbar B_0 \tag{4.5}$$

In the case of hydrogen, the energy E_1 ($m_I = 1/2$) of the magnetic moment in the same direction as the magnetic field is

$$E_1 = -\frac{1}{2} \gamma \hbar B_0 \tag{4.6}$$

and the energy E_2 ($m_I = -1/2$) of the magnetic moment opposite to the magnetic field is

$$E_2 = \frac{1}{2} \gamma \hbar B_0. \tag{4.7}$$

Hydrogen with a magnetic moment opposite to the magnetic field, that is, downward in the z-axis direction, has higher energy. As shown in Figure 4.3(b), the energy difference ΔE is

$$\Delta E = E_2 - E_1$$
$$= \gamma \hbar B_0. \tag{4.8}$$

This splitting is called Zeeman splitting, and the energy level is called the Zeeman level. This energy gap is proportional to the static magnetic field strength B_0.

When a nucleus in such a state is irradiated by an electromagnetic wave with photon energy equal to ΔE from the RF coil, the spin in the lower Zeeman level absorbs the energy and moves to the upper level. The nucleus then emits an electromagnetic wave with the same photon energy and returns to the lower Zeeman level. This electromagnetic wave is detected by the RF coil and is recorded as an NMR signal. The photon energy is proportional to the frequency ω_0 of the electromagnetic wave. The frequency corresponding to the energy ΔE is expressed as

$$\omega_0 = \frac{\Delta E}{\hbar} = \gamma B_0 \tag{4.9}$$

and is called the Larmor frequency or magnetic resonance frequency. Because of the above gyromagnetic ratio, when a 1-T static magnetic field is applied externally to hydrogen, resonance occurs because of irradiation with electromagnetic waves with a frequency of 42.58 MHz.

In a state of thermal equilibrium, the ratio of the number of spins in the two Zeeman levels is the Boltzmann distribution. Here, if the number of spins with energy E is $N(E)$, the Boltzmann distribution is given by

$$N(E) \propto \exp\left(-\frac{E}{kT}\right) \tag{4.10}$$

where k is the Boltzmann constant (1.38×10^{-23} J/K), and T is the temperature. Considering that $kT \gg \gamma \hbar B_0$ at room temperature, if the ratio of the number N_1 of spins in the lower level (same direction as the magnetic field) and the number N_2 of spins in the higher level (opposite the magnetic field) is calculated from Equations (4.8) and (4.10), then

$$\frac{N_2}{N_1} = \exp\left(-\frac{\gamma \hbar B_0}{kT}\right) \approx 1 - \frac{\gamma \hbar B_0}{kT}. \tag{4.11}$$

At $B_0 = 1$ T and $T = 300$ K, $\gamma \hbar B_0/kT = 6.8 \times 10^{-6}$, and low-energy level N_1 is slightly larger than high-energy level N_2.

From the above discussion, the magnetic moments at the lower and upper levels are $+\gamma \hbar/2$ and $-\gamma \hbar/2$, respectively. When N_s hydrogen nuclei are contained in the unit volume of a measured object such as a human body and a static magnetic field of strength B_0 is applied in the z direction, the z component of the magnetic moment per unit volume can be obtained as follows:

$$
\begin{aligned}
M_0 &= \frac{\left(+\frac{\gamma \hbar}{2}\right)N_1 + \left(-\frac{\gamma \hbar}{2}\right)N_2}{N_1 + N_2} N_s \\
&= \frac{\gamma \hbar}{2} \cdot \frac{N_1 - N_2}{N_1 + N_2} N_s \\
&\approx \frac{\gamma^2 \hbar^2 B_0 N_s}{2kT} \cdot \frac{N_1}{N_1 + N_2} \\
&\approx \frac{\gamma^2 \hbar^2 B_0 N_s}{4kT}
\end{aligned}
\tag{4.12}
$$

Here, while obtaining the second to third rows by equation (4.11), the fourth row was obtained using the fact that the difference between N_1 and N_2 is very small, such that $N_1 + N_2 \approx 2N_1$. When the external magnetic field is in the z direction, Zeeman splitting does not occur in the x and y directions, so the sum of the magnetic moments in the state of thermal equilibrium becomes zero.

The magnetic moment per unit volume is called macroscopic magnetization or simply magnetization and is represented by the vector M. In a state of thermal equilibrium

$$M = M_0 e_z \approx \frac{\gamma^2 \hbar^2 B_0 N_s}{4kT} e_z. \tag{4.13}$$

4.2.3 Relaxation of Magnetization

Immediately after the strength of the external magnetic field applied to the nuclear spin changes abruptly or when the nuclear spin moves between Zeeman levels due to NMR, the magnetization temporarily deviates from the state of thermal equilibrium. However, as time passes, magnetization approaches the value determined by Equation (4.13) based on the external magnetic field. This process is called relaxation. The rate of relaxation and the mechanism are known to differ between the process in which the z component of magnetization approaches M_0 and the process in which the x and y components approach zero.

The process in which the z component approaches M_0 is called spin-lattice relaxation or T_1 relaxation. As can be seen from the discussion in the previous section, the z component of magnetization is determined by the ratio of the low-energy upward magnetic moment and the high-energy downward magnetic moment. T_1 relaxation is a process in which energy is exchanged between nuclear spin and the surrounding environment. Here, the surrounding environment also has an energy level difference equal to the energy of the photon emitted by the nuclear spin and absorbs the energy. The concept of an environment that absorbs such energy is called a lattice. When the z component M_z of magnetization is M_1 at time $t = 0$, its approach to a state of thermal equilibrium M_0 is expressed by the following equation:

$$M_z = M_0 + \left(M_1 - M_0\right) e^{-\frac{t}{T_1}} \tag{4.14}$$

The time constant T_1 representing the speed of relaxation is called the T_1 relaxation time. At time $t = T_1$, $(M_z - M_0)/(M_1 - M_0) = 1/e = 0.368$, and the difference between the z component of the magnetization M_z and the thermal equilibrium state M_0 at that time is 36.8% of $M_1 - M_0$, which is the value at $t = 0$.

T_1 relaxation time is related to the speed of thermal motion of the molecule to which the nucleus belongs, and when the representative frequency of thermal motion is close to the Larmor frequency, energy exchange becomes active and the T_1 relaxation time becomes short. For medium-sized biomolecules and water molecules in some bound state with biomolecules, the T_1 relaxation time is relatively short because the thermal motion frequency is close to the Larmor frequency. For molecules that are small and fast, such as pure water, or for macromolecules that move very slowly, the T_1 relaxation time is longer because of the slower energy exchange.

The process by which the x and y components approach zero is called spin–spin relaxation or T_2 relaxation. Spins not only interact with the externally applied static magnetic field B_0 but also interact with local magnetic fields created by surrounding spins. The local magnetic field strength is different for each nucleus and varies with time. This acts in a direction to disassemble the orientation of the magnetic moments of the nuclei. When the x component M_x or the y component M_y of the magnetization are M_1 at time $t = 0$, the process where this approaches the thermal equilibrium state zero is expressed by the following equation:

$$M_x = M_1 e^{-\frac{t}{T_2}}$$

$$\tag{4.15}$$

$$M_y = M_1 e^{-\frac{t}{T_2}}$$

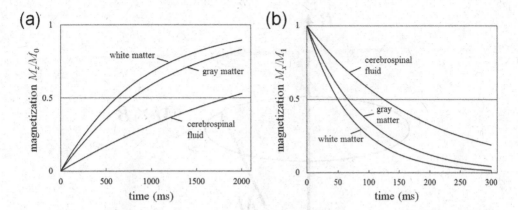

FIGURE 4.4 (a) T_1 relaxation and (b) T_2 relaxation of nuclear magnetizations in biological tissues.

The time constants for the x and y components are equal, and this is called the T_2 relaxation time. At time $t = T_2$, $M_x/M_1 = 1/e = 0.368$, and magnetization attenuates to 36.8% of the initial value.

For soft tissues, typical T_1 relaxation times are several hundred milliseconds and T_2 relaxation times are tens of milliseconds. Figures 4.4(a) and (b) show the curves for T_1 relaxation and T_2 relaxation times, respectively, for gray matter ($T_1 = 1124$ ms, $T_2 = 95$ ms), white matter ($T_1 = 884$ ms, $T_2 = 72$ ms), and cerebrospinal fluid ($T_1 = 2650$ ms, $T_2 = 180$ ms). In MRI measurements, in addition to T_2 relaxation, there is also the effect of magnetostatic inhomogeneity on the macroscopic scale, which causes magnetization to attenuate faster than pure T_2 relaxation. This is called T_2^* relaxation.

4.2.4 BLOCH EQUATION

What is observed using the RF coil is not the magnetic moment of individual nuclei, but the movement of magnetization given by the nuclei as a group. Magnetization generates a magnetic field in the surrounding space, and when the surrounding magnetic field changes with time because of the movement of magnetization, the magnetic flux linked to the RF coil placed in the vicinity also changes. Then, a voltage is generated in the RF coil according to the law of electromagnetic induction, and this is recorded as an MRI signal. Therefore, analyzing the motion of magnetization is very important for understanding the physics of MRI. The motion of magnetization is described by the following equation:

$$\frac{d\boldsymbol{M}}{dt} = \gamma \boldsymbol{M} \times \boldsymbol{B} - \frac{M_x \boldsymbol{e}_x + M_y \boldsymbol{e}_y}{T_2} + \frac{(M_0 - M_z)\boldsymbol{e}_z}{T_1} \tag{4.16}$$

where $\boldsymbol{M} = (M_x, M_y, M_z)$ is magnetization; $\boldsymbol{B}$ is the magnetic field applied; $\boldsymbol{e}_x$, $\boldsymbol{e}_y$, and $\boldsymbol{e}_z$ are unit vectors in the x, y, and z directions, respectively; M_0 is the value of M_z in a state of thermal equilibrium; and T_1 and T_2 are the relaxation times. Equation (4.16) takes its name from the physicist who proposed it and it is called the Bloch equation. This forms the basis for NMR and MRI. Under conditions where relaxation can be ignored, just the first terms on the left and right sides can be extracted and analyzed.

$$\frac{d\boldsymbol{M}}{dt} = \gamma \boldsymbol{M} \times \boldsymbol{B} \tag{4.17}$$

$d\boldsymbol{M}/dt$ is the velocity vector of $\boldsymbol{M}$ and refers to the direction in which $\boldsymbol{M}$ moves. This means that $\boldsymbol{M}$ keeps moving in the direction normal to the surface of magnetization $\boldsymbol{M}$ and magnetic field $\boldsymbol{B}$ (direction $\boldsymbol{M} \times \boldsymbol{B}$). As shown in Figure 4.5, this motion is a rotational motion around an axis given

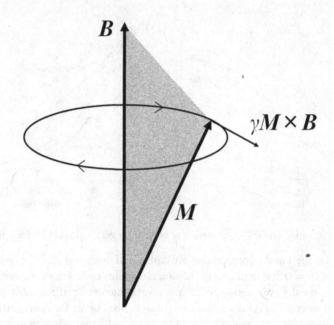

FIGURE 4.5 Motion of magnetization following the Bloch equation.

by magnetic field **B**. The classical physics interpretation is that the magnetization with spin angular momentum, which corresponds to rotation, precesses like a spinning top around the axis of **B**. The angular velocity of the precession is $\gamma|B|$, which is consistent with the magnetic resonance frequency expressed by Equation (4.9).

The second and third terms on the right side of Equation (4.16) represent the effect of relaxation. For confirmation, differentiating Equation (4.14) and Equation (4.15) gives

$$\frac{dM_z}{dt} = -\frac{M_z - M_0}{T_1}$$

$$\frac{dM_x}{dt} = -\frac{M_x}{T_2} \tag{4.18}$$

$$\frac{dM_y}{dt} = -\frac{M_y}{T_2}$$

Multiplying the first to third formulas by e_z, e_x, and e_y, respectively, and consolidating gives

$$\frac{dM}{dt} = -\frac{M_x e_x + M_y e_y}{T_2} + \frac{(M_0 - M_z)e_z}{T_1} \tag{4.19}$$

Combining Equations (4.17) and (4.19) gives Equation (4.16).

There are three magnetic fields applied in MRI: the main field, gradient field, and RF field. The main field is applied with strength B_0 in the z direction and is expressed as $\boldsymbol{B}_0 = B_0\boldsymbol{e}_z$. The main field has a very high degree of uniformity and can be considered to have almost the same direction and magnitude at any position within the imaging field.

A set of three coils that generates a magnetic field inclined in the x, y, and z directions generates the gradient field. In MRI, a magnetic field distribution is generated by adding the gradient magnetic field $\boldsymbol{\beta}$ to the main magnetic field $\boldsymbol{B}_0$ as a vector. Figure 4.6 shows the magnetic field distribution when the x gradient coil is operated. The magnitude and direction of the gradient magnetic field $\boldsymbol{\beta}$

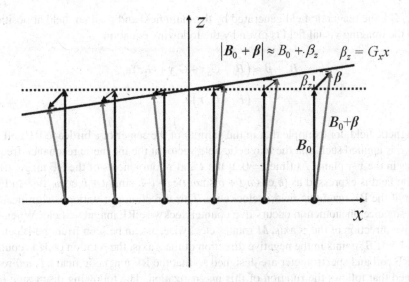

FIGURE 4.6 Vectorial summation of the main field and gradient field.

vary spatially, and the x gradient coil is designed so that the z component of β, that is, β_z is proportional to the value of the x coordinate. That is, with G_x as constant, $\beta_z = G_x x$ holds true. The strength of the gradient magnetic field is very weak compared with that of the main field. In Figure 4.6, β is drawn larger for easy viewing, but β actually has a magnitude 1/1000 that of B_0. At this time, the vector $B_0 + \beta$, which is the sum of both magnetic fields, points almost in the z direction (direction of B_0). Using the law of cosines and the approximation $(1 + x)^{1/2} \approx 1 + x/2$, when $x \ll 1$, the magnitude of $B_0 + \beta$ is expressed as follows:

$$|B_0 + \beta| = |B_0| \sqrt{1 + \frac{2B_0 \cdot \beta}{|B_0|^2} + \frac{|\beta|^2}{|B_0|^2}}$$

$$\approx |B_0| \left[1 + \frac{B_0 \cdot \beta}{|B_0|^2} + \frac{|\beta|^2}{2|B_0|^2} \right] \qquad (4.20)$$

$$\approx |B_0| + \frac{B_0 \cdot \beta}{|B_0|}$$

$$= B_0 + \beta_z$$

In the change from the second to the third row, the fact that the third term $|\beta|^2/2|B_0|^2$ is very small compared with the second term $B_0 \cdot \beta/|B_0|^2$ is used. From the above, the combined magnetic field of the main magnetic field and the magnetic field generated by the x gradient coil can be expressed as

$$B_0 + \beta \approx |B_0 + \beta| e_z$$

$$\approx (B_0 + \beta_z) e_z \qquad (4.21)$$

$$= (B_0 + G_x x) e_z$$

Similarly, the y and the z gradient coils generate magnetic field distributions such that $\beta_z = G_y y$ and $\beta_z = G_z z$. Thus, when the gradient field generated by the three-axis gradient coil set is expressed as

$G = (G_x, G_y, G_z)$, the magnetic field generated by the main field and gradient field at position $r = (x, y, z)$ within the imaging visual field is given by the following equation:

$$B_0 + \beta \approx \left(B_0 + G_x x + G_y y + G_z z \right) e_z$$

$$= \left(B_0 + G \cdot r \right) e_z \tag{4.22}$$

The RF magnetic field, for example that in the vicinity of the center of a birdcage RF coil as shown in Figure 4.7, is applied such that the magnetic field vector at the magnetic resonance frequency ω_0 = γB_0 rotates in the x–y plane. At time $t = 0$, if the x and y components of the RF magnetic field are b_x and b_y, the field is expressed as $(b_x \cos\omega_0 t + b_y \sin\omega_0 t)e_x + (-b_x \sin\omega_0 t + b_y \cos\omega_0 t)e_y$. In Figure 4.7, the rotation of the RF magnetic field is clockwise. This direction of rotation is significant, and no magnetic resonance phenomenon occurs in a counterclockwise RF magnetic field. When B_0 points in the positive direction of the z- axis, M rotates clockwise, as can be seen from the Bloch equation and Figure 4.5. If B_0 points in the negative direction of the z axis, the rotation of M is counterclockwise. The RF coil and spectrometer are designed so that the RF magnetic field has a direction and rotation speed that follows the rotation of this magnetization. The following discussion holds true for such cases.

From the above, the external magnetic field in the Bloch equation is given by the following equation:

$$B = \left(b_x \cos \omega_0 t + b_y \sin \omega_0 t \right) e_x + \left(-b_x \sin \omega_0 t + b_y \cos \omega_0 t \right) e_y + \left(B_0 + G \cdot r \right) e_z \tag{4.23}$$

To proceed with the analysis of such a magnetic field, it is often more convenient to present the magnetization and magnetic field components in a frame that rotates around the z axis at the frequency

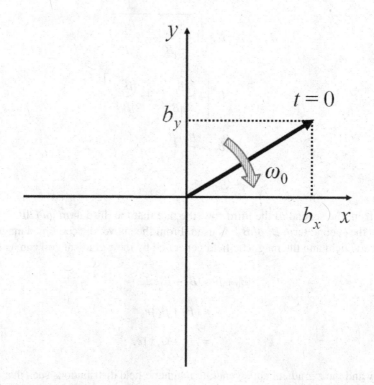

FIGURE 4.7 RF magnetic field.

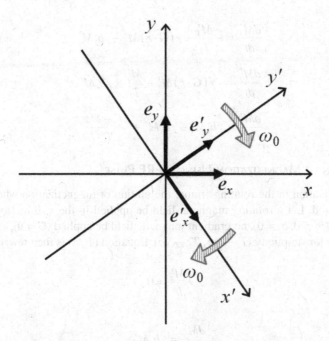

FIGURE 4.8 Bloch equation is described on a rotating frame with the angular frequency of ω_0.

ω_0. As shown in Figure 4.8, if the axes of the rotating frame are x', y', and z', the following relationship holds true for the unit vector of each axis.

$$e'_x = e_x \cos \omega_0 t - e_y \sin \omega_0 t$$

$$e'_y = e_x \sin \omega_0 t + e_y \cos \omega_0 t \tag{4.24}$$

$$e'_z = e_z$$

Using these unit vectors, Equation (4.23) can be written as follows:

$$B = b_x e'_x + b_y e'_y + \left(B_0 + G \cdot r \right) e'_z \tag{4.25}$$

In addition, each component of magnetization is transformed as follows:

$$M = M_x e_x + M_y e_y + M_z e_z = M'_x e'_x + M'_y e'_y + M'_z e'_z \tag{4.26}$$

Noting that e'_x, e'_y, and e'_z are functions of time, as shown in Equation (4.24), differentiation is carried out as follows:

$$\frac{dM}{dt} = \frac{dM'_x}{dt} e'_x + \frac{dM'_y}{dt} e'_y + \frac{dM'_z}{dt} e'_z + M'_x \frac{de'_x}{dt} + M'_y \frac{de'_y}{dt} + M'_z \frac{de'_z}{dt}$$

$$= \left(\frac{dM'_x}{dt} + \omega_0 M'_y \right) e'_x + \left(\frac{dM'_y}{dt} - \omega_0 M'_x \right) e'_y + \frac{dM'_z}{dt} e'_z \tag{4.27}$$

By substituting Equations (4.25), (4.26), and (4.27) into Equation (4.16) and rearranging them by component, the Bloch equation in the rotating frame can be obtained.

$$\frac{dM'_x}{dt} = -\frac{M'_x}{T_2} + \gamma\left(\boldsymbol{G} \cdot \boldsymbol{r}\right)M'_y - \gamma b_y M'_z$$

$$\frac{dM'_y}{dt} = -\gamma\left(\boldsymbol{G} \cdot \boldsymbol{r}\right)M'_x - \frac{M'_y}{T_2} + \gamma b_x M' \qquad (4.28)$$

$$\frac{dM'_z}{dt} = \gamma b_y M'_x - \gamma b_x M'_y - \frac{M'_z - M_0}{T_1}$$

4.2.5 EXCITATION OF MAGNETIZATION USING AN RF PULSE

Using the Bloch equation in the rotating frame, the motion of magnetization when an RF field is irradiated is analyzed. Let a rotating magnetic field be applied in the x' direction of the rotating coordinate system ($b_x \neq 0$, $b_y = 0$), no gradient magnetic field be applied ($\boldsymbol{G} = \boldsymbol{0}$), and the relaxation effect be negligible for simplicity ($T_1 \to \infty$, $T_2 \to \infty$). Equation (4.28) is then rewritten as follows:

$$\frac{dM'_x}{dt} = 0 \qquad (4.29)$$

$$\frac{dM'_y}{dt} = \gamma b_x M'_z$$

$$\frac{dM'_z}{dt} = -\gamma b_x M'_y$$

The first equation shows that the x' component of magnetization does not change with time. Multiplying the third equation by an imaginary unit i and adding it to the second equation gives

$$\frac{d}{dt}\left(M'_y + iM'_z\right) = -i\gamma b_x\left(M'_y + iM'_z\right) \qquad (4.30)$$

Given the thermal equilibrium state $M'_y + iM'_z = iM_0$ as the initial condition, the solution to this equation becomes

$$M'_y + iM'_z = iM_0 \exp\left(-i\gamma b_x t\right) \qquad (4.31)$$

If the real part and the imaginary part are written separately

$$M'_y = M_0 \sin\left(\gamma b_x t\right) \qquad (4.32)$$

$$M'_z = M_0 \cos\left(\gamma b_x t\right)$$

As shown in Figure 4.9, the motion of magnetization at this time is such that the magnetization that was initially oriented in the z direction rotates around the x' axis so that it falls over time. The angular velocity of this rotational motion is γb_x. When an RF field is applied in a pulse of time width τ, the angle α at which the magnetization falls is given as follows:

$$\alpha = \tan^{-1}\frac{M'_y}{M'_z} = \gamma b_x \tau \qquad (4.33)$$

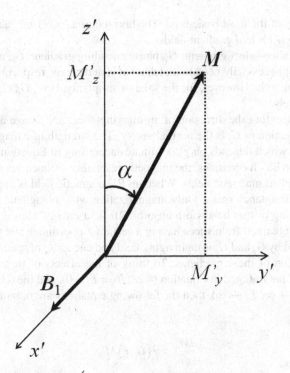

FIGURE 4.9 Motion of magnetization caused by an application of the RF field.

The angle α is called the flip angle, the RF pulse with $\alpha = 90°$ is called the 90° pulse, and the RF pulse with $\alpha = 180°$ is called the 180° pulse.

When the RF pulse irradiation is stopped, the nuclear spin relaxes according to Equation (4.18) while releasing energy and returns to a state of thermal equilibrium.

4.2.6 PULSE SEQUENCE AND IMAGE RECONSTRUCTION

During MRI acquisition, RF fields and gradient fields are applied with precisely designed timing and waveform. Diagramming of these actions results in a pulse sequence. Among many types of pulse sequences developed through previous studies, Figure 4.10 shows a gradient echo pulse

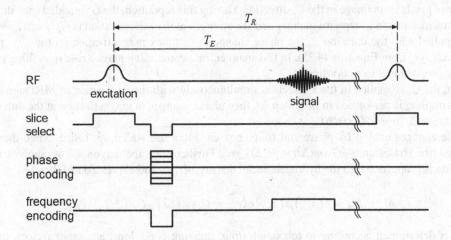

FIGURE 4.10 Pulse sequence diagram of gradient echo.

sequence, which is one of the most basic type. The horizontal axis is time, and the vertical axis is the intensity of the RF fields and gradient fields.

The directions of a slice-select gradient, G_s, phase encoding gradient, G_p, and frequency encoding gradient, G_r, do not necessarily need to match an x, y, and z axis, respectively, and they may be inclined to a coordinate axis. However, for the sake of simplicity, here, G_r, G_p, and G_s match an x, y, and z axis, respectively.

G_s plays a role in deciding the direction of an imaging slice, and image acquisition occurs for slices in which the direction of G_s is the normal vector. The strength of a magnetic field applied to the object is $B_0 + G_s z$, which depends on z coordinates according to Equation (4.22). As described in formula (4.9) and the Bloch equations, the magnetic resonance frequencies are $\omega(z) = \gamma(B_0 + G_z z)$ proportional to the applied magnetic fields. When an RF magnetic field is applied, the pulse has a finite bandwidth on a frequency space. Only magnetization with a magnetic resonance frequency in a range corresponding to this bandwidth absorbs RF field energy. Hence, magnetic resonance signals are selectively obtained from slices having a certain z-coordinate and thickness.

An x-y plane defined by G_r and G_p is an imaging field and encoding of spatial information occurs because of the operation of these gradients. To think of the effects of these gradients in simple terms, if an RF field is not radiated in Equation (4.28) ($b_x = b_y = 0$), and the relaxation effect is considered negligible ($T_1 \to \infty$, $T_2 \to \infty$), then the following equations are obtained from the first and second equations:

$$\frac{dM'_x}{dt} = \gamma \left(\boldsymbol{G} \cdot \boldsymbol{r} \right) M'_y \tag{4.34}$$

$$\frac{dM'_y}{dt} = -\gamma \left(\boldsymbol{G} \cdot \boldsymbol{r} \right) M'_x$$

If complex magnetization is composed assuming the x component of magnetization to be the real part, the y component to be the imaginary part, and initial magnetization to be the complex number M_{1c}, then the following solution is obtained according to considerations that are similar to deriving solution (4.31) from Equation (4.29):

$$M'_x + iM'_y = M_{1c} \exp\left[-i\gamma \left(\boldsymbol{G} \cdot \boldsymbol{r} \right) t \right] \tag{4.35}$$

In the gradient echo imaging, the pulse sequence in Figure 4.10 is repeated as many times as the number of pixels in an image in the G_p direction. During this repetition, the G_p is added such that the ΔG_p units increase in a stepwise manner and the strength at the nth operation is $G_p = n\Delta G_p$. When G_p is applied with the duration τ_p, the phase change in complex magnetization on the x'-y' plane is $-\gamma n(\Delta G_p)y\tau_p$ from Equation (4.35). In this manner, phase encoding plays a role in making phase changes proportional to y coordinates.

Next, the G_r is applied in the x direction simultaneously with the acquisition of MRI signals. If signal sampling is performed in time step Δt, then phase changes in magnetization at the mth sampling time $m(\Delta t)$ are $-\gamma G_r x m(\Delta t)$.

Phase changes caused by phase and frequency encoding are added up; hence, when they are extracted, the phases are $-\gamma G_r x m(\Delta t) - \gamma n(\Delta G_p)y\tau_p$. Furthermore, the magnitude of complex magnetization is proportional to the hydrogen nuclei density $\rho(x, y)$, and expressed as follows:

$$M_x + iM_y = A\left(x,y \right)\rho\left(x,y \right)\exp\left[-i\gamma G_r x m\left(\Delta t \right) - i\gamma n\left(\Delta G_p \right)y\tau_p \right] \tag{4.36}$$

$A(x, y)$ is determined according to relaxation time, imaging conditions, and other factors, and is a function that determines contrast (image darkness and brightness). Magnetic resonance signals measured with an RF coil are proportional to the integral of magnetization over all objects. If the parameters are simplified by $\gamma G_r m(\Delta t) = k_x$ and $\gamma n(\Delta G_p)\tau_p = k_y$, the result is as follows:

$$S(k_x, k_y) \propto \iint A(x, y)\rho(x, y)\exp(-ik_x x - ik_y y)\,dxdy$$

$$= \int A(r)\rho(r)\exp(-ik \cdot r)\,dr \qquad (4.37)$$

This formula means that the acquired signal $S(k_x, k_y)$ is expressed by a two-dimensional Fourier transform of $A(x, y)\rho(x, y)$. Consequently, if a recorded signal is processed by a two-dimensional inverse Fourier transform, it is possible to reconstruct hydrogen atom density $\rho(x, y)$ with a contrast function $A(x, y)$ applied.

$$A(x, y)\rho(x, y) \propto \iint S(k_x, k_y)\exp(ik_x x + ik_y y)\,dk_x dk_y$$

$$= \int S(k)\exp(ik \cdot r)\,dk \qquad (4.38)$$

In this manner, extracted data are arranged on a plane made by k_x and k_y, and this is called the k space. Figure 4.11 shows the relationship between k space and a reconstructed image. MRI signals are waveforms to time t. The data in k space are MRI signals obtained using the pulse sequence arranged as digital data sampled through analog–digital (A–D) conversion. Figure 4.12 is the relationship between (a) part of the pulse sequence and (b) k space. The G_r gradient pulse causes a displacement in the k_x axis, and the G_p gradient pulse causes a displacement in the k_y axis. The position on k space is shifted by simultaneously applying the G_r and G_p gradient pulses. Signals are arrayed in k space on the line along the k_x direction by collecting echo signals with a simultaneously applied G_r gradient. It is possible to cover the k space completely with the data array by varying the k_y position of the data line, according to G_p gradient pulse strength, per signal collection. In Figure 4.12(b), the position in the k space is displaced to the target k_y value by the application of G_p gradient pulse

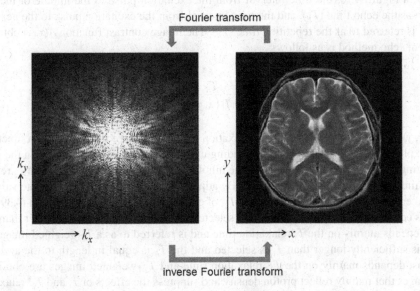

FIGURE 4.11 Signals stored on a k space and a reconstructed image of the human brain. Since the signals in k space are given by a Fourier transform of the distribution of magnetization, the reconstruction is performed by an inverse Fourier transform.

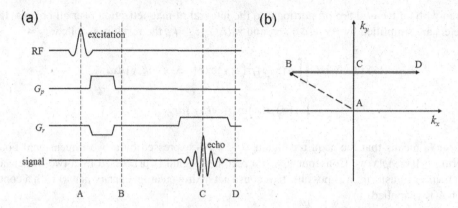

FIGURE 4.12 Storing an echo signal on a k space. A gradient pulse in (a) causes a displacement on a k space in (b). An echo signal is stored along a line parallel to k_x.

and the first G_r gradient pulse, and echo signals are recorded while applying the second G_r gradient pulse. Images are obtained through inverse Fourier transform of the collected k space data array.

k space data do not directly show the magnetization space distribution, but they show the magnitude and phase of various spatial frequency components constituting the magnetization space distribution. This means that k space positions (k_x, k_y) correspond to each spatial frequency and the data are considered to be the magnitude and phase of frequency components. These stored values show complex numbers having a real part and an imaginary part. The low-frequency region around the center of k-space data includes information on signal strength and contrast of a measured object, and the data of high-frequency region far from the center includes information on the boundaries and edges. Such k space characteristics are useful when considering techniques of pulse sequence design and data acquisition.

4.2.7 CONTRAST

As shown in Figure 4.10, the time interval from the excitation pulse to the middle of the echo is referred to as the echo time (T_E), and the time interval from the excitation pulse to the next excitation pulse is referred to as the repetition time (T_R). The image contrast function $A(x, y)$ obtained by the gradient echo method is as follows:

$$A(x,y) = \left\{1 - \exp\left(-\frac{T_R}{T_1(x,y)}\right)\right\} \exp\left(-\frac{T_E}{T_2^*(x,y)}\right)$$

(4.39)

Here, $T_1(x, y)$ and $T_2^*(x, y)$ are the object's relaxation times, which are assumed to be a function of x and y to explicitly show that they differ depending upon their location in the object. In the gradient echo imaging, the T_2 contrast is affected by the inhomogeneity of B_0, and the effective T_2 relaxation time containing this effect is denoted by T_2^*. By adjusting the T_R and T_E to appropriate values, it is possible to enhance the differences in T_1 and T_2^* of each tissue according to Equation (4.39). When a T_R that is equal in length to the average T_1 is selected and the T_E is sufficiently shorter than T_2^*, the contrast depends mainly on the T_1 relaxation time and is referred to as a T_1-weighted image. When a T_R that is sufficiently longer than T_1 is selected and the T_E is equal in length to the average T_2^*, the contrast depends mainly on the T_2^* relaxation time and T_2^*-weighted images are obtained. To obtain images that mainly reflect proton density and suppress the effects of T_1 and T_2^* relaxation as much as possible, a sufficiently long T_R and short T_E are selected. Spin echo imaging, another major pulse sequence, allows us to have a contrast depending on T_2 instead of T_2^*. Figure 4.13 shows a T_1-weighted image and T_2-weighted image acquired by the spin echo method.

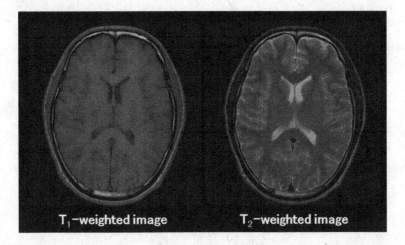

FIGURE 4.13 T_1-weighted image and T_2-weighted image of the human brain obtained using a spin echo sequence.

4.3 NUMERICAL ANALYSES OF THE BLOCH EQUATION

In what follows, all magnetization and magnetic field components are expressed in a rotating frame. Assuming magnetization $M = (M_x', M_y', M_z') = (m_x, m_y, m_z)$, $M_0 = m_0$, $\omega_x = -\gamma b_x$, $\omega_y = -\gamma b_y$, $\omega_G = -\gamma G \cdot r$, $R_1 = 1/T_1$, and $R_2 = 1/T_2$, the Bloch equation in a rotating frame can be represented by the following matrix form:

$$\begin{pmatrix} \dot{m}_x \\ \dot{m}_y \\ \dot{m}_z \end{pmatrix} = \underbrace{\begin{pmatrix} -R_2 & -\omega_G & \omega_y \\ \omega_G & -R_2 & -\omega_x \\ -\omega_y & \omega_x & -R_1 \end{pmatrix}}_{\Omega} \begin{pmatrix} m_x \\ m_y \\ m_z \end{pmatrix} + \begin{pmatrix} 0 \\ 0 \\ R_1 m_0 \end{pmatrix} \tag{4.40}$$

To make it simple, the analysis is divided into two cases: when the RF magnetic field is irradiated and when it is not.

In a typical pulse sequence, the RF pulse width is often very short compared with the T_1 and T_2 relaxation times of the object. In this case, as the effect of relaxation is small when the RF magnetic field is irradiated, the analysis can be approximated to $R_1 = R_2 = 0$. In the case where the RF magnetic field is not irradiated and the effect of relaxation is taken into account, let $\omega_x = \omega_y = 0$.

In the case where the RF magnetic field is irradiated, the following equation must be solved:

$$\begin{pmatrix} \dot{m}_x \\ \dot{m}_y \\ \dot{m}_z \end{pmatrix} = \underbrace{\begin{pmatrix} 0 & -\omega_G & \omega_y \\ \omega_G & 0 & -\omega_x \\ -\omega_y & \omega_x & 0 \end{pmatrix}}_{\Omega_1} \begin{pmatrix} m_x \\ m_y \\ m_z \end{pmatrix} \tag{4.41}$$

As preparation for the analysis, given the matrix Ω_1 and time t, the exponential function of the matrix is

$$\exp(\Omega_1 t) = \sum_{p=0}^{\infty} \frac{1}{p!}(\Omega_1 t)^p \tag{4.42}$$

When Ω_1 is constant, differentiating the above equation term-by-term gives

$$\frac{d}{dt}\exp(\Omega_1 t) = \sum_{p=1}^{\infty}\frac{1}{p!}\Omega_1^p p t^{p-1}$$

$$= \Omega_1 \sum_{p=1}^{\infty}\frac{1}{(p-1)!}\Omega_1^{p-1} t^{p-1} \tag{4.43}$$

$$= \Omega_1 \exp(\Omega_1 t)$$

For an arbitrary constant $c = (c_x, c_y, c_z)$, the general solution to Equation (4.41) is

$$\begin{pmatrix} m_x \\ m_y \\ m_z \end{pmatrix} = \exp(\Omega_1 t)\begin{pmatrix} c_x \\ c_y \\ c_z \end{pmatrix} \tag{4.44}$$

To solve the Bloch equation numerically, we are required to find the equation for calculating the (κ + 1)th magnetization vector $m^{\kappa+1} = (m_x^{\kappa+1}, m_y^{\kappa+1}, m_z^{\kappa+1})$ at time $t^{\kappa+1} = t^{\kappa} + \Delta t^{\kappa}$ from the κth magnetization vector $m^{\kappa} = (m_x^{\kappa}, m_y^{\kappa}, m_z^{\kappa})$ at time t^{κ}, which was Δt^{κ} seconds earlier. Next, starting from the initial magnetization (m_x^0, m_y^0, m_z^0), we simply iterate the equation to calculate the magnetization vector at each time. Eliminating the constant c from these equations

$$m^{\kappa} = \exp(\Omega_1 t^{\kappa})c \tag{4.45}$$

$$m^{\kappa+1} = \exp[\Omega_1(t^{\kappa} + \Delta t^{\kappa})]c \tag{4.46}$$

we obtain

$$\begin{pmatrix} m_x^{\kappa+1} \\ m_y^{\kappa+1} \\ m_z^{\kappa+1} \end{pmatrix} = \exp(\Omega_1 \Delta t^{\kappa})\begin{pmatrix} m_x^{\kappa} \\ m_y^{\kappa} \\ m_z^{\kappa} \end{pmatrix} \tag{4.47}$$

Depending on the programming language used for numerical analysis, it may be necessary to convert $\exp(\Omega_1 \Delta t^{\kappa})$ into a form that is easier to calculate. For a matrix such as Ω_1 in Equation (4.40), the following Rodrigues formula holds:

$$\exp(\Omega_1 \Delta t^{\kappa}) = I + \frac{\sin\alpha}{\alpha}(\Omega_1 \Delta t^{\kappa}) + \frac{1-\cos\alpha}{\alpha^2}(\Omega_1 \Delta t^{\kappa})^2 \tag{4.48}$$

where I is the identity matrix and α is

$$\alpha = \Delta t^{\kappa}\sqrt{\omega_x^2 + \omega_y^2 + \omega_G^2} \tag{4.49}$$

Using this formula, $\exp(\Omega_1 \Delta t^{\kappa})$ can be calculated as follows:

$$\exp(\Omega_1 \Delta t^{\kappa}) = \begin{pmatrix} (1-\cos\alpha)p_x^2 + \cos\alpha & p_x p_y(1-\cos\alpha) - p_z\sin\alpha & p_z p_x(1-\cos\alpha) + p_y\sin\alpha \\ p_x p_y(1-\cos\alpha) + p_z\sin\alpha & (1-\cos\alpha)p_y^2 + \cos\alpha & p_y p_z(1-\cos\alpha) - p_x\sin\alpha \\ p_z p_x(1-\cos\alpha) - p_y\sin\alpha & p_y p_z(1-\cos\alpha) + p_x\sin\alpha & (1-\cos\alpha)p_z^2 + \cos\alpha \end{pmatrix}$$

$$\tag{4.50}$$

provided that

$$\boldsymbol{p} = \begin{pmatrix} p_x \\ p_y \\ p_z \end{pmatrix} = \frac{\Delta t^\kappa}{\alpha} \begin{pmatrix} \omega_x \\ \omega_y \\ \omega_z \end{pmatrix} \tag{4.51}$$

In the case where the RF magnetic field is not irradiated, the following equation must be solved:

$$\begin{pmatrix} \dot{m}_x \\ \dot{m}_y \\ \dot{m}_z \end{pmatrix} = \underbrace{\begin{pmatrix} -R_2 & -\omega_G & 0 \\ \omega_G & -R_2 & 0 \\ 0 & 0 & -R_1 \end{pmatrix}}_{\Omega_2} \begin{pmatrix} m_x \\ m_y \\ m_z \end{pmatrix} + \begin{pmatrix} 0 \\ 0 \\ R_1 m_0 \end{pmatrix} \tag{4.52}$$

The following is the general solution of this equation, which can be confirmed by substituting it into Equation (4.52):

$$\begin{pmatrix} m_x \\ m_y \\ m_z \end{pmatrix} = \exp(\Omega_2 t) \begin{pmatrix} c_x \\ c_y \\ c_z \end{pmatrix} + \begin{pmatrix} 0 \\ 0 \\ m_0 \end{pmatrix} \tag{4.53}$$

Repeating the same reasoning that obtained Equation (4.47) from Equation (4.41), we obtain:

$$\begin{pmatrix} m_x^{\kappa+1} \\ m_y^{\kappa+1} \\ m_z^{\kappa+1} \end{pmatrix} = \exp(\Omega_2 \Delta t^\kappa) \begin{pmatrix} m_x^\kappa \\ m_y^\kappa \\ m_z^\kappa \end{pmatrix} + \left(I - \exp(\Omega_2 \Delta t^\kappa)\right) \begin{pmatrix} 0 \\ 0 \\ m_0 \end{pmatrix} \tag{4.54}$$

Next, $\exp(\Omega_2 \Delta t^\kappa)$ is calculated. By letting λ be the eigenvalue of matrix $\Omega_2 \Delta t^\kappa$, the characteristic equation is

$$\det \begin{pmatrix} -R_2 \Delta t^\kappa - \lambda & -\omega_G \Delta t^\kappa & 0 \\ \omega_G \Delta t^\kappa & -R_2 \Delta t^\kappa - \lambda & 0 \\ 0 & 0 & -R_1 \Delta t^\kappa - \lambda \end{pmatrix} \tag{4.55}$$

$$= -\left(\lambda + R_1 \Delta t^\kappa\right)\left[\left(\lambda + R_2 \Delta t^\kappa\right)^2 + \left(\omega_G \Delta t^\kappa\right)^2\right] = 0$$

The eigenvalue and the corresponding eigenvector of length 1 are obtained as follows:

$$\lambda_1 = -\left(R_2 + i\omega_G\right)\Delta t^\kappa, \quad \lambda_2 = -\left(R_2 - i\omega_G\right)\Delta t^\kappa, \quad \lambda_3 = -R_1 \Delta t^\kappa \tag{4.56}$$

$$\varepsilon_1 = \frac{1}{\sqrt{2}} \begin{pmatrix} 1 \\ i \\ 0 \end{pmatrix}, \quad \varepsilon_2 = \frac{1}{\sqrt{2}} \begin{pmatrix} i \\ 1 \\ 0 \end{pmatrix}, \quad \varepsilon_3 = \begin{pmatrix} 0 \\ 0 \\ 1 \end{pmatrix} \tag{4.57}$$

We then create a matrix of the eigenvectors and find the inverse matrix as follows:

$$P = \frac{1}{\sqrt{2}} \begin{pmatrix} 1 & i & 0 \\ i & 1 & 0 \\ 0 & 0 & \sqrt{2} \end{pmatrix}, \quad P^{-1} = \frac{1}{\sqrt{2}} \begin{pmatrix} 1 & -i & 0 \\ -i & 1 & 0 \\ 0 & 0 & \sqrt{2} \end{pmatrix}$$

The following equation holds for these matrices:

$$\exp\left(P^{-1}\Omega_2\Delta t^\kappa P\right) = \sum_{p=0}^{\infty} \frac{1}{p!}\left(P^{-1}\Omega_2\Delta t^\kappa P\right)^p$$

$$= \sum_{p=0}^{\infty} \frac{1}{p!}P^{-1}\left(\Omega_2\Delta t^\kappa\right)^p P \tag{4.59}$$

$$= P^{-1}\left[\sum_{p=0}^{\infty} \frac{1}{p!}\left(\Omega_2\Delta t^\kappa\right)^p\right]P$$

$$= P^{-1}\exp\left(\Omega_2\Delta t^\kappa\right)P$$

Therefore,

$$\exp\left(\Omega_2\Delta t^\kappa\right) = P\exp\left(P^{-1}\Omega_2\Delta t^\kappa P\right)P^{-1} \tag{4.60}$$

When the calculation is performed using

$$\exp\left(P^{-1}\Omega_2\Delta t^\kappa P\right) = \exp\begin{pmatrix} -\left(R_2+i\omega_G\right)\Delta t^\kappa & 0 & 0 \\ 0 & -\left(R_2-i\omega_G\right)\Delta t^\kappa & 0 \\ 0 & 0 & -R_1\Delta t^\kappa \end{pmatrix}$$

$$= \sum_{p=0}^{\infty} \frac{1}{p!}\begin{pmatrix} -\left(R_2+i\omega_G\right)\Delta t^\kappa & 0 & 0 \\ 0 & -\left(R_2-i\omega_G\right)\Delta t^\kappa & 0 \\ 0 & 0 & -R_1\Delta t^\kappa \end{pmatrix}^p \tag{4.61}$$

$$= \begin{pmatrix} \exp\left[-\left(R_2+i\omega_G\right)\Delta t^\kappa\right] & 0 & 0 \\ 0 & \exp\left[-\left(R_2-i\omega_G\right)\Delta t^\kappa\right] & 0 \\ 0 & 0 & \exp\left(-R_1\Delta t^\kappa\right) \end{pmatrix}$$

we obtain

$$\exp\left(\Omega_2\Delta t^\kappa\right) = \begin{pmatrix} \exp\left(-R_2\Delta t^\kappa\right)\cos\left(\omega_G\Delta t^\kappa\right) & -\exp\left(-R_2\Delta t^\kappa\right)\sin\left(\omega_G\Delta t^\kappa\right) & 0 \\ \exp\left(-R_2\Delta t^\kappa\right)\sin\left(\omega_G\Delta t^\kappa\right) & \exp\left(-R_2\Delta t^\kappa\right)\cos\left(\omega_G\Delta t^\kappa\right) & 0 \\ 0 & 0 & \exp\left(-R_1\Delta t^\kappa\right) \end{pmatrix} \tag{4.62}$$

In the aforementioned discussion, equations were derived for a time-discretized external magnetic field. In the case where the RF field is irradiated, Equations (4.47) and (4.50) are used. In the case where it is not irradiated, Equations (4.54) and (4.62) are used.

The time step interval Δt^κ does not have to be the same for every step. However, to avoid violating the assumptions made when obtaining Equation (4.43), it is necessary to choose an interval that keeps the fluctuation of each element in matrix Ω_1 or Ω_2 sufficiently small during that time step. In practice (for example, when the RF magnetic field is irradiated with a pulse waveform such as a sinc or Gaussian), the fluctuations of ω_x and ω_y at each step are minimized by selecting a time step that is sufficiently short compared with the pulse width. Periods when no RF magnetic field or gradient magnetic field is applied can be combined into a single step, since only R_1 and R_2, which do not change over time, are left in $\boldsymbol{\Omega}_2$.

In the aforementioned analysis, which was divided into two separate cases, the equation was solved by ignoring the relaxation effect in the case where the RF magnetic field was irradiated. However, depending on the pulse sequence and imaging conditions, an RF magnetic field and a gradient magnetic field may be applied simultaneously, and the analysis may need to include the effect of relaxation. Examples of conditions when this may be necessary include when the pulse width of the RF magnetic field is very long, or when the TR is so short that it is on the same order as the RF pulse width. In such cases, the following equation can be used for analysis:

$$
\begin{pmatrix} m_x^{\kappa+1} \\ m_y^{\kappa+1} \\ m_z^{\kappa+1} \end{pmatrix} = \exp\left(\Omega \Delta t^\kappa\right) \begin{pmatrix} m_x^\kappa \\ m_y^\kappa \\ m_z^\kappa \end{pmatrix}
$$

$$
+ \frac{R_1 m_0}{R_1 \omega_G^2 + R_2\left(\omega_x^2 + \omega_y^2\right) + R_1 R_2^2} \left(I - \exp\left(\Omega \Delta t^\kappa\right)\right) \begin{pmatrix} \omega_x \omega_G + \omega_y R_2 \\ \omega_y \omega_G - \omega_x R_2 \\ \omega_G^2 + R_2^2 \end{pmatrix}
$$

(4.63)

4.4 FUNCTIONAL MRI

4.4.1 NEUROVASCULAR COUPLING

Functional MRI is a technique for acquiring a map of brain activity from changes in MRI signals derived from neuronal activity in the brain. This technique was demonstrated by Ogawa et al. [3, 4]. It has been known that a relationship exists between neuronal activity and the blood flow and hemoglobin oxygenation. Oxygen is consumed by neuronal activity, so the blood oxygenation rate decreases over several hundred milliseconds, but the vascular system locally increases blood flow to compensate and increases oxygen supply. In many cases, more oxygen is supplied than that consumed by neurons owing to the strong feedback action, and the hemoglobin oxygenation rate rises over seconds after neuronal activity. Hemoglobin magnetic properties depend upon bonding with oxygen. Deoxyhemoglobin that does not bond with oxygen is paramagnetic, and oxyhemoglobin that bonds with oxygen is diamagnetic. Because of its magnetic moment, paramagnetic deoxyhemoglobin generates a microscopic disturbance in magnetic field distribution and shortens the T_2^* relaxation time. In the tissue in which neuronal activity arises, the oxygenation rate rises, so T_2^* relaxation time is prolonged, and consequently, the signal strength rises. This is the blood oxygenation level dependent (BOLD) effect. In addition to the T_2^* relaxation time being prolonged, the signal strength is increased because of the increase in blood flow; hence, protons that have not been excited flow in from outside the imaging slices. Functional MRI is the capturing, statistical sampling, and imaging of these signal changes.

What makes functional MRI superior to various other functional neuroimaging techniques is that it is noninvasive and has high spatial resolution. Other methods for obtaining brain function images include positron emission tomography (PET) and single photon emission computed tomography (SPECT), but these require radioactive tracer agents. In contrast, functional MRI does not need tracers or contrast agents and detects neuron activity through the body's inherent hemoglobin magnetic properties. In terms of spatial resolution, functional MRI is superior to PET and SPECT in many cases. In addition, electroencephalography (EEG) and magnetoencephalography (MEG) are also noninvasive and detect neuronal electricity activity based on head surface potential and magnetic fields, so they have the advantage of making high temporal resolution on the order of milliseconds possible. However, an inverse problem needs to be solved to estimate neuronal current distribution in the brain from measured potential and magnetic fields on the brain surface. The issue with consistently finding a solution to this inverse problem is that it has difficult mathematical properties. In functional MRI, consistent high-resolution mapping is achievable without such difficulties. Near-infrared spectroscopy (NIRS) is another technology for detecting changes in the hemoglobin oxygenation rate with surface optical sensors, but NIRS can only measure the cerebral cortex, while functional MRI is capable of visualizing even deep brain activity.

Block design, event-related design, and resting state are the three main approaches by which brain activity is induced and then measured. In block design, the blocks that continuously produce brain activity by providing stimulation over several consecutive seconds or minutes to test subjects and the subsequent blocks that do not provide stimulation for several seconds or minutes are alternately arranged. The MRI signals between these two blocks are statistically analyzed, and then, the regions presenting significant differences are extracted. In event-related design, one-time stimulation is delivered, and subsequently, MRI signal responses, including time changes, are acquired. Thereafter, second-time or further stimulation is delivered with sufficient intervals. The advantage of this method is that timely information is also acquired. Resting state is a technique that was established recently. It deals with brain activity that is spontaneously produced without any particular external stimulation. Focusing on periodic fluctuations in individual voxel BOLD signals in places with spontaneously occurring activity enables the extraction of combinations of voxels with high correlation to these periodic fluctuations. A high correlation means some type of coupling between neurons, making it possible to locate neural network structures in the brain from measured results.

4.4.2 APPLICATIONS OF FUNCTIONAL MRI

Functional MRI offers an extremely wide range of applications, including fundamental research into brain function and psychology, and disease diagnosis. It is possible to catch microscopic activity in cerebral functional column units by utilizing spatial resolution, which is a major advantage of functional MRI. In the following section, several application examples will be introduced.

In recent years, it has been known that stimulation using electromagnetic fields sometimes improves symptoms of neurological and psychiatric diseases in which treatment with drugs has been unsuccessful. For example, pain relaxation through repetitive transcranial magnetic stimulation of the primary motor cortex to deal with refractory neuropathic pain has been indicated in clinical trials [5]. There has been accumulating clinical evidence for such treatment, while the mechanism by which electrical stimulation of the brain demonstrates therapeutic effects still presents many unknowns. Thus, research is being conducted to identify brain activity induced by direct stimulation of the primary motor cortex through functional MRI [6]. This has resulted in the identification of the process by which activity is induced in regions related to pain recognition, such as the striatum and thalamus, through stimulation of the primary motor cortex, as shown in Figure 4.14. Such research can provide clues to clarify this for treatment with unknown mechanisms.

In many MRI devices, images are acquired with subjects in a supine position with their heads placed at the center of a magnet. This position is different from peoples' everyday lives, and the fact that the visual field of test subjects is restricted to enable the insertion of their head deeply into a magnet has been an issue. Therefore, an asymmetric magnet design has been proposed; this design is capable of placing the uniform magnetic field regions used in imaging in places close to the edges instead of the center of the magnet [7]. There have been reports of imaging test samples with prototype magnets made for proof of principle. As for future applications, the possibility of diagnosing dementia more objectively and at an earlier stage has been indicated by ascertaining brain activity when dealing with the Benton visual retention test and other measures used in dementia testing.

As described above, general functional MRI is based on the BOLD effect, but the possibility of functional MRI using other contrast mechanisms is being explored. There is a report that diffusion MRI signals increased temporarily and locally in regions of activity within the brain, and it was mentioned that functional MRI using this may be possible [8]. In addition, there are also reports

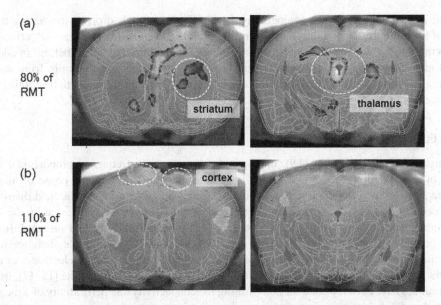

FIGURE 4.14 Functional MRI of the rat brain showing the activations evoked by electrical stimulations to the motor cortex. The intensity of stimulation was (a) 80% of the resting motor threshold (RMT) and (b) 110% of the RMT.

indicating the possibility of detecting weak currents produced from neuronal activity as signal changes in MRI [9]. It is expected that, if such methods are established, they will lead to new forms of functional neuroimaging that will have superior temporal and spatial resolution.

4.5 RECENT ADVANCES IN IMAGING TECHNIQUES AND APPLICATIONS

4.5.1 DIFFUSION

Imaging of diffusion phenomena using MRI has been carried out since the 1980s, mainly for self-diffusion of water molecules contained in living tissue [10]. Clinical applications began to be extensively demonstrated in the 1990s when rapid imaging techniques became available. These clinical studies showed a major advantage of diffusion MRI is in the diagnosis of acute cerebral infarction [11]. In the 1990s, active treatment of cerebral infarction began, and diffusion-weighted images that can depict the lesion in several hours after the onset were rapidly spread. In addition to acute cerebral infarction, diffusion-weighted images are known to show contrasts different from T_1-weighted and T_2-weighted images for tumors, and their usefulness has been established. Furthermore, in the tissue composed of fibrous cells such as the white matter of the brain, it has become possible to visualize the structure of the nerve fiber bundles in a three-dimensional manner by utilizing the property that the diffusion coefficient varies depending on the direction [12].

4.5.2 FLOW AND PRESSURE

It is also possible to visualize fluid dynamics parameters, such as flow velocity and pressure. If protons are repeatedly excited with intervals shorter than T_1, magnetization saturation is produced and the signal strength drops. In contrast, protons that receive excitation only once from a thermal equilibrium state generate strong signals. When this phenomenon is used to excite and measure an imaging volume repeatedly, the blood flowing in from outside the volume generates strong signals. Consequently, it is possible to extract arterial blood selectively with high signals; this principle is then used in magnetic resonance angiography. In addition, as signals drop by irradiating regions outside of a slice with a pulse selectively saturating protons, it is possible to extract selectively protons flowing into a volume with specific timing, which is called arterial spin labeling. In addition, attempts are being made to use MRI to quantify volumetric changes in the brain from elevated intracranial pressure – that is, cerebral compliance [13, 14] – which is expected to be applied while monitoring a brain disease.

4.5.3 ELECTRIC PROPERTIES OF TISSUE

According to Larmor's formula (4.9), magnetization rotates at a frequency proportional to the strength of an applied magnetic field, which is reflected in the frequency and phase of magnetic resonance signals. Focusing upon this, it is possible to find the internal magnetic field distribution of the object from MRI signals. Furthermore, magnetic field distribution is affected by the values of electromagnetic properties, including current, conductivity, permittivity, and susceptibility, and thus it is possible to visualize these electrical properties using MRI. For example, there are reports of techniques in which a pulsed magnetic field is applied to the object through electrodes or a coil, and conductivity is determined based on changes in signals produced at this time [15–17]. In addition, techniques have been reported for determining the conductivity and permittivity of a measured body based on RF field distribution, and clinical applications of this technique are expected based on the need to evaluate specific absorption rate (SAR) in high-field MRI [18–20]. Furthermore, a technique has been proposed using the relationship of the self-diffusion coefficient of water and live tissue conductivity obtained from diffuse MRI to map conductivity, which is characterized by being able to make measurements including conductivity anisotropy [21]. Figure 4.15 shows conductivity images of the human brain obtained using diffusion MRI.

4.5.4 MRI-GUIDED SURGERY AND TREATMENT

MRI is different from computed tomography (CT) and PET in that there is no radiation exposure, so repeated imaging is possible. It is expected to be a means of navigation in surgery and various other treatments. When this happens, the therapeutic instruments used need to be nonmagnetic, but various nonmagnetic tools are already being developed. In addition, in recent years, surgery has been complemented by methods in which the body is treated without cutting by irradiating affected areas with ultrasonic waves, electromagnetic waves, or radiation. High-intensity focused ultrasound (HIFU) is a therapeutic method in which ultrasonic waves of high amplitude are focused on a single spot, and then the disease sites are eliminated by pyrogenicity. By using MRI, it is possible to monitor temperature distribution in real time in addition to anatomical information from affected areas. Consequently, it is possible to use MRI throughout all treatment, including preparation of an HIFU treatment plan through detailed visualization of lesions, monitoring heat generation in the therapeutic process, and evaluating postoperative status. In recent years, there have also reportedly been examples of developing HIFU devices under MRI guides [23, 24].

4.5.5 IMAGE PROCESSING

Image processing using information-processing technology represented by artificial intelligence (AI) has had a significant impact on MRI. Automatic detection of aneurysms using surface extraction and other techniques has been conducted for a long time. Recently, the use of neural networks has made pattern recognition possible for a wider range of subjects. For example, automatic detection of lesions is now possible, and it is becoming useful in reducing the risk for radiologists overlooking lesions while interpreting large images on an everyday basis. In addition, techniques have also been reported that enable a neural network to learn the relationship of images including a lot of

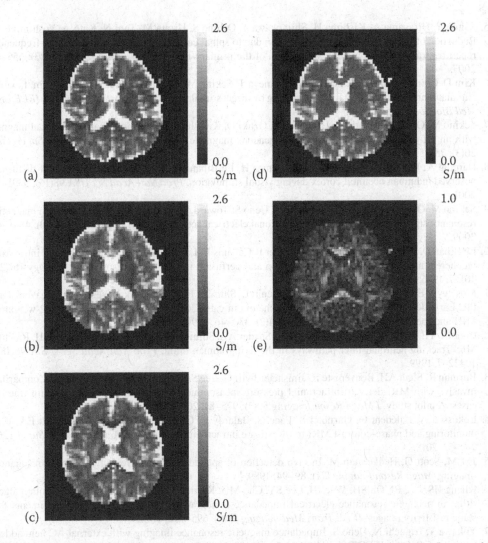

FIGURE 4.15 Conductivity imaging of the human brain using diffusion MRI. (a through c) Images of the estimated conductivities in the anterior-posterior, right-left, and superior-inferior directions, respectively. (d, e) Images of the mean conductivity and anisotropy index [22].

noise to images not including noise and generate images that have reduced noise. The scope of AI application is expected to become increasingly broad, encompassing even application to functional MRI image analysis.

REFERENCES

1. Lauterbur PC: Image formation by induced local interactions: Examples employing nuclear magnetic resonance. *Nature* 242(5394): 190–1, 1973.
2. Wada H, Sekino M, Ohsaki H, Hisatsune T, Ikehira H, Kiyoshi T: Prospect of high-field MRI. *IEEE Trans Appl Supercond* 20(3): 115–22, 2010.
3. Ogawa S, Lee TM, Kay AR, Tank DW: Brain magnetic resonance imaging with contrast dependent on blood oxygenation. *Proc Natl Acad Sci USA* 87(24): 9868–72, 1990.
4. Ogawa S, Tank DW, Menon R, Ellermann JM, Kim SG, Merkle H, Ugurbil K: Intrinsic signal changes accompanying sensory stimulation: Functional brain mapping with magnetic resonance imaging. *Proc Natl Acad Sci USA* 89(13): 5951–5, 1992.

 5. Saitoh Y, Hirayama A, Kishima H, Shimokawa T, Oshino S, Hirata M, Tani N, Kato A, Yoshimine T: Reduction of intractable deafferentation pain due to spinal cord or peripheral lesion by high-frequency repetitive transcranial magnetic stimulation of the primary motor cortex. *J Neurosurg* 107(3): 555–9, 2007.

 6. Kim D, Chin Y, Reuveny A, Sekitani T, Someya T, Sekino M: An MRI-compatible, ultra-thin, flexible stimulator array for functional neuroimaging by direct stimulation of the rat brain. *Conf Proc IEEE Eng Med Biol Soc* 6702–5, 2014.

 7. Sekino M, Ohsaki H, Wada H, Hisatsune T, Ozaki O, Kiyoshi T: Fabrication of an MRI model magnet with an off-centered distribution of homogeneous magnetic field zone. *IEEE Trans Appl Supercond* 20(3): 781–5, 2010.

 8. Darquié A, Poline JB, Poupon C, Saint-Jalmes H, Le Bihan D: Transient decrease in water diffusion observed in human occipital cortex during visual stimulation. *Proc Natl Acad Sci USA* 98(16): 9391–5, 2001.

 9. Sekino M, Ohsaki H, Yamaguchi-Sekino S, Ueno S: Toward detection of transient changes in magnetic resonance signal intensity arising from neuronal electrical activities. *IEEE Trans Magn* 45(10): 4841–4, 2009.

10. Le Bihan D, Breton E, Lallemand D, Grenier P, Cabanis E, Laval-Jeantet M: MR imaging of intravoxel incoherent motions: Application to diffusion and perfusion in neurologic disorders. *Radiology* 161(2): 401–7, 1986.

11. Moseley ME, Cohen Y, Mintorovitch J, Chileuitt L, Shimizu H, Kucharczyk J, Wendland MF, Weinstein PR: Early detection of regional cerebral ischemia in cats: Comparison of diffusion- and T2-weighted MRI and spectroscopy. *Magn Reson Med* 14(2): 330–46, 1990.

12. Conturo TE, Lori NF, Cull TS, Akbudak E, Snyder AZ, Shimony JS, McKinstry RC, Burton H, Raichle ME: Tracking neuronal fiber pathways in the living human brain. *Proc Natl Acad Sci U S A* 96(18): 10422–7, 1999.

13. Burman R, Shah AH, Benveniste R, Jimsheleishvili G, Lee SH, Loewenstein D, Alperin N: Comparing invasive with MRI-derived intracranial pressure measurements in healthy elderly and brain trauma cases: A pilot study. *J Magn Reson Imaging* 50(3): 975–81, 2019.

14. Lokossou A, Balédent O, Garnotel S, Page G, Balardy L, Czosnyka Z, Payoux P, Schmidt EA: ICP monitoring and phase-contrast MRI to investigate intracranial compliance. *Acta Neurochir Suppl* 126: 247–53, 2018.

15. Joy M, Scott G, Henkelman M: In vivo detection of applied electric currents by magnetic resonance imaging. *Magn Reson Imaging* 7(1): 89–94, 1989.

16. Khang HS, Lee BI, Oh SH, Woo EJ, Lee SY, Cho MH, Kwon O, Yoon JR, Seo JK: J-substitution algorithm in magnetic resonance electrical impedance tomography (MREIT): Phantom experiments for static resistivity images. *IEEE Trans Med Imaging* 21(6): 695–702, 2002.

17. Yukawa Y, Iriguchi N, Ueno S: Impedance magnetic resonance imaging with external AC field added to main static field. *IEEE Trans Magn* 35(5): 4121–3, 1999.

18. Ueno S, Iriguchi N: Impedance magnetic resonance imaging: A method for imaging of impedance distributions based on magnetic resonance imaging. *J Appl Phys* 83(11): 6450–2, 1998.

19. Katscher U, Voigt T, Findeklee C, Vernickel P, Nehrke K, Dössel O: Determination of electric conductivity and local SAR via B1 mapping. *IEEE Trans Med Imaging* 28(9): 1365–74, 2009.

20. Sekino M, Tatara S, Ohsaki H: Imaging of electric permittivity and conductivity using MRI. *IEEE Trans Magn* 44(11): 4460–3, 2008.

21. Sekino M, Ohsaki H, Yamaguchi-Sekino S, Iriguchi N, Ueno S: Low-frequency conductivity tensor of rat brain tissues inferred from diffusion MRI. *Bioelectromagnetics* 30(6): 489–99, 2009.

22. Sekino M, Inoue Y, Ueno S: Magnetic resonance imaging of electrical conductivity in the human brain. *IEEE Trans Magn* 41(10): 4203–5, 2005.

23. Verpalen IM, Anneveldt KJ, Nijholt IM, Schutte JM, Dijkstra JR, Franx A, Bartels LW, Moonen CTW, Edens MA, Boomsma MF: Magnetic resonance-high intensity focused ultrasound (MR-HIFU) therapy of symptomatic uterine fibroids with unrestrictive treatment protocols: A systematic review and meta-analysis. *Eur J Radiol* 120: 108700, 2019.

24. Liu X, Ellens N, Williams E, Burdette EC, Karmarkar P, Weiss CR, Kraitchman D, Bottomley PA: High-resolution intravascular MRI-guided perivascular ultrasound ablation. *Magn Reson Med* 83(1): 240–53, 2020.

5 Chemical Exchange Saturation Transfer and Amide Proton Transfer Imaging

Takashi Yoshiura

CONTENTS

5.1 Introduction 101
5.2 Principle of CEST... 102
5.3 Endogenous CEST and Amide Proton Transfer Imaging 103
5.4 APT Imaging Method.. 104
5.5 Quantifying APT Effect 104
5.6 Clinical Applications of APT Imaging.. 105
    5.6.1 Clinical Applications to Brain Tumors.. 105
    5.6.2 Prediction of Malignancy Grade in Diffuse Gliomas 105
    5.6.3 Prediction of Genetic Mutation in Gliomas.. 106
    5.6.4 Differential Diagnosis of Brain Tumors 107
    5.6.5 Discrimination Between Treatment-Related Changes and Viable Tumors.............. 108
    5.6.6 Applications to Other Brain Tumors 108
    5.6.7 APT Imaging of Head and Neck Lesions.. 109
    5.6.8 APT Imaging of Breast, Chest, and Abdominal Lesions.. 109
    5.6.9 Pitfalls and Limitations of APT Imaging in Oncology 110
5.7 CEST Imaging of pH... 111
5.8 CEST Imaging of GAGs.. 113
5.9 CEST Imaging of Glycogen.. 113
5.10 Exogenous CEST Agents.. 113
5.11 Concluding Remarks 115
References... 115

5.1 INTRODUCTION

Magnetic resonance (MR) imaging plays a critical role in modern clinical medicine. In addition to its non-invasiveness, the most eminent aspect of MR imaging is its multi-functionality. While MR imaging can provide exquisite high-resolution morphological images, it can also afford functional information, such as water molecular diffusion, perfusion, and neuronal activities. MR has also shown promise in molecular probing. By taking advantage of subtle differences in nuclear magnetic resonance frequencies, MR spectroscopy (MRS) detects and distinguishes metabolites in biological tissues. However, MRS has an intrinsically low signal-to-noise ratio (SNR); the SNR of MRS is far lower than that of positron emission tomography (PET). Given the low SNR, MRS typically reports an averaged signal within a 1–2 cm voxel, thus seriously limiting its role in clinical settings.

Chemical exchange saturation transfer (CEST) is a novel MR molecular imaging contrast first denominated by Ward and Balaban [1] in 2000. CEST allows for the imaging of low-concentration endogenous and exogenous molecules with spatial resolution comparable with those in conventional MR imaging. In addition, CEST imaging has the potential to provide physiological information, such as pH and glucose metabolism. CEST has recently attracted much interest in the community of clinical imaging. Several pioneering works have demonstrated its feasibility and uses in clinical practice. This chapter illustrates CEST imaging with a focus on its clinical impacts.

5.2 PRINCIPLE OF CEST

CEST is based on saturation transfer from a solute molecule to water molecules via chemical exchange (Figure 5.1). In this transfer, a proton physically moves back and forth between the solute and the solvent (water). The solute proton has a resonance frequency different from that of water but is not detectable by conventional MR imaging due to the very low solute concentration (typically on the order of millimolar). The exchangeable protons of the solute are saturated via a selective saturation radio frequency (RF) pulse at the solute frequency. The saturation is transferred to the bulk water via chemical exchange, resulting in a decrease in the bulk water's magnetization and signal. The saturation transfer continues during irradiation by the saturation pulse, resulting in a build-up of saturation in the bulk water pool. This amplification mechanism allows for the indirect observation of low-concentration solutes.

Figure 5.2 shows the Z-spectrum, which is a plot of the bulk water magnetization along the Z-axis against different saturation pulse frequencies relative to water frequency (offset frequency). The water signal is strongly suppressed around 0 ppm, which is the water resonance frequency, due to direct saturation. The CEST effect is observed as a small signal reduction in the spectrum at the specific frequency of the solute protons (Figure 5.2).

The proton exchange rate must be slow enough to enable the selective saturation of the solute protons but fast enough to induce a detectable water signal reduction. It is also notable that the proton exchange rate is dependent on environmental conditions, such as pH and temperature.

A higher magnetic field strength is generally advantageous in CEST as the shift difference, measured in hertz, increases in proportion to the magnetic field strength.

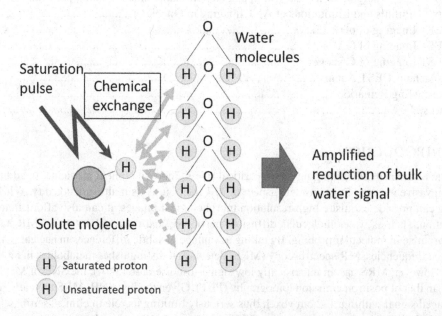

FIGURE 5.1 Principle of CEST. A saturation RF pulse with a specific frequency saturates exchangeable protons on a solute molecule. The saturated protons are transferred to water molecules via chemical exchange. The saturation of water molecules accumulates during continuous irradiation of the saturation pulse, thus amplifying the bulk water signal reduction.

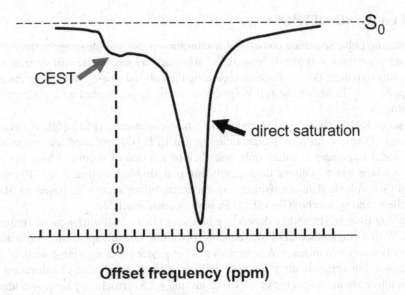

FIGURE 5.2 Z-spectrum. The magnetization of protons on water molecules along the Z-axis is plotted at varying offset frequencies of the saturation pulse. A strong suppression due to direct saturation is present at the water resonance frequency (0 ppm). The CEST effect is observed as a smaller signal reduction at the specific frequency for the solute exchangeable protons (ω ppm).

5.3 ENDOGENOUS CEST AND AMIDE PROTON TRANSFER IMAGING

CEST experiments can be conducted using both endogenous and exogenous molecules. The feasibility of endogenous CEST imaging was first demonstrated in the imaging of urea in the bladder of healthy human subjects [2]. Since then, several endogenous metabolites with exchangeable protons that have optimal exchange properties under physiological conditions have been identified and imaged *in vivo*. These include proteins and peptides with amide (–NH) [3]; glutamate and creatine with amine (–NH$_2$) [4,5]; and glycogen, glycosaminoglycan (GAG), and myo-inositol with hydroxyl (–OH) protons [6–8]. Thus, CEST-based MR imaging has shown promise as a non-invasive, non-ionizing tool for molecular imaging.

Amide proton transfer (APT) imaging is a specific type of endogenous CEST imaging technique proposed by Zhou et al. [3]. This method enables the semi-quantitative measurement of amide protons (–NH) in intrinsic mobile proteins and peptides, which have a specific resonance frequency 3.5 ppm downfield to that of water. Because this resonance frequency is sufficiently remote from that of water, and amide protons are relatively abundant in tissue, especially in tumors, APT imaging is considered the most clinically relevant type of CEST imaging.

The amide proton transfer ratio (APTR), which is the exchange transfer effect per proton, is described as follows [3]:

$$APTR = \frac{k[\text{amide proton}]}{2[\text{H}_2\text{O}]R_{1w}}\left(1 - e^{-R_{1w}t_{sat}}\right)$$

where k is the exchange rate constant, R_{1w} (= $1/T_{1w}$) is the spin–lattice relaxation rate of water, and t_{sat} is the saturation time. As shown by this equation, the APT effect is dependent not only on the fractional concentration of exchangeable amide protons but also on several other parameters, including the T1 of water, T_{1w}, and the saturation pulse length t_{sat}. As previously described, the proton exchange rate constant k is dependent on environmental conditions, such as pH and temperature.

5.4 APT IMAGING METHOD

An APT imaging pulse sequence consists of a saturation pulse and an image acquisition component. The saturation pulse is typically long (0.5–2 s) because a longer pulse will saturate more water protons [9], although there is no consensus regarding the optimal shape, strength, or duration of the saturation pulse [10]. To obtain the full Z-spectrum, imaging is repeated with different saturation pulse frequencies.

Several sequences, such as two-dimensional (2D) fast spin-echo (FSE) [11], 2D gradient-echo (GE) [12], and 2D spin-echo echo planar imaging (SE-EPI) [13], are used for image acquisition. Two-dimensional sequences evaluate only one slice of a lesion or require a long period of time to evaluate a whole lesion. Volume data acquisition is desirable for clinical use. Recently, three-dimensional (3D) volume data acquisition sequences, including sequences based on 3D FSE [14] and 3D gradient- and spin-echo (GRASE) [15], have become available.

The scanning time varies widely depending on the number of saturation pulse frequencies and slice levels. With a single-slice 2D FSE sequence obtaining the full spectrum, the scanning time may be approximately two minutes. A scan with a 3D sequence covering a large section of the brain may be completed in approximately seven to eight minutes if the choices of saturation pulse frequencies are limited to those specific to APT (i.e., around ± 3.5 ppm). In any case, an additional scan may be necessary for voxel-wise correction for B0 inhomogeneity (as explained in the next section).

5.5 QUANTIFYING APT EFFECT

As described earlier, CEST imaging is based on the saturation of solute protons at their specific frequency. However, this process may be hindered by magnetic field inhomogeneities in the magnet, which in turn could produce a shift in the Z-spectrum. Thus, correction for B0 inhomogeneities in each voxel is essential to obtain accurate measurements. Several different correction methods have been proposed; these include realigning the 0 ppm to the minima of the interpolated Z-spectrum [3]; a water suppression shift referencing method, in which direct water saturation is identified using low-power RF pulses [16]; and correction using a separately obtained B0 map [17].

The Z-spectrum not only consists of direct saturation and the CEST (APT) effect but also includes signal reductions by the non-specific magnetization transfer (MT) effect and the nuclear Overhauser effect (NOE) (Figure 5.3). The non-specific MT effect represents saturation transfer via complex mechanisms involving cross-relaxation [18] and chemical exchange [19]. In contrast to CEST, which is based on saturation transfer through proton exchange (chemical exchange) between water and mobile structures with long T2, non-specific MT occurs between semisolid macromolecules with short T2, such as myelin and water. The NOE is another saturation transfer mechanism that is observed upfield from the water frequency [20].

The APT effect is most often quantified as the asymmetry of the MT ratio, MTR_{asym}, at 3.5 ppm:

$$MTR_{asym}\left(3.5\ \mathrm{ppm}\right) = \left(S_{-3.5\ ppm} - S_{3.5\ ppm}\right)/S_0,$$

where $S_{-3.5\ ppm}$ and $S_{3.5\ ppm}$ are the signal intensities at −3.5 and 3.5 ppm, respectively, and S_0 is the unsaturated signal intensity.

This approach effectively eliminates the overlapping direct saturation and non-specific MT effects that occur due to their symmetric natures; however, the upfield NOE signal remains a confounding factor in the quantification of downfield APT. Moreover, the APT signal is not simply overlapped but is also diluted by competing with the direct saturation of water (spillover effect) and semisolid macromolecular MT effect. Nevertheless, the MTR_{asym} value at 3.5 ppm is widely accepted as an approximate index of APT signal intensity and is often called APT-weighted signal intensity (APTWSI). APT-weighted images are obtained by performing a voxel-wise mapping of this index (Figure 5.3).

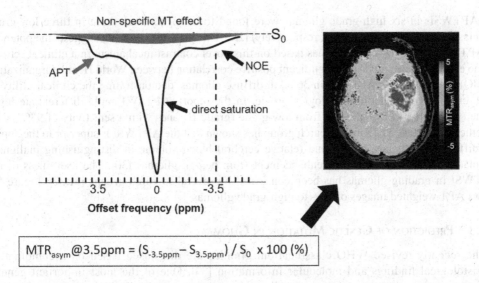

FIGURE 5.3 The APT effect measured as MTR_{asym} at 3.5 ppm, the asymmetry of the MT ratio at 3.5 ppm excluding contributions from symmetric direct saturation, and the non-specific MT effect. The NOE upfield of water remains a confounding factor. APT-weighted images are obtained via the voxel-by-voxel mapping of MTR_{asym} at 3.5 ppm.

To address the above-mentioned issue of MTR_{asym}, several analysis methods have been proposed. The Z-spectrum was fitted with a sum of Lorentzian functions corresponding to direct saturation, MT effect, NOE, and CEST effects, such as those from amide and amine, so that each contributor can be evaluated [21]. Zaiss et al. [22] proposed an index called apparent exchange-dependent relaxation (AREX), which can remove the effect of background MT and correct for T1. Model-based analyses where measured data were fitted with Bloch–McConnell equations have been proposed [23], although they are not readily available for clinical imaging studies. These more sophisticated methods of analysis would allow for more quantitative measurements of APT signals that can better reflect biological functions [22,24].

5.6 CLINICAL APPLICATIONS OF APT IMAGING

5.6.1 CLINICAL APPLICATIONS TO BRAIN TUMORS

APT imaging selectively detects amide protons in mobile proteins and peptides. In contrast to semi-solid structural proteins, mobile proteins primarily consist of cytosolic, endoplasmic reticulum, and secreted proteins. The underlying assumption regarding the APT imaging of tumors is that there is a close relationship between the proliferative activity of the tumor and mobile protein synthesis. Yan et al. [25] used rats implanted with gliosarcomas and showed that compared with healthy brain tissue, tumor tissue contains an increased concentration of cytosolic proteins that contribute to the APT signal despite there being virtually no difference between the total protein concentrations of the two tissues. Moreover, proteomic analyses have shown that there are different expressions between astrocytomas and normal brain tissues as well as among different grades of astrocytomas [26].

5.6.2 PREDICTION OF MALIGNANCY GRADE IN DIFFUSE GLIOMAS

Discrimination between low-grade and high-grade gliomas is critical because, regarding many tumors, the prognoses and thus appropriate therapeutic strategies differ substantially between these two cases [27]. In previous studies, the APTWSI (MTR_{asym} at 3.5 ppm) was found to be 3%–4% higher in tumor tissue than in peritumoral brain tissue for human brain tumors at 3T [28]. Moreover, the APTWSIs in six high-grade gliomas were found to be higher than those in three low-grade gliomas ($2.9 \pm 0.6\%$ vs. $1.2 \pm 0.2\%$, respectively) [28]. This preliminary result suggests the potential of APT imaging for grading gliomas based on this novel contrast mechanism in a clinical setting. Togao et al. [29] have reported a significant positive correlation between World Health Organization (WHO) grades and the APTWSI in 36 adult diffuse gliomas, demonstrating the clinical utility of APT imaging in predicting histological grade. In the report, APTSWI could differentiate high-grade (grade III and IV) gliomas from low-grade (grade II) ones with a sensitivity of 93% and a specificity of 100%. The same research group has shown that the APTWSI is superior to the apparent diffusion coefficient (ADC) and relative cerebral blood volume in distinguishing malignant gliomas without intense contrast enhancement from benign gliomas [30]. The usefulness of the APTWSI in grading gliomas has been confirmed by several research groups [31,32]. Figure 5.4 shows APT-weighted images of low- to high-grade gliomas.

5.6.3 Prediction of Genetic Mutation in Gliomas

In the recently revised WHO classification, gliomas are diagnosed based on a combination of histological findings and molecular information [33]. One of the most important genetic

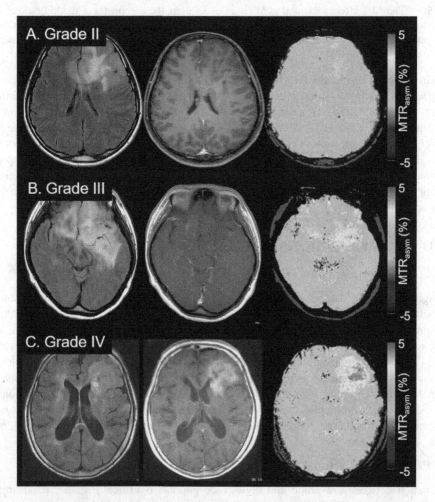

FIGURE 5.4 APT-weighted images of diffuse gliomas: (A) grade II – oligodendroglioma, (B) grade III – anaplastic oligodendroglioma, and (C) grade IV – glioblastoma. Note that the APTWSI (MTR$_{asm}$ at 3.5 ppm) increases as the grade increases.

mutations in diffuse gliomas is the isocitrate dehydrogenase (IDH) mutation. This mutation has been implicated in oncogenesis in several types of tumors and is known to have a major impact on the prognosis of affected patients. That is, the presence of the IDH mutation is recognized as a marker of favorable prognosis in grade II and III astrocytomas and glioblastomas [34,35]. Thus, preoperative detection of this mutation may facilitate the planning of an appropriate therapeutic strategy.

Jiang et al. [36] investigated whether APT imaging is useful for non-invasively predicting the IDH mutation status in grade II gliomas. They reported that IDH-wildtype gliomas tended to have higher APTWSIs than IDH-mutant tumors (IDH-wildtype: $1.39 \pm 0.49\%$ vs. IDH-mutant: $0.93 \pm 0.44\%$, respectively), suggesting the predictive value of APT imaging.

5.6.4 DIFFERENTIAL DIAGNOSIS OF BRAIN TUMORS

Few studies have focused on using APT imaging in the differential diagnosis of brain tumors. Differentiation between high-grade gliomas from primary central nervous system malignant lymphomas (PCNSLs) with conventional MR imaging techniques is often challenging. Jiang et al. [37] have demonstrated the usefulness of APT imaging in differentiating high-grade gliomas from PCNSLs. In their report, PCNSLs had significantly lower maximum APTWSIs ($APTW_{max}$) than high-grade gliomas ($3.38 \pm 1.06\%$ vs. $4.36 \pm 1.30\%$, respectively). This finding was hypothetically attributed to higher nucleus/cytoplasm ratios in PCNSLs, which would reduce the amount of cytoplasmic protein. Moreover, PCNSLs showed significantly lower differences between the maximum and minimum APTWSIs ($APTW_{max-min}$) than high-grade gliomas ($0.76 \pm 0.42\%$ vs. $2.55 \pm 1.20\%$, respectively), indicating a lower heterogeneity. The $APTW_{max-min}$ could differentiate the two tumor types with a sensitivity of 100% and a specificity of 84.6% [37]. Figure 5.5 compares APT-weighted images of a glioblastoma and a PCNSL.

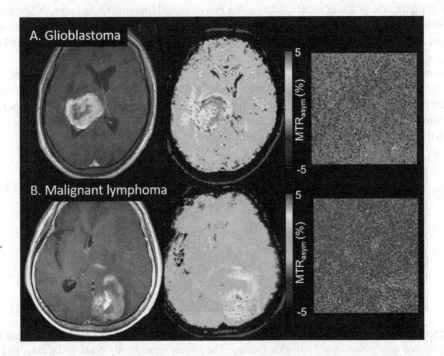

FIGURE 5.5 Comparison of glioblastoma (A) and PCNSL (B) images. Lower APTWSI is observed in PCNSL than in glioblastoma. Hematoxylin and eosin-stained histological sections show more densely aggregated cells with a higher nucleus/cytoplasm ratio in PCNSL than in glioblastoma.

It can also be difficult to differentiate between glioblastoma and solitary brain metastasis (SBM) on conventional MR images. Kamimura et al. [38] compared APTWSI histogram parameters between glioblastoma and SBM in the enhanced areas and peritumoral non-enhanced areas with T2-prolongation, i.e., peritumoral high signal intensity areas (PHAs). They found that the mean and 10th, 25th, 50th, 75th, and 90th percentiles for APTWSI in enhancing areas of glioblastomas were significantly higher than those of SBMs, whereas no significant difference was found in any histogram parameters for PHA. Their results suggest that APTWSI in enhancing areas is an imaging marker that can be used to differentiate between glioblastomas and SBMs.

5.6.5 DISCRIMINATION BETWEEN TREATMENT-RELATED CHANGES AND VIABLE TUMORS

Chemoradiation therapy following surgical resection has been regarded as a standard treatment for high-grade gliomas [39]. When an enhancing lesion emerges during a follow-up MR examination in patients with post-treatment gliomas, discriminating between recurrent tumors and treatment-related changes is crucial but often challenging when using conventional MR techniques [40]. An animal experiment predicted the potential utility of APT as a biomarker for this differentiation [41], and this utility has been verified in several clinical studies [42–45]. In these studies, recurrent tumors showed significantly higher APTWSIs than treatment-related changes; these high APTWSIs may be related to active proliferation. Figure 5.6 shows an APT-weighted image of a case of clinically diagnosed radiation necrosis.

5.6.6 APPLICATIONS TO OTHER BRAIN TUMORS

Meningioma is one of the most common extra-axial brain tumors. A majority of meningiomas are benign (grade I), but approximately 10% of cases are of higher grade (grade II or III) [33]. The latter are associated with higher rates of post-surgical recurrence and worse prognoses [33]. It is difficult to predict the grade of meningioma using conventional imaging techniques. A few previous reports have demonstrated how APT imaging can be used for this purpose. Joo et al. [46] compared 44 benign and 13 atypical (grade II) meningiomas and found that normalized MTR_{asym} (MTR_{asym} in tumor minus MTR_{asym} in normal-appearing white matter) was significantly higher in atypical meningioma than in benign meningioma. However, Yu et al. [47] failed to show a significant difference in the mean value of MTR_{asym} between grade I and II and concluded that the ability of APT to predict the grade of meningiomas is rather limited. Because of the relative rarity of higher grade meningiomas and pathological diversity, further studies with larger populations are necessary to establish the value of APT imaging in characterizing meningiomas.

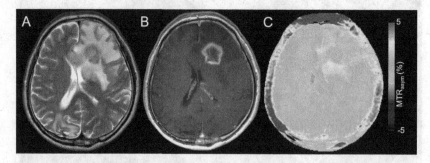

FIGURE 5.6 APT imaging of clinically diagnosed radiation necrosis. A patient with multiple brain metastases was treated with gamma-knife radiation therapy. The patient subsequently developed a mass-like lesion in the left frontal lobe. A T2-weighted image (A) and post-contrast T1-weighted image (B) show a heterogeneously enhancing lesion in the left frontal lobe. Unlike in cases of highly proliferating tumors, the APT-weighted image (C) shows no apparent elevation of APTWSI in the lesion.

5.6.7 APT Imaging of Head and Neck Lesions

The APT imaging of extracranial regions can be very challenging due to magnetic field inhomogeneity, movement caused by respiration and the heartbeat, and contamination from fat signals. However, APT imaging in the head and neck seems to be successful. Several studies have demonstrated its feasibility for obtaining images of lesions in the nasopharynx and salivary glands without major alterations in imaging techniques [48–50]. Law et al. [49] used APT imaging to differentiate benign salivary gland tumors from malignant head and neck tumors, specifically nasopharyngeal undifferentiated carcinomas, squamous cell carcinomas, and non-Hodgkin lymphomas, in 117 patients. They found that the APT signal of malignant tumors quantified by MTR_{asym} at 3.5 ppm was significantly higher than in benign tumors. Notably, APT imaging in combination with ADC derived from diffusion-weighted imaging improved the discriminating performance. The same group reported that early intra-treatment APT imaging may be useful in predicting the six-month outcome following concurrent chemoradiation therapy in patients with nasopharyngeal carcinoma [51]. Figure 5.7 shows an example of APT imaging in the head and neck region.

5.6.8 APT Imaging of Breast, Chest, and Abdominal Lesions

Compared with the upper abdomen, the pelvic region is less affected by respiratory and peristaltic movement, which makes it easier to obtain APT-weighted images with appropriate image quality. Takayama et al. [52] demonstrated the feasibility of APT imaging for endometrial carcinomas of the uterus using a free breathing imaging technique similar to that used for the brain. To minimize motion-related artifacts, the duration of the saturation pulse was shortened to 0.5 s. The authors showed that the APTWSI (MTR_{asym} at 3.5 ppm) correlated with the histological grade of the tumor. Li et al. [53] showed the utility of APT imaging for squamous cell carcinoma in the uterine cervix. It was revealed that the APTWSI is significantly positively correlated with the histological grade of the tumor, which affect the therapeutic strategy and the prognosis. Nishie et al. [54] compared pre-treatment APTWSIs and pathologically determined tumor responses to neoadjuvant chemotherapy in patients with rectal cancer. The mean APTWSI of the low-response group was significantly higher than that of the high-response group. This finding suggests the value of APT imaging in predicting responses to treatment.

APT imaging in the breast and lung appears to be much more challenging. Respiratory movements are apparent obstacles in these regions. Moreover, magnetic field homogeneity can be disturbed in the air-filled lung. Despite those confounding factors, Ohno et al. [55] demonstrated the

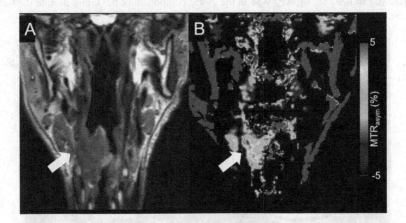

FIGURE 5.7 MR image of a patient with a hypopharyngeal squamous cell carcinoma: T1-weighted image (A) and APT-weighted image (B). The tumor (indicated by the arrows) is clearly visualized as a high APTWSI area.

feasibility of APT imaging in the lung. In the breast, the interference of fat signals with the APT measurement can be an additional problem. Feasibility in the breast has been shown in a few pilot studies [56,57].

5.6.9 Pitfalls and Limitations of APT Imaging in Oncology

Despite previous reports showing the clinical value of APT imaging in the management of brain tumors, the sources of signal contributors in APT imaging is only partially understood. In brain tumor imaging, APT contrast has been primarily attributed to intracellular cytosolic protein. However, clinical and experimental studies have shown that there may be other sources that contribute to increased APTWSIs.

First, it is known that a high APTWSI is often observed in cysts and cavities filled with protein-containing fluid [58]. Figure 5.8 shows an example of a cystic tumor with a high APTWSI. A high APTWSI in cysts appears to be independent of the malignancy grade. Thus, cystic components have been carefully excluded in previous studies regarding the evaluation of gliomas [29,42].

Second, there are several reports of hematoma as a source of high APTWSIs [59,60] (Figure 5.9). Blood includes abundant proteins (such as albumin) and peptides. A previous experimental report showed that porcine whole blood samples showed large MTR_{asym} values greater than 10%, peaking

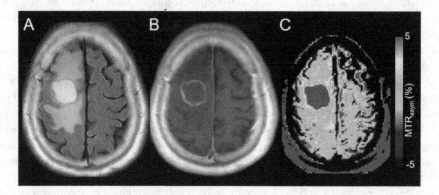

FIGURE 5.8 A patient with a metastatic brain tumor from a lung adenocarcinoma. FLAIR (A) and T1-weighted (B) images show a predominantly cystic mass with an enhancing rim. Note that the APT-weighted image (C) shows a strong APTWSI in the cystic component of the tumor despite the lack of viable tumor cells, presumably due to high protein content.

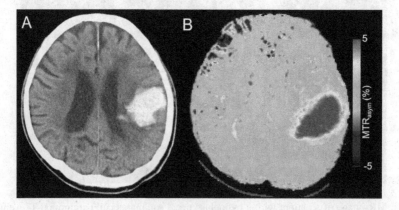

FIGURE 5.9 A patient with an acute cerebral hemorrhage. Brain CT (A) shows hyperdense hematoma in the left parietal lobe. The APT-weighted image (B) shows a markedly high APTWSI in the hematoma.

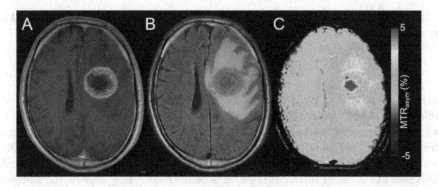

FIGURE 5.10 Brain MR images of a patient with a brain metastasis. The T1-weighted image (A) and FLARI image (B) show a ring-enhancing mass with extensive perifocal edema. Note that the APT-weighted image (C) shows a higher APTWSI in the perifocal edema than in the normal brain parenchyma. A cystic component of the tumor shows a markedly high APTWSI.

at 2.5 to 3 ppm at 3T [61]. In an animal study using a rat model, an increased APTWSI was observed in a hyperacute intracranial hemorrhage [59]. In a clinical report, acute hemorrhages showed high APTWSIs even without cellular contributions [59]. High APTWSIs have also been seen in blood flowing in vessels [61]. Recently, very high APTWSIs (MTR_{asym} values of up to 6.9%) have been reported in a cavernous malformation [62]. Thus, the blood in tumor vessels is a potential source of high APTWSIs in hypervascular tumors. Several studies have revealed that edematous brain tissue can show an increased APTWSI. In an early study by Wen et al. [58], it was found that in patients with gliomas, the APTWSI in edematous areas adjacent to the tumor core was significantly higher than that in normal-appearing white matter in the contralateral hemisphere. Because glioma cells can invade beyond the tumor core, the increased APTWSI may be attributable to either amides in infiltrating tumor cells or edema. More importantly, Yu et al. [63] have reported that the APTWSI was significantly higher in edematous areas surrounding brain metastases than in normal-appearing white matter. Because edematous areas surrounding a brain metastasis represent vasogenic edema and usually include no tumor cells, this result strongly suggests that edema can be associated with increased APTWSI. This finding was attributed to intravascular proteins and peptides penetrating the extravascular space through a leaky blood–brain barrier. Figure 5.10 shows a case of a brain metastasis with an increased APTWSI in the surrounding edema.

Finally, as previously mentioned, the APTWSI can be affected by the local pH and temperature through alterations in the proton exchange rate. Studies have shown that in many tumor types, the intracellular pH is higher than that in normal brain tissue [64]. Because the amide proton exchange rate is base catalyzed in the physiological pH range, the exchange rate increases as pH increases, thus increasing the APT contrast. Nevertheless, pH has not been considered a major source of increased APTWSI in malignant tumors. These should be recognized as possible confounding factors of the interpretation of APT imaging results.

5.7 CEST IMAGING OF PH

pH sensitively reflects tissue metabolism and can be a useful biomarker for detecting pathological conditions. For example, in brain ischemia, a loss of oxygen supply activates anaerobic metabolism, which in turn induces lactate production and lowers pH. Moreover, pH is considered related to hypoxia in rapidly proliferating malignant tumors and abnormal neuronal hyperactivities in psychiatric diseases. As mentioned earlier, the CEST signal can be affected by environmental factors, such as pH and temperature, by changing the proton exchange rate. Zhou et al. [3] showed that noninvasive imaging of pH is feasible based on the CEST of endogenous amide protons, namely APT imaging. In APT, the proton exchange rate is base catalyzed for pH values above five and are thus proportional to the hydroxyl ion concentration and exponentially proportional to pH [3]:

$$k = k_{base}[OH] = k_{base} \times 10^{pH-pKw}$$

where K_w is the ion product of water at a given temperature. Therefore, under assumptions of constant amide proton concentration and water content, it is possible to obtain a pH-weighted image. Using a k_{base} value measured in an animal experiment with 31 P MRS, Zhou et al. [3] further demonstrated the feasibility of pH quantification of focal brain ischemia in a rat model. Sun et al. [65] demonstrated the value of APT imaging in identifying ischemic "penumbra," the hypoperfused region that is still viable but at risk of infarction. In a rat ischemia model, they showed that a pH-weighted image at the hyperacute stage can predict the extent of final infarction more accurately than diffusion or perfusion-weighted imaging. This information may allow the optimal treatment strategy for patients with hyperacute ischemic stroke to be chosen. Nevertheless, to date, the clinical translation of APT-based pH-weighted imaging has not progressed, which is in part due to small signal intensity changes. Figure 5.11 shows a pH-weighted image of a patient with hyperacute infarction.

Previous studies have shown that tumor cells have alkaline intracellular pH as opposed to acidic extracellular pH. The acidic extracellular pH is considered a result of hypoxia in rapidly growing tumors. The acidic extracellular pH has a profound impact on the aggressiveness of the tumor, resulting in increased tumor invasion [66] and angiogenesis [67]. Furthermore, hypoxic tumor tissue with lowered pH is resistant to radiation therapy and chemotherapies. Thus, imaging pH may provide valuable information regarding tumor malignancy grade, the prediction of treatment efficacy, and prognosis. Challenges regarding APT-based pH imaging of tumors include the obvious violation of the assumption of constant amide proton concentration. Another proposed approach for pH-weighted imaging is based on the CEST of amine protons (at 2.0 ppm downfield to water) on proteins, peptides, and amino acids. In contrast to APT, the amine CEST-related signal increases pH decreases within the pH range of 6–7. Harris et al. [68] examined patients with glioblastoma using amine CEST pH-weighted imaging and reported that tumors with acidic signatures on pH-weighted images were associated with an elevated uptake of 18F-fluoro-DOPA, lactate on MRS, and perfusion abnormalities. In addition, they found that patients with acidic tumors at pre-treatment had significantly lower progression-free survival rate. However, this approach does not solve the issue of protein concentration dependence. McVicar et al. [69] proposed a unique solution for pH-weighted endogenous CEST imaging. Their method, called amine/amide concentration-independent detection (AACID), utilizes the ratio of CEST-mediated contrast from amine and amide protons to measure absolute pH independently from the protein concentration. In an animal experiment at 9.4T, they showed its feasibility in acute cerebral ischemia [69]. Furthermore, the same

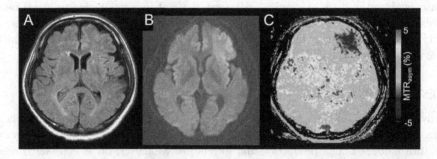

FIGURE 5.11 pH-weighted imaging of hyperacute stroke. No obvious signal intensity change is observed in the FLAIR image (A). The diffusion-weighted image at a b value of 1000 s/mm² (B) shows faint hyperintensity in the left frontal lobe. The APT-weighted image (C) shows a decreased APTWSI in the affected area, which reflects a lowered pH due to anaerobic metabolism and lactate production.

group demonstrated pH change in a tumor model induced by Cariporide, a sodium–proton exchange inhibitor [70]. When measuring the pH of tumors using the AACID technique, there is an assumption that the amine/amide exchangeable proton concentration ratio is constant across tumors, which in reality may be violated. In addition, clinical translation requires validation at lower magnetic fields (e.g., 3T), especially regarding the detection of amine CEST signals.

5.8 CEST IMAGING OF GAGS

GAGs are long, linear, unbranched polysaccharides consisting of repeating disaccharide units that are widely distributed in connective tissues, such as joint cartilage and intervertebral disks, where GAGs act as lubricant or shock absorbers. GAGs are classified into six categories: chondroitin sulfate, dermatan sulfate, keratan sulfate, heparin, heparan sulfate, and hyaluronan. GAG loss is characteristically seen in osteoarthritis, a chronic degenerative disorder of joint cartilage most often seen in the knees of aged women as well as in intervertebral disk degeneration. Since osteoarthritis is a disease of very high prevalence and social impact, imaging methods to detect the early degeneration of joint cartilage have been vigorously pursued. The current gold-standard imaging technique is called delayed gadolinium-enhanced MRI contrast (dGEMRIC), which allows for the indirect evaluation of GAG loss through the diffusion of $(DTPA)^{2-}$, which has a repulsive force against negatively charged GAGs. Other available imaging techniques include the mapping of T2 and T1ρ and 23Na imaging with ultra-high field MR imaging. However, none of these allow for the direct measurement of GAGs in the cartilage. Thus, imaging techniques for directly observing GAGs have been anticipated. Ling et al. [7] first proposed the CEST imaging of GAGs (gagCEST). The gagCEST is based on the chemical exchange at 3-hydroxyl groups (–OH, 1.0 ppm downfield to water frequency) in each GAG unit. An early *in vivo* study on healthy young subjects showed somewhat discouraging results; the amplitude of the gagCEST signal (MTR_{asym}) was up to 6% at 7T, whereas it was negligible at clinically accessible 3T [71]. However, a more recent study has shown the feasibility of gagCEST with clinical 3T MRI by indicating a significant correlation between gagCEST signal amplitude and GAG concentration *in vivo* [72]. As for the intervertebral disks, which have higher GAG concentrations and larger volumes than join cartilages, gagCEST imaging appears to be more clinically feasible. A few studies have demonstrated significant correlations between gagCEST-related signal intensity and the morphological grade of intervertebral disk degeneration [73,74].

5.9 CEST IMAGING OF GLYCOGEN

Glycogen is a storage form of glucose and plays an essential role in maintaining glucose homeostasis. Glycogen is a large polymer of glucose residues. When the body does not need glucose, the glucose surplus is stored as a form of glycogen in the liver and muscles. When energy is needed, glycogen can be broken down to yield glucose molecules. The CEST imaging of glycogen (glycoCEST) [6] is based on proton exchange between hydroxyl protons (–OH, 1.0 ppm downfield to water) and water protons. The initial observations were glycogen-containing phantoms at 4.7T and 9.4T and *in vivo* perfused mouse livers at 4.7T. In the *in vivo* experiment, a decline in glycoCEST signal was observed following the administration of glucagon, which induces glycogen breakdown. To date, there are only a few reports of translation into living humans. Deng et al. [75] used healthy volunteers to demonstrate that reproducing glycoCEST measurements of the liver is feasible with a clinical 3T MRI scanner and showed that compared with post-meal measurement, glycoCEST signals decreased following overnight fasting. Thus far, no study regarding clinical patients has been reported.

5.10 EXOGENOUS CEST AGENTS

Since very few exogenous CEST contrast agents are currently available for human studies, we will discuss them very briefly. Some of the non-ionic iodine contrast materials for X-ray computerized tomography (CT) and angiography have exchangeable protons, and they have been extensively tested as candidates for CEST contrast agents. They can have an immediate impact on clinical medicine since they are FDA approved and directly applicable to the human body. Iopamidol is one of the first and most representative of said contrast materials [76]. Iopamidol has two amide groups (–NH) that resonate at different offset frequencies (4.2 and 5.5 ppm downfield to water), allowing for a ratiometric approach to pH measurement. This approach has been successfully used to evaluate pH changes in acute kidney injuries [77] and tumors [78] in animal models. Another exogenous CEST contrast agent that is applicable to the human body is glucose. Glucose has five proton exchange sites (hydroxyl groups) that resonate at 0.8–2.2 ppm downfield to the water frequency. Therefore, non-radioactive glucose can be used as a CEST contrast agent. Glucose-enhanced CEST imaging, glucoCEST, was first proposed by Chan et al. [79], where an animal model implanted with a malignant tumor graft underwent the intravenous injection of glucose. The accumulation of glucose in the tumor graft was visualized with CEST MR after the injection. This and one other report [80] attracted much attention in the MR community, and its application to human brain tumors was also reported [81]. However, a critical limitation of glucoCEST is the fact that in cells, glucose is rapidly metabolized and intracellular glucose contributes little to the CEST signal. Rather, the observed CEST-related signals in glucoCEST mainly represent glucose in extracellular spaces, including tumor blood vessels. Thus, glucoCEST is considered suboptimal for imaging glucose metabolism. Nevertheless, it is worth noting that natural glucose can be a contrast agent for perfusion and vascular permeability as a safe alternative to gadolinium-based contrast agents, especially for patients with contraindications. Fluorodeoxyglucose (FDG) is a glucose analog, and when labeled with an 18F radioisotope can be used as a PET tracer for glucose metabolism. Unlike glucose, once taken into a cell, FDG remains a form of FGD-6 phosphate without being further metabolized (metabolic trapping), allowing for the imaging of glucose metabolism. Similar to glucose, FGD has multiple exchangeable hydroxyl protons. Thus, non-radioactive FDG could be a candidate for a CEST contrast agent. However, due to its toxicity, FDG is not applicable for human CEST imaging, which requires a much larger amount than PET. Ongoing research projects are pursuing CEST-based glucose metabolic imaging. Recently, CEST-active compounds with lower toxicity, such as 3-O-methyl glucose, have been tested for the purpose of glucose metabolic imaging for humans [82].

One group of CEST contrast agents possess paramagnetic lanthanide ions (such as La^{3+}, Gd^{3+}, and Lu^{3+}) within the chelate. These are called paraCEST agents [83]. Compared with the diamagnetic CEST (diaCEST) contrast agents mentioned previously, the resonance frequencies of the exchangeable protons bound to the paraCEST agents are shifted far away from the water resonance frequency (~50 ppm). This large chemical shift is advantageous because it allows paraCEST agents to exhibit faster chemical exchange rates than diaCEST agents while staying below the intermediate exchange rate limit. A large chemical shift also allows for the saturation of bound protons without the direct saturation of bulk water protons. Moreover, the presence of paramagnetic ions provides an additional exchange mechanism through the molecular exchange of water molecules between the internal and external areas of the paraCEST agents' coordinate spheres. These characteristics give paraCEST agents opportunities to better visualize their CEST effects. A variety of paraCEST agents have been proposed and used in *in vitro* and *in vivo* animal models to probe pH and temperature based on changes in the proton (or molecular) exchange rate [84,85]. The limitations of the paraCEST approach include the potential toxicity of lanthanide ions and the high agent uptake necessary for proper detection, especially in *in vivo* experiments.

lipoCEST is another class of CEST contrast agent that was originally developed to enhance the sensitivity of CEST contrast agents [86]. lipoCEST agents contain a liposome that encapsulates a paramagnetic lanthanide-based shift reagent. The paramagnetic shift reagent induces a chemical shift on the water molecules inside the liposomal cavity. As a result, the entire ensemble of water molecules contained in a liposome becomes an exchangeable pool of protons. Due to the huge number of water molecules inside the liposome and the fact that the effect spreads beyond the liposome, a dramatic sensitivity increase can be achieved. Researchers have attempted to use the lipoCEST approach for various unique applications, including reporting temperature [87] and targeting specific epitopes expressed on the surfaces of diseased cells [88].

5.11 CONCLUDING REMARKS

A number of studies have shown the feasibility and potential of CEST imaging as a tool for preclinical and clinical investigations. In the past ten years, the clinical translation of APT imaging has been progressing; especially regarding brain tumors, and is currently expanding to include head and neck, abdomen, and chest pathologies. CEST imaging of pH could have a significant impact on the clinical management of acute brain infarction and malignant tumors but requires further technical refinement toward clinical use. Exogenous CEST contrast agents, particularly those used to image glucose metabolism, although extremely fascinating, will take time to become clinically available.

REFERENCES

1. Ward KM, Aletras AH, Balaban RS. A new class of contrast agents for MRI based on proton chemical exchange dependent saturation transfer (CEST). *J Magn Reson* 2000;143(1):79–87.
2. Dagher A, Aletras A, Choyke P, Balaban R. Imaging of urea using chemical exchange-dependent saturation transfer at 1.5 T. *J Magn Reson Imaging* 2000;12(5):745–748.
3. Zhou J, Payen JF, Wilson DA, Traystman RJ, van Zijl PC. Using the amide proton signals of intracellular proteins and peptides to detect pH effects in MRI. *Nat Med* 2003;9(8):1085–1090.
4. Cai K, Haris M, Singh A, Kogan F, Greenberg JH, Hariharan H, Detre JA, Reddy R. Magnetic resonance imaging of glutamate. *Nat Med* 2012;18(2):302–306.
5. Haris M, Nanga RP, Singh A, Cai K, Kogan F, Hariharan H, Reddy R. Exchange rates of creatine kinase metabolites: Feasibility of imaging creatine by chemical exchange saturation transfer MRI. *NMR Biomed* 2012;25(11):1305–1309.
6. van Zijl PC, Jones CK, Ren J, Malloy CR, Sherry AD. MRI detection of glycogen in vivo by using chemical exchange saturation transfer imaging (glycoCEST). *Proc Natl Acad Sci U S A* 2007;104(11):4359–4364.
7. Ling W, Regatte RR, Navon G, Jerschow A. Assessment of glycosaminoglycan concentration in vivo by chemical exchange-dependent saturation transfer (gagCEST). *Proc Natl Acad Sci U S A* 2008;105(7):2266–2270.
8. Haris M, Cai K, Singh A, Hariharan H, Reddy R. In vivo mapping of brain myo-inositol. *Neuroimage* 2011;54(3):2079–2085.
9. Togao O, Hiwatashi A, Keupp J, Yamashita K, Kikuchi K, Yoshiura T, Yoneyama M, Kruiskamp MJ, Sagiyama K, Takahashi M, Honda H. Amide proton transfer imaging of diffuse gliomas: Effect of saturation pulse length in parallel transmission-based technique. *PLoS One* 2016;11(5):e0155925.
10. Jones KM, Pollard AC, Pagel MD. Clinical applications of Chemical exchange saturation transfer (CEST) MRI. *J Magn Reson Imaging* 2018;46(1):11–27.
11. Keupp J, Eggers H. Intrinsic field homogeneity correction in fast spin echo based amide proton transfer MRI. In *Proceedings of the 20th Annual Meeting of ISMRM*, Melbourne, Australia, 2012. Abstract 4185.
12. Dixon WT, Hancu I, Ratnakar SJ, Sherry AD, Lenkinski RE, Alsop DC. A multislice gradient echo pulse sequence for CEST imaging. *Magn Reson Med* 2010;63(1):253–256.
13. Ng MC, Hua J, Hu Y, Luk KD, Lam EY. Magnetization transfer (MT) asymmetry around the water resonance in human cervical spinal cord. *J Magn Reson Imaging* 2009;29(3):523–528.
14. Keupp J, Doneva M, Senegas J, Hey S, Eggers H. 3D fast spin-echo amide proton transfer MR with intrinsic field homogeneity correction for neuro-oncology applications. In *Proceedings of the 22nd Annual Meeting of ISMRM*, Milan, Italy, 2014. Abstract 3150.
15. Jones CK, Polders D, Hua J, Zhu H, Hoogduin HJ, Zhou J, Luijten P, van Zijl PC. In vivo three-dimensional whole-brain pulsed steady-state chemical exchange saturation transfer at 7 T. *Magn Reson Med* 2012;67(6):1579–1589.
16. Kim M, Gillen J, Landman B, Zhou JY, van Zijl PC. Water saturation shift referencing (WASSR) for chemical exchange saturation transfer (CEST) experiments. *Magn Reson Med* 2009;61(6):1441–1450.
17. Sun PZ, Farrar CT, Sorensen AG. Correction for artifacts induced by B0 and B1 field inhomogeneities in pH-sensitive chemical exchange saturation transfer (CEST) imaging. *Magn Reson Med* 2007;58(6):1207–1215.

18. Bryant RG. The dynamics of water–protein interactions. *Annu Rev Biophys Biomol Struct* 1996;25:29–53.

19. van Zijl PC, Zhou J, Mori N, Payen JF, Wilson D, Mori S. Mechanism of magnetization transfer during on-resonance water saturation: A new approach to detect mobile proteins, peptides, and lipids. *Magn Reson Med* 2003;49(3):440–449.

20. van Zijl PC, Yadav NN. Chemical exchange saturation transfer (CEST): What is in a name and what isn't? *Magn Reson Med* 2011;65(4):927–948.

21. Cai K, Singh A, Poptani H, Li W, Yang S, Lu Y, Hariharan H, Zhou XJ, Reddy R. CEST signal at 2ppm (CEST@2ppm) from Z-spectral fitting correlates with creatine distribution in brain tumor. *NMR Biomed* 2015;28(1):1–8.

22. Zaiss M, Xu J, Goerke S, Khan IS, Singer RJ, Gore JC, Gochberg DF, Bachert P. Inverse Z-spectrum analysis for spillover-, MT-, and T1 -corrected steady-state pulsed CEST-MRI--Application to pH-weighted MRI of acute stroke. *NMR Biomed* 2014;27(3):240–252.

23. Woessner DE, Zhang S, Merritt ME, Sherry AD. Numerical solution of the Bloch equations provides insights into the optimum design of PARACEST agents for MRI. *Magn Reson Med* 2005;53(4):790–799.

24. Tee YK, Harston GW, Blockley N, Okell TW, Levman J, Sheerin F, Cellerini M, Jezzard P, Kennedy J, Payne SJ, Chappell MA. Comparing different analysis methods for quantifying the MRI amide proton transfer (APT) effect in hyperacute stroke patients. *NMR Biomed* 2014;27(9):1019–1029.

25. Yan K, Fu Z, Yang C, Zhang K, Jiang S, Lee DH, Heo HY, Zhang Y, Cole RN, van Eyk JE, Zhou J. Assessing amide proton transfer (APT) MRI contrast origins in 9 L gliosarcoma in the rat brain using proteomic analysis. *Mol Imaging Biol* 2015;17(4):479–487.

26. Li J, Zhuang Z, Okamoto H, Vortmeyer AO, Park DM, Furuta M, Lee YS, Oldfield EH, Zeng W, Weil RJ. Proteomic profiling distinguishes astrocytomas and identifies differential tumor markers. *Neurology* 2006;66(5):733–736.

27. Clarke J, Butowski N, Chang S. Recent advances in therapy for glioblastoma. *Arch Neurol* 2010;67(3):279–283.

28. Zhou J, Blakeley JO, Hua J, Kim M, Laterra J, Pomper MG, van Zijl PC. Practical data acquisition method for human brain tumor amide proton transfer (APT) imaging. *Magn Reson Med* 2008;60(4):842–849.

29. Togao O, Yoshiura T, Keupp J, Hiwatashi A, Yamashita K, Kikuchi K, Suzuki Y, Suzuki SO, Iwaki T, Hata N, Mizoguchi M, Yoshimoto K, Sagiyama K, Takahashi M, Honda H. Amide proton transfer imaging of adult diffuse gliomas: Correlation with histopathological grades. *Neuro Oncol* 2014;16(3):441–448.

30. Togao O, Hiwatashi A, Yamashita K, Kikuchi K, Keupp J, Yoshimoto K, Kuga D, Yoneyama M, Suzuki SO, Iwaki T, Takahashi M, Iihara K, Honda H. Grading diffuse gliomas without intense contrast enhancement by amide proton transfer MR imaging: Comparisons with diffusion- and perfusion-weighted imaging. *Eur Radiol* 2017;27(2):578–588.

31. Choi YS, Ahn SS, Lee SK, Chang JH, Kang SG, Kim SH, Zhou J. Amide proton transfer imaging to discriminate between low- and high-grade gliomas: Added value to apparent diffusion coefficient and relative cerebral blood volume. *Eur Radiol* 2017;27(8):3181–3189.

32. Sakata A, Fushimi Y, Okada T, Arakawa Y, Kunieda T, Minamiguchi S, Kido A, Sakashita N, Miyamoto S, Togashi K. Diagnostic performance between contrast enhancement, proton MR spectroscopy, and amide proton transfer imaging in patients with brain tumors. *J Magn Reson Imaging* 2017;46(3):732–739.

33. Louis DN, Ohgaki H, Wiestler OD, Cavenee WK *WHO Classification of Tumours of the Central Nervous System*. Revised 4th ed. Lyon: IARC, 2016.

34. Eckel-Passow JE, Lachance DH, Molinaro AM, Walsh KM, Decker PA, Sicotte H, Pekmezci M, Rice T, Kosel ML, Smirnov IV, Sarkar G, Caron AA, Kollmeyer TM, Praska CE, Chada AR, Halder C, Hansen HM, McCoy LS, Bracci PM, Marshall R, Zheng S, Reis GF, Pico AR, O'Neill BP, Buckner JC, Giannini C, Huse JT, Perry A, Tihan T, Berger MS, Chang SM, Prados MD, Wiemels J, Wiencke JK, Wrensch MR, Jenkins RB. Glioma groups based on 1p/19q, IDH, and TERT promoter mutations in tumors. *N Engl J Med* 2015;372(26):2499–2508.

35. The Cancer Genome Atlas Research Network. Comprehensive, integrative genomic analysis of diffuse lower-grade gliomas. *N Engl J Med* 2015;372(26):2481–2498.

36. Jiang S, Zou T, Eberhart CG, Villalobos MAV, Heo HY, Zhang Y, Wang Y, Wang X, Yu H, Du Y, van Zijl PCM, Wen Z, Zhou J. Predicting IDH mutation status in grade II gliomas using amide proton transfer-weighted (APTw) MRI. *Magn Reson Med* 2017;78(3):1100–1109.

37. Jiang S, Yu H, Wang X, Lu S, Li Y, Feng L, Zhang Y, Heo HY, Lee DH, Zhou J, Wen Z. Molecular MRI differentiation between primary central nervous system lymphomas and high-grade gliomas using endogenous protein-based amide proton transfer MR imaging at 3 Tesla. *Eur Radiol* 2016;26(1):64–71.

38. Kamimura K, Nakajo M, Yoneyama T, Fukukura Y, Hirano H, Goto Y, Sasaki M, Akamine Y, Keupp J, Yoshiura T. Histogram analysis of amide proton transfer-weighted imaging: Comparison of glioblastoma and solitary brain metastasis in enhancing tumors and peritumoral regions. *Eur Radiol* 2019;29(8):4133–4140.

39. Stupp R, Mason WP, van den Bent MJ, Weller M, Fisher B, Taphoorn MJ, Belanger K, Brandes AA, Marosi C, Bogdahn U, Curschmann J, Janzer RC, Ludwin SK, Gorlia T, Allgeier A, Lacombe D, Cairncross JG, Eisenhauer E, Mirimanoff RO, European Organisation for Research and Treatment of Cancer Brain Tumor and Radiotherapy Groups; National Cancer Institute of Canada Clinical Trials Group. Radiotherapy plus concomitant and adjuvant temozolomide for glioblastoma. *N Engl J Med* 2005;352(10):987–996.

40. Kanda T, Wakabayashi Y, Zeng F, Ueno Y, Sofue K, Maeda T, Nogami M, Murakami T. Imaging findings in radiation therapy complications of the central nervous system. *Jpn J Radiol* 2018;36(9):519–527.

41. Zhou J, Tryggestad E, Wen Z, Lal B, Zhou T, Grossman R, Wang S, Yan K, Fu DX, Ford E, Tyler B, Blakeley J, Laterra J, van Zijl PC. Differentiation between glioma and radiation necrosis using molecular magnetic resonance imaging of endogenous proteins and peptides. *Nat Med* 2011;17(1):130–134.

42. Park JE, Kim HS, Park KJ, Kim SJ, Kim JH, Smith SA. Pre- and posttreatment glioma: Comparison of amide proton transfer imaging with MR spectroscopy for biomarkers of tumor proliferation. *Radiology* 2016;278(2):514–523.

43. Park KJ, Kim HS, Park JE, Shim WH, Kim SJ, Smith SA. Added value of amide proton transfer imaging to conventional and perfusion MR imaging for evaluating the treatment response of newly diagnosed glioblastoma. *Eur Radiol* 2016;26(12):4390–4403.

44. Mehrabian H, Desmond KL, Soliman H, Sahgal A, Stanisz GJ. Differentiation between radiation necrosis and tumor progression using chemical exchange saturation transfer. *Clin Cancer Res* 2017;23(14):3667–3675.

45. Park JE, Lee JY, Kim HS, Oh JY, Jung SC, Kim SJ, Keupp J, Oh M, Kim JS. Amide proton transfer imaging seems to provide higher diagnostic performance in post-treatment high-grade gliomas than methionine positron emission tomography. *Eur Radiol* 2018;28(8):3285–3295.

46. Joo B, Han K, Choi YS, Lee S-K, Ahn SS, Chang JH, Kang S-G, Kim SH, Zhou J. Amide proton transfer imaging for differentiation of benign and atypical meningiomas. *Eur Radiol* 2018;28(1):331–339.

47. Yu H, Wen X, Wu P, Chen Y, Zou T, Wang X, Jiang S, Zhou J, Wen Z. Can amide proton transfer-weighted imaging differentiate tumor grade and predict Ki-67 proliferation status of meningioma? *Eur Radiol* 2019;29(10):5298–5306.

48. Yuan J, Chen S, King AD, Zhou J, Bhatia KS, Zhang Q, Yeung DKW, Wei J, Mok GSP, Wang Y-X. Amide proton transfer weighted imaging of the head and neck at 3T: A feasibility study on healthy human subjects and patients with head and neck cancer. *NMR Biomed* 2014;27(10):1239–1247.

49. Law BKH, King AD, Ai Q-Y, Poon DM, Chen W, Bhatia KS, Ahuja AT, Ma BB, Yeung DK-W, Mo FKF, Wang Y-X, Yuan J. Head and neck tumors: Amide proton transfer MRI. *Radiology* 2018;288(3):782–790.

50. Liu R, Jiang G, Gao P, Li G, Nie L, Yan J, Jiang M, Duan R, Zhao Y, Luo J, Yin Y, Li C. Non-invasive amide proton transfer imaging and ZOOM diffusion-weighted imaging in differentiating benign and malignant thyroid micronodules. *Front Endocrinol* 2018;9:747.

51. Qamar S, King AD, Ai Q-Y, Law BKH, Chan JSM, Poon DM, Tong M, Mo FKF, Chen W, Bhatia KS, Ahuja AT, Ma BB, Yeung DK-W, Wang Y-X, Yuan J. Amide proton transfer MRI detects early changes in nasopharyngeal carcinoma: Providing a potential imaging marker for treatment response. *Eur Arch Oto-Rhino-Laryngol* 2019;276(2):505–512.

52. Takayama Y, Nishie A, Togao O, Asayama Y, Ishigami K, Ushijima Y, Okamoto D, Fujita N, Sonoda K, Hida T, Ohishi Y, Keupp J, Honda H. Amide proton transfer MR imaging of endometrioid endometrial adenocarcinoma: Association with histologic grade. *Radiology* 2018;286(3):909–917.

53. Li B, Sun H, Zhang S, Wang X, Guo Q. Amide proton transfer imaging to evaluate the grading of squamous cell carcinoma of the cervix: A comparative study using 18F FDG PET. *J Magn Reson Imaging* 2019;50(1):261–268.

54. Nishie A, Asayama Y, Ishigami K, Ushijima Y, Takayama Y, Okamoto D, Fujita N, Tsurumaru D, Togao O, Sagiyama K, Manabe H, Oki E, Kubo Y, Hida T, Hirahashi-Fujiwara M, Keupp J, Honda H. Amide proton transfer imaging to predict tumor response to neoadjuvant chemotherapy in locally advanced rectal cancer. *J Gastroenterol Hepatol* 2019;34(1):140–146.

55. Ohno Y, Yui M, Koyama H, Yoshikawa T, Seki S, Ueno Y, Miyazaki M, Ouyang C, Sugimura K. Chemical exchange saturation transfer MR imaging: Preliminary results for differentiation of malignant and benign thoracic lesions. *Radiology* 2016;279(2):578–589.

56. Dula A, Arlinghaus LR, Dortch RD, Dewey BE, Whisenant JG, Ayers GD, Yankeelov TE, Smith SA. Amide proton transfer imaging of the breast at 3T: Establishing reproducibility and possible feasibility assessing chemotherapy response. *Magn Reson Med* 2013;70(1):216–224.
57. Krikken E, Khlebnikov V, Zaiss M, Jibodh RA, van Diest PJ, Luijten PR, Klomp DWJ, van Laathoven HWM, Wijnen JP. Amide chemical exchange saturation transfer at 7T: A possible biomarker for detecting early response to neoadjuvant chemotherapy in breast cancer patients. *Breast Cancer Res* 2018;20(1):51.
58. Wen Z, Hu S, Huang F, Wang X, Guo L, Quan X, Wang S, Zhou J. MR imaging of high-grade brain tumors using endogenous protein and peptide-based contrast. *Neuroimage* 2010;51(2):616–622.
59. Wang M, Hong X, Chang CF, Li Q, Ma B, Zhang H, Xiang S, Heo HY, Zhang Y, Lee DH, Jiang S, Leigh R, Koehler RC, van Zijl PCM, Wang J, Zhou J. Simultaneous detection and separation of hyperacute intracerebral hemorrhage and cerebral ischemia using amide proton transfer MRI. *Magn Reson Med* 2015;74(1):42–50.
60. Jeong HK, Han K, Zhou J, Zhao Y, Choi YS, Lee SK, Ahn SS. Characterizing amide proton transfer imaging in haemorrhage brain lesions using 3T MRI. *Eur Radiol* 2017;27(4):1577–1584.
61. Zheng S, van der Bom IM, Zu Z, Lin G, Zhao Y, Gounis MJ. Chemical exchange saturation transfer effect in blood. *Magn Reson Med* 2014;71(3):1082–1092.
62. Bohara M, Kamimura K, Nakajo M, Yoneyama T, Yoshiura T. Amide proton transfer imaging of cavernous malformation in the cavernous sinus. *Magn Reson Med Sci* 2018;18(2):109–110.
63. Yu H, Lou H, Zou T, Wang X, Jiang S, Huang Z, Du Y, Jiang C, Ma L, Zhu J, He W, Rui Q, Zhou J, Wen Z. Applying protein-based amide proton transfer MR imaging to distinguish solitary brain metastases from glioblastoma. *Eur Radiol* 2017;27(11):4516–4524.
64. Ross BD, Higgins RJ, Boggan JE, Knittel B, Garwood M. 31P NMR spectroscopy of the in vivo metabolism of an intracerebral glioma in the rat. *Magn Reson Med* 1988;6(4):403–417.
65. Sun PZ, Zhou J, Sun W, Huang J, van Zijl PCM. Detection of the ischemic penumbra using pH-weighted MRI. *J Cereb Blood Flow Metab* 2007;27(6):1129–1136.
66. Martínez-Zaguilán R, Seftor EA, Seftor RE, Chu YW, Gillies RJ, Hendrix MJ. Acidic pH enhances the invasive behavior of human melanoma cells. *Clin Exp Metastasis* 1996;14(2):176–186.
67. Shi Q, Le X, Wang B, Abbruzzese JL, Xiong Q, He Y, Xie K. Regulation of vascular endothelial growth factor expression by acidosis in human cancer cells. *Oncogene* 2001;20(28):3751–3756.
68. Harris RJ, Cloughesy TF, Liau LM, Prins RM, Antonios JP, Li D, Yong WH, Pope WB, Lai A, Nghiemphu PL, Ellington BM. pH-weighted molecular imaging of glioma using amide chemical exchange saturation transfer MRI. *Neuro Oncol* 2015;17(11):1514–1524.
69. McVicar N, Li AX, Goncalves DF, Bellyou M, Meakin SO, Prado MAM, Bartha R. Quantitative tissue pH measurement during cerebral ischemia using amine and amide concentration-independent detection (AACID) with MRI. *J Cereb Blood Flow Metab* 2014;34(4):690–698.
70. Albatany M, Li A, Meakin S, Bartha R. In vivo detection of acute intracellular acidification in glioblastoma multiforme following a single dose of cariporide. *Int J Clin Oncol* 2018;23(5):812–819.
71. Singh A, Haris M, Cai K, Kassey VB, Kogan F, Reddy D, Hariharan H, Reddy R. Chemical exchange saturation transfer magnetic resonance imaging of human knee cartilage at 3T and 7T. *Magn Reson Med* 2012;68(2):588–594.
72. Wei W, Lambach B, Jia G, Kaeding C, Flanigan D, Knopp MV. A phase I clinical trial of the knee to assess correlation of gagCEST MRI, delayed gadolinium-enhanced MRI of cartilage and T2 mapping. *Eur J Radiol* 2017;90:220–224.
73. Haneder S, Apprich SR, Schmitt B, Michaely HJ, Schoenberg SO, Friedrich K, Trattnig S. Assessment of glycosaminoglycan content in intervertebral discs using chemical exchange saturation transfer at 3.0 Tesla: Preliminary results in patients with low-back pain. *Eur Radiol* 2013;23(3):861–868.
74. Wada T, Togao O, Tokunaga C, Funatsu R, Yamashita Y, Kobayashi K, Nakamura Y, Honda H. Glycosaminoglycan chemical exchange saturation transfer in human lumbar intervertebral dics: Effect of saturation pulse and relationship with low back pain. *J Magn Reson Imaging* 2017;45(3):863–871.
75. Deng M, Chen S-Z, Yuan J, Chan Q, Zhou J, Wang Y-X. Chemical exchange saturation transfer (CEST) MR technique for liver imaging at 3.0 Tesla: An evaluation of different offset number and an after-meal and over-night-fast comparison. *Mol Imaging Biol* 2016;18(2):274–282.
76. Aime S, Calabi L, Biondi L, Miranda MD, Ghelli S, Paleari L, Rebaudengo C, Terreno E. Iopamidol: Exploring the potential use of well-established X-ray contrast agent for MRI. *Magn Reson Med* 2005;53(4):830–834.

77. Longo DL, Busato A, Landarzo S, Antico F, Aime S. Imaging the pH evolution of an acute kidney injury model by means of Iopamodol, a MRI-CEST pH-responsive contrast agent. *Magn Reson Med* 2013;70(3):859–864.

78. Jones KM, Randtke EA, Yoshimaru ES, Howison CM, Chalasani P, Klein RR, Chambers SK, Kuo P, Pagel MD. Clinical translation of tumor acidosis measurement with acidoCEST MRI. *Mol Imaging Biol* 2017;19(4):617–625.

79. Chan KW, McMahon MT, Kato Y, Liu G, Bulte JW, Bhujwalla ZM, Artemov D, van Zijl PC. Natural D-glucose as a biodegradable MRI contrast agent for detecting cancer. *Magn Reson Med* 2012;68(6):1764–1773.

80. Walker-Samuel S, Ramasawmy R, Trrealdea F, Rega M, Rajkumar V, Johnson SP, Richardson S, Concalves M, Parkers HG, Arstad E, Thomas DL, Pedley RB, Lythgoe MF, Golay X. In vivo imaging of glucose uptake and metabolism in tumors. *Nat Med* 2013;19(8):1067–1072.

81. Xu X, Yadav NN, Knutsson L, Kalyani R, Hall E, Laterra J, Blakeley J, Strowd R, Pomper M, Barker P, Chan KWY, Liu G, McMahon MT, Stevens RD, van Zijl PCM. Dynamic glucose-enhanced (DGE) MRI: Translation to human scanning and first results in glioma patients. *Tomography* 2015;1(2):105–114.

82. Rivlin M, Tsarfaty I, Navon G. Functional molecular imaging of tumors by chemical exchange saturation transfer MRI of 3-O-methyl-D-glucose. *Magn Reson Med* 2014;72(5):1375–1380.

83. Soesbe TC, Wu Y, Sherry AD. Advantages of paramagnetic chemical exchange saturation transfer (CEST) complexes having slow to intermediate water exchange properties as responsive agents. *NMR Biomed* 2013;26:829–838.

84. McVicar N, Li AX, Suchy M, Hudson RHE, Menon RS, Bartha R. Simultaneous in vivo pH and temperature mapping using a PARACEST-MRI contrast agent. *Maggn Reson Med* 2013;70(4):1016–1025.

85. Wu Y, Zhang S, Soesbe TC, Yu J, Vinogradov E, Lenkinski RE, Sherry AD. pH imaging of mouse kidney in vivo using a frequency-dependent paraCEST agent. *Magn Reson Med* 2016;75(6):2432–2441.

86. Ferrauto G, Castelli DD, Gregorio ED, Terreno E, Aime S. LipoCEST and cellCEST imaging agents: Opportunities and challenges. *WIREs Nanomed Nanobiotechnol* 2016;8:602–618.

87. Langereis S, Keupp J, Van Velthoven JLJ, De Roos IHC, Burdinski D, Pikkemaat AJA, Grull H. A Temperature-sensitive liposomal 1H CEST and 19F contrast agent for MR image-guided drug delivery. *J Am Chem Soc* 2009;131(4):1380–1381.

88. Flament J, Geffroy F, Medina C, Robic C, Mayer J-F, Meriaux S, Valette J, Robert P, Port M, Le Bihan D, Lethimonnier F, Boumezbeur F. In vivo CEST MR imaging of U87 mice brain tumor angiogenesis using targeted lipoCEST contrast agent at 7T. *Magn Reson Med* 2013;69(1):179–187.

6 Diffusion Magnetic Resonance Imaging in the Central Nervous System

Kouhei Kamiya, Yuichi Suzuki, and Osamu Abe

CONTENTS

6.1 Introduction 121
6.2 Basics of dMRI 122
    6.2.1 Fick's Laws 122
    6.2.2 Diffusion Propagator 123
    6.2.3 Random Walk and Central Limit Theorem 124
    6.2.4 Free, Hindered, and Restricted Diffusion 124
    6.2.5 Bloch–Torrey Equation 125
    6.2.6 Stejskal–Tanner Pulsed Gradient Spin–Echo Sequence 125
    6.2.7 q-Space 126
    6.2.8 The Effect of Diffusion Time 127
6.3 Diffusion-Weighted Image and Diffusion Tensor Imaging 129
    6.3.1 Diffusion-Weighted Image 129
    6.3.2 Anisotropic Diffusion 130
    6.3.3 Diffusion Tensor Imaging 131
6.4 Going Beyond DTI 133
    6.4.1 Diffusion Kurtosis Imaging 133
    6.4.2 Model-Based Approaches 134
        6.4.2.1 Biophysical Models of White Matter 134
        6.4.2.2 Models for Clinical Studies 136
    6.4.3 Limitations 137
6.5 Tractography 139
    6.5.1 DTI Tractography 139
    6.5.2 Higher-Order Methods 139
    6.5.3 Applications 141
    6.5.4 Limitations 143
6.6 Current Trends and Perspectives 144
    6.6.1 Diffusion Time Dependence 144
    6.6.2 Multiple Diffusion Encoding 144
6.7 Conclusion 146
References 146

6.1 INTRODUCTION

Diffusion magnetic resonance imaging (dMRI) measures the diffusion of water molecules along different dimensions The pattern of diffusion in biological tissues is determined by the geometry and organization of the sample at the micrometer scale – orders of magnitude below the size of the typical image voxel. The measurement was first introduced by Stejskal and Tanner in 1965 (Stejskal

& Tanner, 1965), and became available in clinical medicine approximately 20 years later (Le Bihan et al., 1986; Wesbey, Moseley, & Ehman, 1984). The unique contrast of diffusion-weighted images, and their sensitivity to tissue microstructure, had significant impact on clinical practice, as first discovered in acute stroke (Moseley et al., 1990). Anisotropy of diffusion in the nervous tissue led to the development of another clinically important tool called tractography, a non-invasive method for visualizing white matter bundles *in vivo* that is now indispensable for presurgical planning. This chapter is devoted to dMRI in the central nervous system, focusing on methods available in clinical settings and their theoretical backgrounds. First, we introduce the basic concepts of diffusion. Then, we move to applications, starting with the conventional diffusion-weighted image and diffusion tensor imaging (DTI). We next introduce methods going beyond DTI: diffusion kurtosis imaging (DKI) and biologically inspired compartment models. We also describe methods to estimate fiber orientation from dMRI, which forms the basis of tractography. Finally, we introduce emerging applications of non-conventional dMRI sequences and future perspectives. To keep the size of this chapter reasonable, we only touch on the basics. For more in-depth discussions on the theory and basic concepts of diffusion, the readers are referred to Price (2009), Jones et al. (2013), Kiselev (2017), and Novikov et al. (2019). Also, several recent reviews focus on recent topics such as microstructure models (Alexander et al., 2019; Jelescu & Budde, 2017), as well as tractography and its application in network analysis (Jeurissen et al., 2019; Sotiropoulos & Zalesky, 2019).

6.2 BASICS OF DMRI

6.2.1 FICK'S LAWS

Adolf Fick (Fick, 1855) described the flux of particles in gases or liquids driven by a concentration gradient as:

$$F(\mathbf{r},t) = -D\nabla C(\mathbf{r},t),$$ (6.1)

where F is the rate of transfer of the diffusing particles passing through a unit area of the sample (i.e., flux), C is the concentration of the particles, and $\mathbf{r}$ is the position (a three-dimensional column vector) (Figure 6.1). D is referred to as the diffusion coefficient or diffusivity. Between 10°C and body temperature, D_{water} ranges from ~1 to 3 µm²/ms. The negative sign indicates that the flow is in the opposite direction to that of increasing concentration. Equation (6.1) is called Fick's first law of diffusion. Because the conservation of the total number of particles requires that the change of

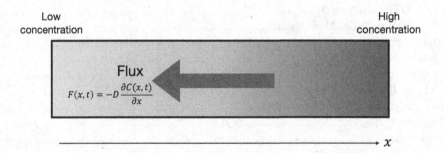

FIGURE 6.1 Illustration of Fick's laws, represented in one dimension with the concentration gradient along the x direction. The flux is driven by the concentration gradient (Fick's first law of diffusion). The conservation of the total number of the particles requires that the change of concentration in the small volume Δx over short time interval Δt is the net inflow: $\dfrac{\partial C(x,t)}{\partial t} = -\dfrac{\partial F(x,t)}{\partial x}$. Substituting this into the first law leads to Fick's second law of diffusion.

concentration is related to the local flux divergence $\dfrac{\partial C(\mathbf{r},t)}{\partial t} = -\nabla F(\mathbf{r},t)$, Fick's second law of diffusion can be derived:

$$\frac{\partial C(\mathbf{r},t)}{\partial t} = D\nabla^2 C(\mathbf{r},t). \tag{6.2}$$

6.2.2 Diffusion Propagator

In biological tissues, macroscopic concentration gradients are not the driving force behind diffusion. Rather, the process is driven by the local imbalance of exterior collisions with surrounding water molecules under random thermal motion (Brownian motion). Albert Einstein (Einstein, 1905) first showed the link between diffusion and Brownian motion. He showed that a probability density function (PDF) called the diffusion propagator $P(\mathbf{r}_0|\mathbf{r}_1,t)$, which represents the conditional probability that a particle starting at position $\mathbf{r}_0$ at time zero will move to $\mathbf{r}_1$ after a time t, obeys Fick's law in the same manner as the particle concentration. As we will see below, what we can actually measure is the ensemble average of net displacements, $\mathbf{R} = \mathbf{r}_1 - \mathbf{r}_0$. The voxel-averaged PDF of net displacements $\mathbf{R}$ over time t is defined as

$$\bar{P}(\mathbf{R},t) = \int P(\mathbf{r}_0) P(r_0 \mid r_0 + \mathbf{R}, t) dr_0, \tag{6.3}$$

where $P(\mathbf{r}_0)$ is the probability of finding the spin at starting position $\mathbf{r}_0$. Usually, we consider homogenous $P(\mathbf{r}_0) = \dfrac{1}{V}$, where V is the voxel volume. $\bar{P}(\mathbf{R},t)$ is termed the average or mean propagator.

$P(\mathbf{r}_0|\mathbf{r}_1,t)$ is the solution of the diffusion equation (Einstein, 1905),

$$\frac{\partial}{\partial t} P(\mathbf{r}_0|\mathbf{r}_1,t) = D\nabla^2 P(\mathbf{r}_0|\mathbf{r}_1,t). \tag{6.4}$$

Equation (6.4) is analogous to Fick's second law of diffusion. In the absence of barriers (i.e., free diffusion), given the initial condition that all particles start from position $\mathbf{r}_0$ at time $t=0$, the solution to Equation (6.4) is Gaussian

$$P(\mathbf{r}_0|\mathbf{r}_1,t) = (4\pi Dt)^{-3/2} \exp\left(-\frac{(\mathbf{r}_1 - \mathbf{r}_0)^2}{4Dt}\right). \tag{6.5}$$

Equation (6.5) indicates that $P(\mathbf{r}_0|\mathbf{r}_1,t)$ does not depend on the initial position $\mathbf{r}_0$ but only on the net displacement, $\mathbf{R} = \mathbf{r}_1 - \mathbf{r}_0$. The mean propagator

$$\bar{P}(\mathbf{R},t) = (4\pi Dt)^{-3/2} \exp\left(-\frac{\mathbf{R}^2}{4Dt}\right) \tag{6.6}$$

implies that the distribution of the particle mean displacement at any time t is Gaussian, and the width of distribution (mean squared displacement) increases linearly with t:

$$\langle \mathbf{R}^2 \rangle = 6Dt. \tag{6.7}$$

In one dimension, this is

$$\langle X^2 \rangle = 2Dt. \tag{6.8}$$

Here, $<\cdot>$ denotes the ensemble average over the voxel. For a certain property w, it is defined as $\langle w \rangle = \int wp(w)dw$, where $p(w)$ is the PDF of value w. The range of diffusion time t explored in typi-

cal dMRI experiments of the brain is roughly from 20 to 100 ms. Assuming that the average diffusion coefficient in the brain is ~1 $\mu m^2/ms$, dMRI probes a length scale on the order of 5 to 20 μm. This implies that dMRI is a powerful tool for probing subcellular structures of neural tissue, such as axons and cellular compartments, on the micrometer scale.

6.2.3 RANDOM WALK AND CENTRAL LIMIT THEOREM

The diffusion process can be best described as a random walk of particles (Kiselev, 2017). Here, we consider the case of free diffusion as the simplest example. The path of a particle can be split into very many small steps, taking place during a short time interval. In dMRI, we are interested in time intervals on the millisecond scale, which is much longer than the characteristic timescale of thermal motion, and therefore we can expect each step to be random and uncorrelated. This implies that the net displacement is the sum of very many short, statistically independent steps, thus satisfying the condition for the central limit theorem and indicting that the net displacement has a Gaussian distribution.

6.2.4 FREE, HINDERED, AND RESTRICTED DIFFUSION

Diffusion in nervous tissue deviates from simple Gaussian distribution because of the presence of barriers such as cell membranes and myelin. In this setting, the diffusion propagator is influenced not only by the intrinsic diffusivity but also by the microstructural characteristics of the barriers. To express diffusion properties in biological tissues, it is often useful to distinguish between "restricted" and "hindered" diffusion (Jones et al., 2013). Restricted diffusion refers to particles residing inside small compartments, for example, inside impermeable membranes. Such particles cannot displace beyond the barriers. In hindered diffusion, the movement of the particles is impeded but not confined within a limited space. Typically, the extracellular space, the space between densely packed cells and axons, is regarded as hindered diffusion. Figure 6.2 illustrates the relationships between time and the mean square displacement for free, hindered, and restricted diffusion.

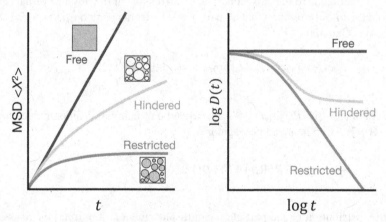

FIGURE 6.2 Illustration of the mean squared displacement (MSD) and diffusion coefficient (D) as function of diffusion time (t) for the case of free, restricted, and hindered diffusion. For free diffusion (dark gray), MSD increases linearly with t, and D does not depend on t. Hindered diffusion (light gray) shows similar behavior to free diffusion for very small t where only a small portion of particles can reach a barrier. Increasing t deviates it from the case of free diffusion. For a very long t, it becomes indistinguishable from a uniform medium with the diffusion coefficient somewhat reduced as compared with the intrinsic diffusivity. Restricted diffusion (middle gray) refers to particles residing inside small compartments, which cannot displace beyond the barriers. In this case, when increasing t, MSD shows a plateau and D approaches zero.

6.2.5 Bloch–Torrey Equation

Another approach to the dMRI signal is based on the Bloch–Torrey equation (Torrey, 1956). The transverse magnetization $m(\mathbf{r}, t)$ (a two-dimensional vector in the plane transverse to the B_0 field, represented by a complex number) is represented by:

$$\frac{\partial m(\mathbf{r}, t)}{\partial t} = -i\gamma \left(B_0 + \mathbf{g}(t) \cdot \mathbf{r} \right) m(\mathbf{r}, t) - \frac{m(\mathbf{r}, t)}{T_2} + \nabla \cdot D(\mathbf{r}) \cdot \nabla m(\mathbf{r}, t). \tag{6.9}$$

Here, $\mathbf{r}$ is the spin position, T_2 is the transverse relaxation time, and $\mathbf{g}(t)$ is the time-dependent diffusion-sensitizing gradient. For simplicity, we assumed a magnetically homogenous sample and homogenous relaxation rate in Equation (6.9). The above two approaches, the diffusion propagator and Bloch–Torrey equation, are tightly connected.

6.2.6 Stejskal–Tanner Pulsed Gradient Spin–Echo Sequence

Stejskal and Tanner (1965) developed the theory and methodology to measure diffusion with MRI. Figure 6.3 illustrates the Stejskal–Tanner pulsed gradient spin–echo (PGSE) sequence, the most commonly used dMRI sequence. A 90° radiofrequency (RF) pulse is applied at $t = 0$ to excite

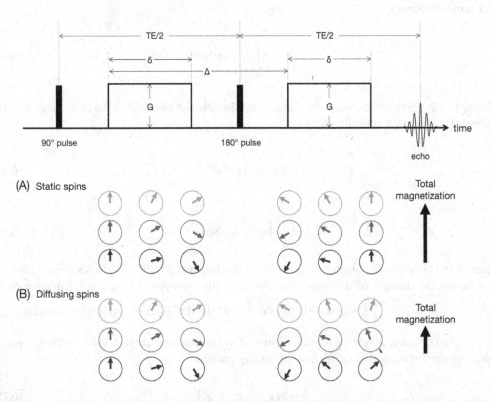

FIGURE 6.3 The effect of diffusion weighting in a Stejskal and Tanner pulsed gradient spin–echo (PGSE) sequence. (A) and (B) show two ensembles of spins, static (A) and diffusing (B). Different spins are represented by different grays. For the static spins, the dephasing induced by the first and the second lobes of the diffusion gradient are completely balanced, which means no signal loss occurs other than that caused by T2 relaxation. In contrast, the diffusing spins will experience different magnetic fields depending on the positions at each moment, and hence the dephasing induced by the first lobe cannot be reversed by the second one, leading to extra loss of signal amplitude as compared with the static spins.

the sample, and a 180° RF pulse is applied at $t = TE/2$ to reverse the phase of the spins. Diffusion-sensitizing gradients are applied on both sides of the 180° pulse. For static spins, dephasing before and after the 180° pulse is exactly reversed and therefore cancels because each spin experiences a constant magnetic field. In contrast, for diffusing molecules, each spin experiences a different magnetic field depending on its position. The dephasing cannot be completely reversed, resulting in a reduced signal amplitude (i.e., extra signal loss compared with the signal obtained without the diffusion-sensitizing gradient). In typical dMRI experiments, we are concerned with the effect of a spatially homogenous gradient **g**, defined as the gradient of the magnetic field in the direction of $\mathbf{B}_0$

$$\mathbf{g} = \left(\frac{\partial B_z}{\partial x} \quad \frac{\partial B_z}{\partial y} \quad \frac{\partial B_z}{\partial z} \right)^{\mathrm{T}}. \tag{6.10}$$

For the spin–echo sequence, we define $\mathbf{g}(t)$ as the "effective" gradient, accounting for the phase reversals due to the 180° pulse. Under the presence of such gradient, the magnetic field at point **r** is

$$B(\mathbf{r}) = B_0 + \mathbf{g} \cdot \mathbf{r}, \tag{6.11}$$

and the phase shift at time t of a spin following a path $\mathbf{r}(t)$ is given by the temporal integration of the Larmor frequency

$$\varphi(t) = \gamma \int_0^t \mathbf{g}(t') \cdot \mathbf{r}(t') dt', \tag{6.12}$$

where γ is the gyromagnetic ratio. To characterize the gradient imposed in the pulse sequence, the following notations are introduced:

$$\mathbf{q}(t) = \gamma \int_0^t \mathbf{g}(t') dt', \tag{6.13}$$

$$b = \int_0^{TE} \mathbf{q}(t') \cdot \mathbf{q}(t') dt'. \tag{6.14}$$

Equation (6.14) defines the b-value, or the b-factor (Le Bihan et al., 1986), an index commonly used to represent the strength of diffusion weighting. For the example in Figure 6.3, Equation (6.14) yields $b = \gamma^2 G^2 \delta^2 \left(\Delta - \frac{\delta}{3} \right)$, where G is the strength of the diffusion-sensitizing gradient (see also Figure 6.4). Diffusion weighting is often expressed by combination of the b-value and a unit vector indicating the direction of the diffusion-sensitizing gradient

$$\mathbf{b} = b \left(g_x \quad g_y \quad g_z \right)^{\mathrm{T}}. \tag{6.15}$$

6.2.7 q-SPACE

To appreciate the relationship between the signal and the diffusion propagator, it is useful to consider the short gradient pulse (SGP) limit ($\delta \ll \Delta$) (Figure 6.4). The transverse magnetization of a

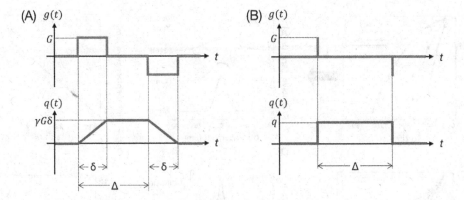

FIGURE 6.4   The effective gradient $g(t)$ (top row) and the corresponding $q(t)$ (bottom row). (A) Stejskal and Tanner PGSE sequence and (B) short gradient pulses.

spin can be described with a complex number $e^{-i\varphi}$, where φ is the phase acquired in the magnetic field. According to Equation (6.12), the net phase shift of a spin which was at position $\mathbf{r}_0$ during the first gradient pulse and at $\mathbf{r}_1$ during the second is $\mathbf{q} \cdot (\mathbf{r}_1 - \mathbf{r}_0)$. Thus, the normalized signal intensity (Stejskal & Tanner, 1965) is

$$\frac{S(\mathbf{q},t)}{S_0} = \int \exp\left(-i\mathbf{q} \cdot (\mathbf{r}_1 - \mathbf{r}_0)\right) P(\mathbf{r}_0) P(\mathbf{r}_0|\mathbf{r}_1,t) d\mathbf{r}_0 d\mathbf{r}_1, \tag{6.16}$$

where S_0 is the reference signal that would be obtained without the diffusion-sensitizing gradient. Note that $\mathbf{q}$ here is different from the time-dependent function $\mathbf{q}(t)$ (Figure 6.4).

Using the mean propagator, Equation (6.16) can be rewritten as

$$\frac{S(\mathbf{q},t)}{S_0} = \int \exp\left(-i\mathbf{q} \cdot \mathbf{R}\right) \bar{P}(\mathbf{R},t) d\mathbf{R}. \tag{6.17}$$

Equation (6.17) implies that the dMRI signal is the Fourier transform of the mean propagator (Figure 6.5). This leads to the formalism of q-space imaging (QSI) (Assaf et al., 2000), which aims to recover the mean propagator by measurements varying $\mathbf{q}$.

In clinical scanners equipped with a maximal gradient strength of 40 to 80 mT/m, the high q-values necessary for QSI cannot be achieved without increasing δ; hence the SGP condition is inevitably violated. In addition, in biological tissues, because QSI measures the displacement probability averaged over the voxel, the contributions from multiple compartments (e.g., intra- and extra-cellular compartments) are entangled.

6.2.8   THE EFFECT OF DIFFUSION TIME

The mean propagator $\bar{P}(\mathbf{R},t)$ is a function of $\mathbf{R}$ and t, therefore implying that at least two parameters ($\mathbf{q}$ and t) are needed to characterize dMRI experiments (Kiselev, 2017; Novikov et al., 2019). The b-value alone does not characterize a measurement, because two measurements with the same b but different q and t will result in different signals. The exception to this is the case of free diffusion. In the above example of the SGP limit, substituting Equation (6.6) into Equation (6.17) results in

$$\frac{S}{S_0} = \exp(-bD), \tag{6.18}$$

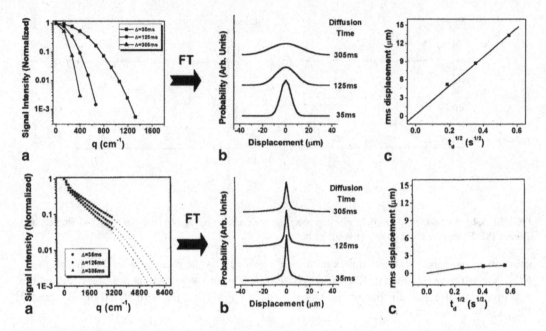

FIGURE 6.5 The diffusion MRI signal is the Fourier transform (FT) of the mean propagator. The top row shows the signal obtained from t-butanol solution (free diffusion), and the bottom row shows the signal obtained from the bovine optic nerve (restricted diffusion). From left to right: (a) the normalized signal attenuation as a function of q for different diffusion times (Δ), (b) the mean displacement obtained by FT of the data shown in (a), and (c) the root mean square displacement as a function of $t^{1/2}$. (Reprinted with permission from Assaf et al. 2000.)

where $b = q^2 t$. As we will see in the following sections, the b-value is indeed useful for experimental conditions that can be implemented on clinical scanners, and hence is widely used. Typically, we estimate the diffusivity using Equation (6.18) in a low q-value regime. Here, we emphasize that estimated diffusivity is a function of the diffusion time t (Figure 6.2), although it does not explicitly appear in Equation (6.18). We will return to the relationship between diffusivity and the diffusion propagator in Section 6.4.1.

Along the t axis, at least three characteristic regimes have been described in the dMRI literature (Kiselev, 2017; Novikov et al., 2019; Sen, 2004). These regimes are defined relative to the correlation length, l_c, the typical distance over which the environment of each local microstructure changes (cell size, cell packing, axon undulation, caliber change, etc.). The correlation length sets the timescale $t_c = l_c^2 / D$, which is the typical time for molecules to diffuse over the distance l_c.

1) Short time limit ($t \ll t_c$)

   Within a very short diffusion time, only a small portion of particles can reach a barrier, and the other particles do not experience the effect of the barrier. The relevant parameter here is the net surface-to-volume ratio (S/V) of the barriers. The diffusion coefficient in this regime is described (Mitra et al., 1992) as

$$D(t) \approx D_0 \left(1 - \frac{4}{3d\sqrt{\pi}} \frac{S}{V} \sqrt{D_0 t} \right), \tag{6.19}$$

where D_0 is the intrinsic diffusivity and d is the number of dimensions along which molecules can diffuse. For biological tissues, meeting this short time limit by using PGSE is practically impossible. Oscillating gradient spin echo (OGSE) is better suited to probe diffusion times shorter than PGSE (Does et al., 2003; Novikov & Kiselev, 2011). So far, the short time limit has been probed mainly in preclinical studies using animal models of tumors that have relatively large cell sizes (Reynaud, 2017).

2) Approaching the long time limit ($t \to \infty$)

$D(t)$ decreases with t, approaching the limit $D_\infty \equiv D(t \to \infty)$ in a medium-specific manner

(de Swiet & Sen, 1996; Novikov et al., 2014). Novikov et al. (2014) demonstrated that the behavior approaching the long time limit depends on the distinct types of long-range spatial correlations of the structure, for example, regular lattices, or short-range disorder where the barriers are uncorrelated or exhibit correlation over a finite length. This theory has been shown to give a good explanation for the diffusion time dependence observed in the human brain within the range of diffusion time relevant for clinical studies (Fieremans et al., 2016).

3) Long time limit ($t \to \infty$)

For sufficiently long diffusion times, Gaussian distribution becomes the effective description for connected pores (hindered diffusion) (Kiselev, 2017) (Figure 6.2). In biological tissues, the premise for the central limit theorem (Section 6.2.3) does not always hold because the presence of barriers results in correlations between adjacent steps of particle motion. However, such correlation is only effective over the timescale $t_c = l_c^2 / D$. For sufficiently long times, the steps become uncorrelated, justifying the use of the central limit theorem. Thus, at long times, the effective diffusion propagator becomes Gaussian. In other words, with increasing diffusion time, the effects of small local environments are "gradually forgotten" (Novikov et al., 2019), and their effects on measured signal finally become indistinguishable from that of a uniform medium, with the diffusion coefficient D_∞ somewhat reduced as compared with the intrinsic diffusivity D_0. If the sample consists of multiple non-exchanging compartments, the model for the long time limit becomes the sum of Gaussians. We will see such models in Section 6.4.2.

6.3 DIFFUSION-WEIGHTED IMAGE AND DIFFUSION TENSOR IMAGING

6.3.1 DIFFUSION-WEIGHTED IMAGE

As we have seen, we estimate the diffusion coefficient by using Equation (6.18). The estimated diffusion coefficient depends on many experimental factors, especially the diffusion time, and thus is often called the apparent diffusion coefficient (ADC). Because Equation (6.18) has two unknowns (S_0 and D), at least two measurements with distinct b-values are required. In addition, it was shown very early that ADC in the nervous tissue is strongly dependent on direction (Moseley et al., 1990) (Figure 6.6). The simplest way to remove dependence on direction is to average ADC values obtained along three orthogonal axes:

$$\text{ADC} = \frac{\text{ADC}_x + \text{ADC}_y + \text{ADC}_z}{3}.$$

(6.20)

This defines ADC, which is widely used clinically. ADC approximates the mean diffusivity, which is introduced in Section 6.3.3. When using a visual image to make a diagnosis, it is useful to create an isotropic diffusion-weighted image by multiplying the three images obtained along the three axes and taking the cube root, so that $S = S_0\exp(-b \cdot \text{ADC})$. The term "-weighted" implies that factors other than diffusion contribute to the signal. In a typical PGSE sequence, the overall signal attenuation is given by

$$A = \left(1 - \exp\left(-\frac{TR}{T_1}\right)\right)\exp\left(-\frac{TE}{T_2}\right)\exp(-bD).$$

(6.21)

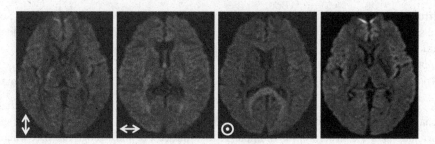

FIGURE 6.6 The diffusion MRI signal in the nervous tissue is strongly dependent on direction. The first three images from left to right: the images obtained by applying a diffusion-sensitizing gradient in the anterior-posterior, left-right, and superior-inferior directions, respectively. In each image, the signal is lower in the brain region containing fibers running in the direction parallel to the diffusion-sensitizing gradient. In the rightmost position is the isotropic diffusion-weighted image, where the dependency on direction has been cancelled out by taking the geometric mean of the three images.

Hence, higher signal intensity on diffusion-weighted images is not always due to smaller D. For example, tissue with a very large T_2 value can show a high signal on a diffusion-weighted image ["T2 shine through"].

Although the formulation is simple, ADC and diffusion-weighted images dramatically improved clinical image-based diagnoses of neurological diseases (Citton et al., 2012; O'Connor et al., 2013). They provide information that directly alters clinical management by identifying lesions that would otherwise have been undetected or by narrowing the list of differential diagnoses. First, in 1990, a profound decrease in ADC and hyperintensity on diffusion-weighted images were discovered to reflect acute cerebral ischemia (Moseley et al., 1990). The diffusion abnormalities can be seen approximately one hour after the insult – hours before computed tomography or other MR sequences show changes. Since then, ADC and diffusion-weighted images have been shown to be useful in characterizing brain lesions, including abscess, encephalitis, prion disease, tumors, secondary degeneration, demyelination, metabolic, drug- or toxin-induced encephalopathies, and status epilepticus, among others. This list is still evolving, with continuing discoveries and establishment of previously unrecognized disease entities (Konno et al., 2018; Sone et al., 2016; Takanashi, 2009).

The microstructural underpinnings behind the diffusion abnormalities differ among diseases and are not completely elucidated yet. In tumors, correlations between radiology and pathology findings suggest that smaller ADC values correlate with higher tumor cellularity (Gauvain et al., 2001). This concept is useful to narrow the differential diagnosis and is widely adopted in clinical image interpretation, although it is of course not a one-to-one relationship and there are many other factors contributing to ADC.

Maple syrup urine disease (MSUD) is another example where dMRI has yielded insightful clinical images. In MSUD, reduction of ADC is located specifically in the brain regions undergoing active myelination (Sakai et al., 2005). The presence of myelin, especially unstable myelin, is considered essential for the development of the diffusion abnormality in MSUD, and it is speculated to result from myelin splitting, with subsequent entrapment of water between the myelinic lamellae (intra-myelinic edema), which is observed in an animal model of MSUD (Harper et al., 1990).

6.3.2 ANISOTROPIC DIFFUSION

The organization of biological tissues, and therefore the diffusion process within them, is anisotropic, i.e., dependent on direction. The directional properties of diffusion can be represented using a rank-2 tensor, which can now be expressed as a 3×3 matrix:

$$D = \begin{pmatrix} D_{xx} & D_{xy} & D_{xz} \\ D_{yx} & D_{yy} & D_{yz} \\ D_{zx} & D_{zy} & D_{zz} \end{pmatrix}. \tag{6.22}$$

D is symmetric ($D_{ij} = D_{ji}$), so there are six independent elements. Because any symmetric matrix can be diagonalized by using an orthogonal matrix, D can be represented as

$$D = \begin{pmatrix} e_1 & e_2 & e_3 \end{pmatrix} \begin{pmatrix} \lambda_1 & 0 & 0 \\ 0 & \lambda_2 & 0 \\ 0 & 0 & \lambda_3 \end{pmatrix} \begin{pmatrix} e_1 & e_2 & e_3 \end{pmatrix}^T, \tag{6.23}$$

where e_1, e_2, and e_3 are the three eigenvectors of D (represented as 3×1 column vectors), whereas λ_1, λ_2, and λ_3 are the corresponding eigenvalues ($\lambda_1 \geq \lambda_2 \geq \lambda_3$), the diffusivities in the corresponding directions. e_1, e_2, and e_3 are orthogonal and unitary vectors. A diffusion tensor can be visualized with an ellipsoid, whose axes are along the directions of the eigenvectors (Figure 6.7).

6.3.3 Diffusion Tensor Imaging

Diffusion tensor imaging (DTI) is the extension of Equation (6.18), and it describes three-dimensional diffusion in anisotropic homogenous media. By using the b-matrix defined as

$$B = \int_0^{TE} q(t')q(t')^T dt', \tag{6.24}$$

the signal can be represented (Basser, Mattiello, & Le Bihan, 1994b) as

$$S = S_0 \exp\left(-\sum_i \sum_j B_{ij}D_{ij}\right) = S_0 \exp\left(-b\sum_i \sum_j g_i g_j D_{ij}\right). \tag{6.25}$$

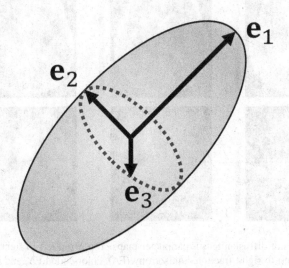

FIGURE 6.7 Anisotropic Gaussian diffusion can be visually represented by using an ellipsoid where the three principal directions correspond to the eigenvectors of the diffusion tensor.

Because **D** has six independent elements, estimating **D** requires at least six measurements with different **B** in addition to one without diffusion weighting ($b=0$). To estimate **D**, log S is modeled as $y = X\beta + \epsilon$, where **y** is the column vector of log S, **X** is the design matrix directly related to the b-matrix, **β** is the column vector of parameters (S_0 and the independent elements of **D**), and ϵ is the column vector of error terms. A basic way to solve such a system is to use linear least squares (LLS) often represented by matrix pseudoinversion: $\beta = \left(X^T X\right)^{-1} X^T y$. In the presence of noise, optimal estimation can be obtained by weighted LLS (WLLS): $\beta = \left(X^T w X\right)^{-1} X^T w y$, where the weight is estimated by using either the observed noisy signals (Basser, Mattiello, & Le Bihan, 1994a; Koay et al., 2006) or the preliminary LLS estimates (Salvador et al., 2005).

Once **D** is obtained, one can derive several rotationally invariant scalar metrics that describe the size and shape of the diffusion tensor (Basser, 1995). These parameters are useful for summarizing the diffusion properties of a given voxel and are used as quantitative metrics in clinical studies. Here we introduce the most commonly reported metrics. Figure 6.8 shows examples of parameter maps in a normal human brain.

1) Mean diffusivity (MD): the average of the eigenvalues. $\text{MD} = \dfrac{\lambda_1 + \lambda_2 + \lambda_3}{3} = \dfrac{tr(\mathbf{D})}{3}$.

2) Axial diffusivity (AD) and radial diffusivity (RD): the diffusion coefficients parallel and perpendicular to the principal direction. $\text{AD} = \lambda_1$, $\text{RD} = \dfrac{\lambda_2 + \lambda_3}{2}$.

3) Fractional anisotropy (FA): a dimensionless index that describes the shape of the diffusion tensor, normalized to have a value between 0 (isotropic diffusion) and 1.

$$\text{FA} = \sqrt{\frac{3}{2}} \sqrt{\frac{\left(\lambda_1 - \text{MD}\right)^2 + \left(\lambda_2 - \text{MD}\right)^2 + \left(\lambda_3 - \text{MD}\right)^2}{\lambda_1^2 + \lambda_2^2 + \lambda_3^2}}. \tag{6.26}$$

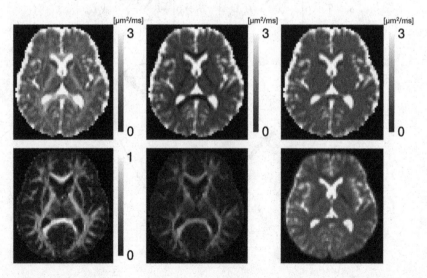

FIGURE 6.8 Examples of diffusion tensor parameter maps. Top, from left to right: axial/radial/mean diffusivity. Bottom, from left to right: fractional anisotropy (FA), color-coded FA, and non-diffusion-weighted image. Color-coded FA represents direction of the first eigenvector, by assigning red, green, and blue to left-right, anterior-posterior, and superior-inferior directions, respectively.

The applications of DTI in clinical studies have shown that DTI metrics are sensitive to disease-related changes in the brain and correlate with disease severity in many neurological diseases, for example, multiple sclerosis (Rovaris et al., 2005), Alzheimer's disease (Sexton et al., 2011), Parkinson's disease (Atkinson-Clement et al., 2017), and amyotrophic lateral sclerosis (Li et al., 2012).

One of the limitations of DTI is that interpretation in terms of specific tissue properties is highly ambiguous. Basically, DTI metrics can be sensitive to anything that affects the microstructural environments (e.g., size and shape of cells, axonal density, axon diameter, axon undulation, fiber orientation, degree of myelination, membrane permeability) (Beaulieu, 2002) and are not specific to any particular factor (Jones et al., 2013). FA has been used erroneously as a marker of "white matter integrity" (Jones et al., 2013), presumably because it shows smaller values in the context of most neurological diseases than in healthy controls (Goveas et al., 2015; Horsfield & Jones, 2002). However, by design, FA is not meant to correlate with axonal integrity. FA is confounded by many other factors, particularly fiber orientation, which can easily mask other effects and sometimes leads to greater values of FA than would be expected according to actual pathology (Assaf et al., 2006; Douaud et al., 2011).

6.4 GOING BEYOND DTI

6.4.1 DIFFUSION KURTOSIS IMAGING

Diffusion kurtosis imaging (DKI) (Jensen & Helpern, 2010; Jensen et al., 2005) can be seen as an extension of DTI into the fourth order term of the cumulant expansion. Thorough discussion of cumulant expansion can be found in Kiselev (2013). For simplicity, we first consider it in one dimension. It has been empirically shown that the dMRI signal is predicted well by Equation (6.18) with a b-value up to ~1000 s/mm^2 and that deviation is observed for greater b-values. It is then straightforward to consider the Taylor expansion of $\log S$ around $b=0$:

$$\log S = \log S_0 + \frac{d \log S}{db}\bigg|_{b=0} b + \frac{1}{2}\frac{d^2 \log S}{db^2}\bigg|_{b=0} b^2 + \cdots \tag{6.27}$$

Although the Taylor expansion is a general mathematical framework, this representation is also connected to the diffusion propagator. Under the SGP approximation, expansion of the exponential factor in the right-hand side of Equation (6.17) (Kiselev, 2013) yields

$$\log \frac{S}{S_0} = -\frac{1}{2!}q^2 \langle X^2 \rangle_c + \frac{1}{4!}q^4 \langle X^4 \rangle_c + \cdots = -\frac{1}{2!}q^2 \langle X^2 \rangle + \frac{1}{4!}q^4 \left(\langle X^4 \rangle - 3\langle X^2 \rangle^2\right) + \cdots \tag{6.28}$$

Here, X denotes the displacement along the direction of the diffusion-sensitizing gradient. The expansion (6.28) is called the cumulant expansion. $\langle X^n \rangle$ and $\langle X^n \rangle_c$ are called the nth order moment and cumulant, respectively. In the absence of bulk flow, the odd order moments and cumulants are zero due to the symmetry of the diffusion propagator. Equation (6.28) implies that the initial decay yields the ensemble-averaged mean squared displacement. In other words, the effective or apparent diffusion coefficient, $D = \langle X^2 \rangle / 2t$, can be derived from the low q limit of the signal attenuation. Introducing excess kurtosis $K = \frac{\langle X^4 \rangle}{\langle X^2 \rangle^2} - 3$, a unitless parameter that describes deviation from Gaussian distribution ($K=0$ for Gaussian),

$$\log S = \log S_0 - bD + \frac{1}{6}b^2 D^2 K + \cdots \tag{6.29}$$

Thus, we can access D and K with dMRI. Extension to three dimensions and truncating after b^2 (Jensen & Helpern, 2010; Jensen et al., 2005) yields

$$\log S(b,\mathbf{g}) = \log S_0 - b\sum_{i,j} g_i g_j D_{ij} + \frac{1}{6} b^2 \bar{D}^2 \sum_{j,j,k,l} g_i g_j g_k g_l W_{ijkl}, \qquad (6.30)$$

where $\bar{D}$ is the MD. One can see that DTI is a truncation of the same series after b. $\mathbf{W}$ is a rank-4 tensor called the kurtosis tensor. $\mathbf{W}$ is fully symmetric and therefore has 15 independent elements. As in DTI, $\mathbf{W}$ can be estimated basically with weighted LLS (Veraart et al., 2013), although it is challenging because it is more subject to noise. Several scalar metrics can be derived from the kurtosis tensor (Glenn et al., 2015; Hansen et al., 2013; Jensen & Helpern, 2010). DKI has several advantages over DTI. Kurtosis expresses deviations from the Gaussian distribution, which are indicators of the presence of barriers and are therefore expected to be meaningful in biological tissues. DKI also provides a more robust estimation of the diffusivity, in that the dependence of estimated diffusivity on the choice of b-values is reduced (Veraart et al., 2011). Applications of DKI in clinical studies typically use 30 to 60 directions of diffusion-sensitizing gradient over a minimum of two non-zero b-values (shells) with a maximum of $b = 2000$–3000 s/mm^2.

DKI has been shown to be potentially useful for studying microstructure in normal development and aging (Das et al., 2017; Falangola et al., 2008), as well as in diseases, including tumors (Raab et al., 2010; Van Cauter et al., 2012), Alzheimer's disease (Falangola et al., 2013), Parkinson's disease (Kamagata et al., 2014), and epilepsy (Bonilha et al., 2015). These studies reported that DKI provides information complementary to DTI that is of clinical relevance, for example, discriminating patients from controls, evaluating tumor grade, and correlating with neurocognitive functions. In particular, several studies reported that DKI provides improved sensitivity to diseases than DTI in relatively isotropic regions such as the gray matter (Bonilha et al., 2015) and white matter regions with prominent crossing fibers (Kamagata et al., 2014), where the usefulness of FA might be limited.

Because the Taylor expansion is a general mathematical framework, DTI and DKI are valid without requiring any assumptions regarding the microstructure of the sample, provided that b is sufficiently close to 0. This is one of the strengths of DTI and DKI, but at the same time is linked to their limitation. The strength is that DTI/DKI does not rely on specific assumptions regarding the underlying tissue and therefore can be applied to any organs or lesions. However, this also means that DTI/DKI metrics are not specific to certain tissue properties. In addition, the expansion is valid only to a certain maximal value of b (convergence radius) (Kiselev, 2013). Thus, DTI/DKI metrics are dependent on the choice of the maximal b-value (Hutchinson et al., 2017) (Figure 6.9). The convergence radius depends on microstructure geometries, but one convenient criterion is $b \leq 3/DK$ (Jensen & Helpern, 2010). For *in vivo* measurement of the brain, empirical evidence indicates maximum b-values of approximately 1000 s/mm^2 for DTI and of 2000 to 3000 s/mm^2 for DKI (Jensen & Helpern, 2010). The behavior of the dMRI signal outside the convergence radius cannot be predicted by DTI/DKI, and therefore is expected to provide additional information.

6.4.2 MODEL-BASED APPROACHES

6.4.2.1 Biophysical Models of White Matter

To derive parameters that will give specific information about the tissue microstructure, biology-inspired tissue models have been proposed. In these, one assumes a simplified geometry as a model of the underlying tissue and then estimates a set of parameters characterizing that geometry by fitting the model to dMRI data. Such approaches can potentially provide specific, biologically meaningful parameters if, and only if, the assumptions are met so that the model accurately describes the relevant microstructural features that have effects to the measurements. Here, we focus on models for the white matter, which has been the tissue most extensively investigated so far.

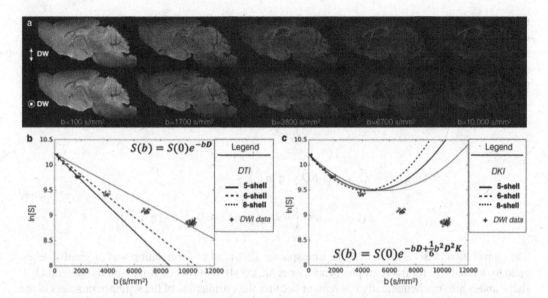

FIGURE 6.9 Diffusion MRI signal in a fixed mouse brain. (a) Diffusion-weighted images in a sagittal slice across a range of b-values ($b = 100–10,000$ s/mm²) with gradient directions along the dorsal-ventral axis (top row) and left-right axis (bottom row). (b) and (c) The mean signal value within a region of gray matter is plotted against the b-value. The lines represent fits to the DTI (b) and DKI (c) signal representation, respectively. The fitting curve using five (up to $b = 1700$ s/mm²), six (up to $b = 3800$ s/mm²), and eight (full data) different b-values are shown, demonstrating dependence of the DTI/DKI fit on the choice of b-value range. DKI = diffusion kurtosis imaging; DTI = diffusion tensor imaging. (Reprinted with permission from Hutchinson et al. 2017.)

Since first proposed by Stanisz et al. (1997) in the study of bovine optic nerve, innumerable versions of white matter models have been developed (Alexander et al., 2010; Assaf & Basser, 2005; Fieremans et al., 2011; Jespersen et al., 2007; Kaden et al., 2016; Novikov et al., 2018; Pasternak et al., 2009; Reisert et al., 2017; Zhang et al., 2012). These models usually combine two or three simple geometries. The typical components used to build models are described in Panagiotaki et al. (2012).

Most white matter models currently used in clinical studies rely on a common assumption that the axons can be described as infinitely thin cylinders ("sticks") and that the exchange between the stick and the extra-axonal space is negligible. In other words, diffusion in the intra-axonal compartment is assumed to follow a one-dimensional Gaussian pattern, along the length of the axon. The dMRI signal in currently available clinical scanners is insensitive to signal attenuation caused by intra-axonal diffusion in the perpendicular direction (Nilsson et al., 2017), because the typical axonal diameter in the brain is too small (~1 μm) (Innocenti et al., 2015). That means our measurement is not sensitive to axon diameter. The analogy of axons as sticks was recently validated in the white matter in a living human brain by observing $1/\sqrt{b}$ scaling of the signal at very high b-values (either sensitivity to axonal diameter or non-negligible exchange would destroy the scaling) (McKinnon et al., 2017; Veraart, Fieremans, & Novikov, 2019). Building on this analogy, the voxel is further assumed to consist of two or three, non-exchanging, Gaussian compartments. The first compartment includes axons, and possibly dendrites, described as sticks. We denote the intra-axonal diffusivity as $D_{a,\parallel}$. The diffusivity in the perpendicular direction is fixed to zero ($D_{a,\perp} = 0$). The second compartment is the extra-axonal space and includes everything except for axons, i.e., extracellular matrix, cell somas, and glial cells. This compartment is assumed to produce anisotropic Gaussian diffusion, with parallel and perpendicular diffusivities, $D_{e,\parallel}$ and $D_{e,\perp}$. Some models further include a third compartment that is isotropic Gaussian, with a diffusivity D_{iso}, to represent either freely diffusing water such as the cerebrospinal fluid or water with negligible diffusivity in all directions (often referred to as "still water"). The dMRI signal is represented as a convolution of the signal model from a single perfectly aligned fiber population pointing in direction **n** (also called the kernel, or the response function) and the orientation distribution function (ODF) $P(\mathbf{n})$. Recently, Reisert et al. (2017) and Novikov et al. (2018) put all such models under the same umbrella. For the case of two-compartment models, the dMRI signal is expressed as

$$S_{\mathbf{g}}(b) = S_0 \int K(b, \mathbf{g} \cdot \mathbf{n}) P(\mathbf{n}) d\mathbf{n}, \tag{6.31}$$

$$K(b, \mathbf{g} \cdot \mathbf{n}) = f_{\text{intra}} \exp\left(-bD_{a,\parallel}(\mathbf{g} \cdot \mathbf{n})^2\right)$$
$$+ (1 - f_{\text{intra}}) \exp\left(-bD_{e,\parallel}(\mathbf{g} \cdot \mathbf{n})^2 - bD_{e,\perp}\left(1 - (\mathbf{g} \cdot \mathbf{n})^2\right)\right). \tag{6.32}$$

The convolution (6.31) is done over the unit sphere. This kind of formulation was originally developed to derive ODF for tractography (Tournier et al., 2004). It allows dealing with the general ODF shape and is also mathematically convenient because the estimations of the scalar parameters in the kernel and of the ODF can be separated by working in the spherical harmonics basis.

6.4.2.2 Models for Clinical Studies

We describe two models that are popular in clinical studies: 1) neurite orientation dispersion and density imaging (NODDI) (Zhang et al., 2012) and 2) white matter tract integrity (WMTI) (Fieremans et al., 2011). For both methods, the demand on acquisition is similar to that of DKI.

1) NODDI

NODDI is a three-compartment model described by two volume fractions ($f_{\text{intra}}, f_{\text{iso}}$), and four diffusivities ($D_{a,\parallel}$, $D_{e,\parallel}$, $D_{e,\perp}$, D_{iso}). The orientation is modeled by a Watson distribution, which can be represented by a single concentration parameter κ. In NODDI, the third isotropic compartment represents free water, $D_{\text{iso}} = 3.0 \ \mu m^2/ms$. In addition, the extra-axonal compartment is placed outside the convolution [Equation (6.31)], so that the model consists of sticks with dispersion (intra-axonal space), a diffusion tensor ellipsoid (extra-axonal space), and an isotropic compartment (free water). For the sake of fitting stability, NODDI further imposes the following constraints:

$$D_{a,\parallel} = D_{e,\parallel} = 1.7 \ \mu m^2/ms \tag{6.33}$$

$$D_{e,\perp} = (1 - f_{\text{intra}}) D_{e,\parallel}. \tag{6.34}$$

Thus, NODDI actually estimates three free parameters, $f_{\text{intra}}, f_{\text{iso}}$, and κ. NODDI rapidly became popular in clinical research and has been applied to studies of normal brain development (Genc et al., 2017), multiple sclerosis (Granberg et al., 2017), Alzheimer's disease (Slattery et al., 2017), and Parkinson's disease (Andica et al., 2018), among others. For example, Slattery et al. (2017) showed that white matter changes in Alzheimer's disease are likely a combination of axonal loss and smaller orientation dispersion. These two phenomena affect FA in opposite directions, which means that decomposing these factors may increase sensitivity to pathology (Slattery et al., 2017). Studies in a postmortem human spinal cord (Grussu et al., 2017) and in monkey brains (Schilling et al., 2018) showed that NODDI's f_{intra} and κ correlate with axon density and dispersion measured on histology, respectively. However, the use of strong constraints in NODDI has also received criticisms (Lampinen et al., 2017; Novikov et al., 2018). Fixing $D_{a,\parallel}$ and $D_{e,\parallel}$ to a certain value means that the output of NODDI depends on our choice of the prescribed value (Hutchinson et al., 2017). In addition, there is no guarantee that $D_{a,\parallel} = D_{e,\parallel}$. In pathology with substantial changes in diffusivities, the abnormalities in f_{intra}, f_{iso}, or κ are not assured to accurately reflect the tissue properties they claim. The constraints (6.33) and (6.34) impose a direct connection between f_{intra} and $D_{e,\perp}$, so that f_{intra} is affected not only by the change in actual neurite density but also by the change of diffusivities (Lampinen et al., 2017).

2) WMTI

WMTI is a two-compartment model that assumes that the voxel is described by a sum of two diffusion tensors corresponding to the intra- and extra-axonal spaces. The model is described by four parameters (f_{intra}, $D_{a,\parallel}$, $D_{e,\parallel}$, $D_{e,\perp}$). WMTI assumes that axons are perfectly aligned. Under this condition, Fieremans et al. (2011) showed that the model parameters can be derived directly from the diffusion and kurtosis tensors. In that process, two possible solutions arise, where either $D_{a,\parallel} > D_{e,\parallel}$ or $D_{a,\parallel} < D_{e,\parallel}$. In the original work on WMTI, the authors chose $D_{a,\parallel} < D_{e,\parallel}$, although this choice of inequalities is still under active research (Dhital et al., 2019; Kunz et al., 2018). The assumption of perfectly aligned axons is expected to work as an approximation for fiber dispersion up to a small angle (Fieremans et al., 2011), and clinical studies have applied WMTI in white matter regions with high FA values. WMTI has been applied to various human diseases, including acute stroke (Hui et al., 2012), multiple sclerosis (de Kouchkovsky et al., 2016), Alzheimer's disease (Benitez et al., 2014), and traumatic brain injury (Chung et al., 2018), as well as to normal development (Jelescu et al., 2015) and aging (Benitez et al., 2018). Notably, studies of acute stroke (Hui et al., 2012) and traumatic brain injury (Chung et al., 2018) reported a decrease of $D_{a,\parallel}$, which seems consistent with the prominent axonal beading or varicosities (caliber variations along the axons) known to occur in ischemia (Li & Murphy, 2008) as well as in mechanical stretch injury (Tang-Schomer et al., 2012). WMTI has been tested extensively in animal models of demyelination (Guglielmetti et al., 2016; Jelescu, Zurek et al., 2016). Jelescu, Zurek et al. (2016) demonstrated that WMTI's f_{intra} and $D_{e,\perp}$ correlate with axon fraction and g-ratio (the ratio between the inner and the outer diameter of the myelin sheath) on histology, respectively. Guglielmetti et al. (2016) reported that $D_{a,\parallel}$ decreased in the acute inflammatory phase in a mouse model of multiple sclerosis, whereas a decrease in f_{intra} was observed during the remyelination and recovery period.

6.4.3 LIMITATIONS

The validity of model assumptions is fundamental for proper interpretation, but so far, general consensus regarding the "appropriate" model has not been reached. Building a model requires identifying which microstructural features have measurable effects and which are irrelevant (Alexander et al., 2019; Novikov, Kiselev, & Jespersen, 2018), and this is still an ongoing process. First, the multiple Gaussian compartment model gives a picture for a sufficiently long diffusion time ($t \rightarrow \infty$), where no diffusion time dependence is observed any longer. However, diffusion time dependence was observed in the normal human brain in the range of t typically probed in clinical studies, in both parallel and perpendicular directions (Fieremans et al., 2016). The observation means that diffusion is not Gaussian in at least one compartment (Novikov et al., 2019). The dependence on diffusion time in the direction parallel to the axons is thought to arise from the morphological heterogeneity along the axons and dendrites (e.g., varicosities, spines, undulations, changes of myelin thickness), which are on the scale of 10 μm (Giacci et al., 2018; Lee et al., 2019; Tang-Schomer et al., 2012) (Figure 6.10). In the perpendicular direction, the effect of random packing geometry of fibers on the extra-axonal diffusion has been suggested as a candidate source of dependence on diffusion time (Burcaw et al., 2015; Fieremans et al., 2016). Second, it is not very clear how many compartments should be considered. In two-compartment models, everything except for the axons are put into a single

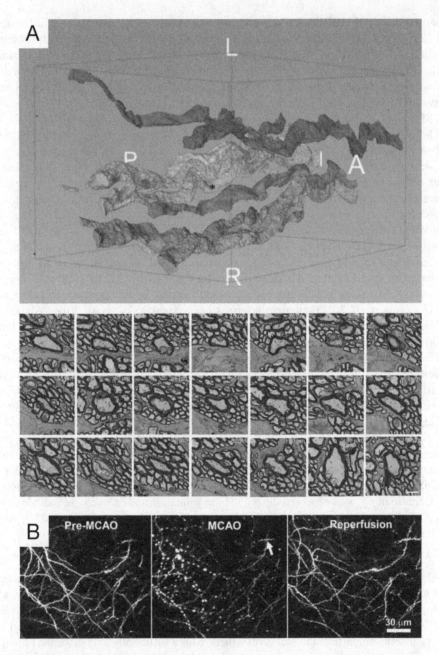

FIGURE 6.10 (A) Complex morphology of axons. Top: three-dimensional rendering of ten randomly selected axons in the optic nerve of a normal rat. Bottom: variability in axonal morphology in a single axon, at 5 μm intervals along its length. Scale bar = 2 μm. (Reprinted with permission from Giacci. 2018.) (B) Two-photon imaging of local changes in blood flow and dendritic structure before, during, and after middle cerebral artery occlusion (MCAO) in a mouse. In the middle panel (52 min after MCAO), extensive dendritic blebbing is observed. A white arrow shows a region in which dendrites were relatively spared, possibly because of residual blood. In the right panel, the animal was reperfused and a significant recovery of dendritic structure was observed. (Reprinted with permission from Li and Murphy. 2008, Copyright [2008] Society for Neuroscience.)

Gaussian compartment, assuming they are all in very fast exchange. However, *in vivo* measurement of exchange is extremely challenging, and this assumption is hard to validate. Furthermore, pathologies may lead to violation of model assumptions, for example, by a change of membrane permeabilities or by the appearance of compartments that do not exist in the healthy brain.

There is also another, rather technical problem in microstructure models: the estimation of model parameters by solving the inverse problem. The works by Jelescu et al. (2016) and Novikov et al. (2018) revealed that parameter estimation of the model [Equations (6.31) and (6.32)] with conventional PGSE measurements is degenerate, i.e., there are multiple sets of model parameters that yield exactly the same signal. This is why methods in current use, like NODDI and WMTI, inevitably introduce some additional constraints to the model. Recently, several proposals have been made to resolve this degeneracy by adding independent measurements, for example, variable TE (Veraart et al., 2018) and multiple diffusion encodings (Coelho et al., 2018; Reisert et al., 2019).

6.5 TRACTOGRAPHY

The observation of anisotropic diffusion in the white matter led to the idea of estimating the directions of white matter fiber bundles and visualizing the fiber trajectories (Basser, 1998). Here, the two main challenges are 1) estimation of the local fiber orientation and 2) reconstruction of the fiber trajectories by tracking the estimated directions (derived from either the principal eigenvector for DTI or peaks in ODF). The latter process is often called fiber tracking or fiber tractography, and the resulting picture is referred to as a tractogram. Here we focus on the former process, that is, methods to obtain information regarding orientation. For comprehensive information and in-depth discussions including tracking methods and further downstream applications (e.g., connectome analysis), the readers are referred to Jones (2010), Jeurissen et al. (2019), and Sotiropoulos and Zalesky (2019).

6.5.1 DTI TRACTOGRAPHY

The first generation of tractography algorithms used the first eigenvector of the diffusion tensor to define the local trajectory at each step of tracking. In principle, the tractogram is generated starting from a certain voxel or voxels (seed) and propagates according to the estimated orientation until the tracts are terminated due to a violation of prescribed criteria (e.g., FA value below threshold, bending with angle greater than threshold). Anatomical knowledge is used as a constraint to reconstruct specific white matter bundles. Typical procedures to reconstruct the major white matter bundles can be found, for example, in Wakana et al. (2007) and Catani and Thiebaut de Schotten (2008). A major limitation of DTI tractography is that it can detect only a single direction per voxel. In the human brain, and with the spatial resolution available on current scanners, it has been demonstrated (Jeurissen et al., 2013) that as much as 90% of white matter voxels contain crossing fibers.

6.5.2 HIGHER-ORDER METHODS

To account for complex fiber geometry, a range of so-called "higher-order" methods have been proposed. These methods have been shown to provide biologically more plausible tractograms even in regions with complex fiber geometries. Figure 6.11 demonstrates their advantage, where the DTI tractography results in limited delineation of the corticospinal tract that can be clinically misleading (Farquharson et al., 2013). Higher-order methods can be grouped into two categories; the first group aims to recover orientation information of the diffusion propagator and the second estimates the fiber ODF by relying on signal models of a single fiber population.

1) Orientation Information of the Diffusion Propagator
   Diffusion spectrum imaging (DSI) (Wedeen et al., 2005) is the direct application of q-space formalism in three dimensions. Although DSI is a powerful and general

framework, the high demand on scanner hardware and acquisition time (515 acquisitions, with maximum b-value of 17000 s/mm^2) makes it difficult to use in clinical practice. To reduce the acquisition requirements, several methods were proposed to work on moderate samplings spread evenly on a sphere in q-space (a shell). q-Ball imaging (Tuch, 2004) approximates the diffusion propagator by Funk–Radon transform of the signals measured on a single shell. Later refinements of q-ball imaging (Descoteaux et al., 2007) further improved the efficiency by working on the basis of spherical harmonics, where representation is more compact and analytical expression of the Funk–Radon transform is available. Typical acquisition for q-ball imaging is 60 directions with a single b-value around 3000 s/mm^2.

2) Spherical Deconvolution

Methods in this category assume that the signal is a convolution of fiber ODF and a signal model of a single fiber population (the kernel, or the response function). One of the most popular methods is constrained spherical deconvolution (CSD) (Tournier et al., 2007; Tournier et al., 2004). In CSD, the kernel is estimated by taking an average over voxels that are assumed to consist of single fiber populations. Then, the fiber ODF is estimated by deconvolving the signal in every voxel with this kernel. This can be done efficiently by working on a spherical harmonics basis (Tournier et al., 2004). Demand on acquisition is similar to that of q-ball imaging. Recently, CSD was extended to use multi-shell acquisition to estimate multi-tissue (white matter, gray matter, and cerebrospinal fluid) kernels (Jeurissen et al., 2014). One of the limitations of CSD is the use of the same kernel for all of the voxels in the brain. As proposed (Jensen et al., 2016; Kaden et al., 2016; Novikov et al., 2018), use of voxel-specific kernels, if possible, will enable more accurate reconstruction of fiber ODF. A recent study comparing histology-based ODF and dMRI showed a need for voxel-specific kernels (Schilling et al., 2019).

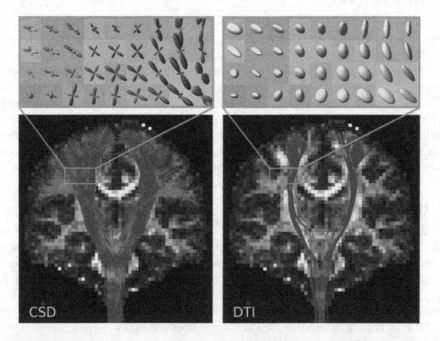

FIGURE 6.11 Corticospinal tract reconstructed using CSD (left) and DTI (right) in a healthy subject. The magnified regions show the fiber orientation estimates within individual voxels. CSD=constrained spherical deconvolution; DTI=diffusion tensor imaging. (Reprinted with permission from Farquharson et al. 2013.)

6.5.3 APPLICATIONS

The unique ability of dMRI tractography to visualize white matter fiber pathways non-invasively made it useful both for clinical medicine and neuroscience research. Despite its limitations (Section 6.5.4), dMRI tractography has at least enabled virtual white matter dissection (Catani & Thiebaut de Schotten, 2008) (Figure 6.12) that roughly matches the results of postmortem studies and has provided good explanations for syndromes caused by disconnection between particular brain regions (Thiebaut de Schotten et al., 2015). Currently, dMRI tractography is an indispensable tool for mapping before neurosurgery (Calabrese, 2016; Voets et al., 2017). The use of higher-order methods has been reported to better predict functional outcomes of surgery for brain tumors by improving the delineation of tracts passing through a peri-tumoral region (Caverzasi et al., 2016).

Another modern application of tractography is the field of connectomics, i.e., the study of the brain as a network (Hagmann et al., 2008; Sporns, Tononi, & Kötter, 2005). Terms used for brain network analysis are derived from graph theory, a field of mathematics with a long history that deals with complex networks. Graph theory describes a network by its nodes and edges. In the brain network, nodes represent cortical and subcortical gray matter regions, whereas edges typically represent the white matter fiber bundles that connect pairs of regions. Thus, building a connectome with dMRI basically consists of three steps: 1) defining nodes; 2) mapping edges; and 3) quantifying edges (Figure 6.13). First, nodes are defined by either registering pre-defined atlases to individual brains or by data-driven approaches (Glasser et al., 2016). Then, dMRI tractography is used to estimate edges (the connections between nodes). Finally, an N-by-N connectivity matrix, where N is the number of nodes, is reconstructed. If we are after a binary matrix, a certain threshold is applied so that each cell of the matrix contains either 0 or 1 (absence/presence of an edge). Alternatively, one may reconstruct a weighted matrix, where each cell contains a measure of the relative weight of that edge (e.g., streamline count). The connectivity matrix can be either analyzed directly (e.g., comparing the weight of each edge between patients and controls) or further fed into a graph-theory framework to derive

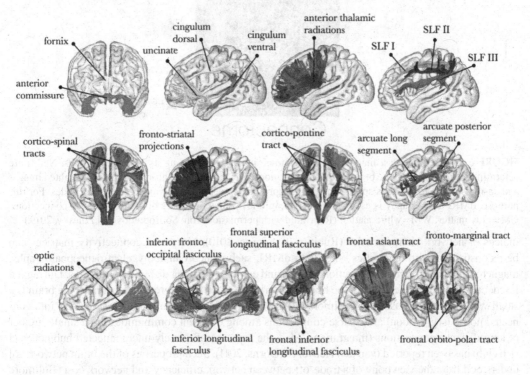

FIGURE 6.12 Virtual dissection of the major fiber pathways in the human brain. SLF = superior longitudinal fasciculus. (Reprinted with permission from de Schotten et al. 2015.)

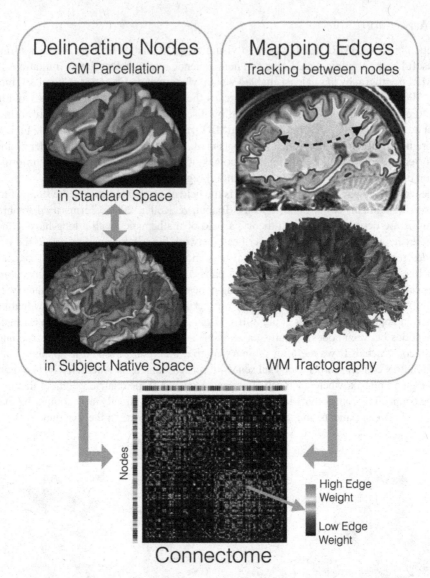

FIGURE 6.13 Generating a macroscale connectome involves estimating its nodes and edges. Nodes are determined by parcellating the brain into distinct regions, typically using high-resolution T1-weighted images and co-registration to the atlas space. The edges represent connection pathways between the nodes. For this purpose, dMRI tractography is used assuming streamlines connecting pairs of nodes represent connections. GM = gray matter; WM = white matter. (Reprinted with permission from Sotiropoulos and Zalesky, 2019.)

indices of network characteristics (Rubinov & Sporns, 2010). Moreover, connectivity matrices can be reconstructed with modalities other than dMRI, such as neuroanatomy, electroencephalography, magnetoencephalography, and functional MRI, and compared among different modalities. During the last decade, interesting features of the brain network have been discovered. For example, the brain is a small-world network (Bassett & Bullmore, 2006), characterized by communities of densely interconnected nodes (segregation) and sparse connections among different communities via a small number of long-distance connections (integration). Also, the presence of densely interconnected hub nodes (a rich club) has been reported (van den Heuvel & Sporns, 2011). Such properties of the brain network are understood from the viewpoint of a trade-off between network efficiency and network cost (Bullmore & Sporns, 2012). Studies comparing patients and controls have provided intriguing results demonstrating alterations of these network properties in diseases [for reviews, see Fornito et al. (2015)].

6.5.4 LIMITATIONS

It cannot be emphasized enough that the tractograms are merely virtual visualizations, or approximations, and are related only indirectly to the real nerve fibers in the brain (Jeurissen et al., 2019; Jones, 2010; Jones et al., 2013). Due to the low signal-to-noise ratio and averaging over the voxel, which are intrinsic characteristics of dMRI, there is a lower limit of crossing angles that can be resolved (around 40°–60°) (Schilling et al., 2018; Tournier et al., 2008). Also, it is impossible to distinguish between different fiber configurations, for example, bending, fanning, and crossing (Figure 6.14) (Jbabdi & Johansen-Berg, 2011). Many different algorithms and implementations have been proposed, and resulting tractograms depend on the choice of algorithms and parameters, as well as the acquisition scheme (Bastiani et al., 2012; Tournier, Calamante, & Connelly, 2012) Such dependence on acquisition and postprocessing methods demonstrates why tractograms should not be confused with biological reality. In a macaque study that compared dMRI tractography with a tracer study, the sensitivity and specificity of tractography were not very satisfactory, regardless of the algorithms used (Thomas et al., 2014).

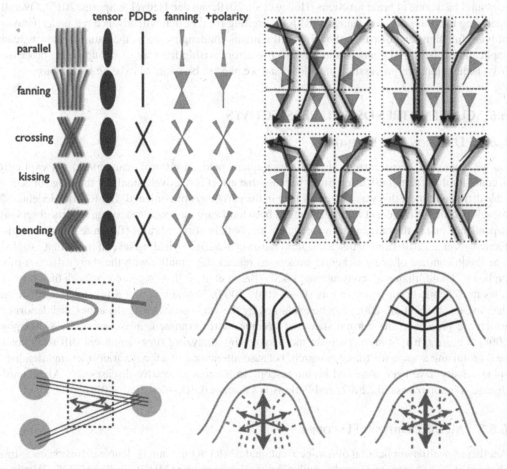

FIGURE 6.14 Illustrations of ambiguities in reconstructing the ODF and their consequences in tractograms. Top left: different fiber geometries can lead to similar ODFs. Methods in current use typically assume symmetric fiber ODF. However, the ground truth geometry might require accounting for the polarity (asymmetry) (the rightmost), which is extremely challenging. Top right: inability to model fanning polarity accurately can result in different tracking results. Bottom left: illustration of kissing fibers. The local fiber ODF is indistinguishable from that of crossing fibers, which can cause false positives. Bottom right: ambiguities near the cortex. The fiber ODF (and hence the tracking results) by dMRI is the same for the two fiber configurations, which have very different biological meanings. ODF = orientation distribution function; PDD = principal diffusion direction. (Reprinted with permission from Jbabdi and Johansen-Berg, 2011.)

Standardization of acquisition and analysis remains a task for the future, and the "optimal" choice may differ depending on the tract of interest. Another limitation intrinsic to dMRI tractography is the lack of information regarding the direction of signal transmission (afferent or efferent connection).

Such ambiguity in tractography reconstruction becomes even more problematic in connectomics, where we wish to have quantification, not just qualitative pictures. From the very beginning, how to define "connectivity" with dMRI has been unclear (Jeurissen et al., 2019; Jones, 2010; Jones et al., 2013; Sotiropoulos & Zalesky, 2019). As a measure of connectivity, studies so far have typically used metrics such as streamline counts of tractography, FA or other DTI metrics, and parameters derived from models like NODDI. However, as we have seen, these are not guaranteed to represent the strength of the connection. Reconstructing a connectivity matrix further involves many methodological choices (definition of nodes, tractography algorithms, thresholding, normalization of weights, etc.), all of which have non-negligible effects on the results (Bastiani et al., 2012; Qi et al., 2015). Despite all these limitations, the connectivity matrices reconstructed with dMRI have correlated with those from other modalities like functional MRI, indicating their promising ability to capture the structural backbone of brain functions (Honey et al., 2010; van den Heuvel & Sporns, 2013). Overall, analyzing the brain network using dMRI is an exciting frontier that will advance our understanding of brain functions and diseases, though it still remains challenging. As is the nature of downstream applications, the dMRI connectome is subject to errors arising from all of the upstream processes (e.g., acquisition, pre-processing, modeling) and we need to be aware of existing limitations.

6.6 CURRENT TRENDS AND PERSPECTIVES

6.6.1 DIFFUSION TIME DEPENDENCE

As we have seen, most current applications implemented on clinical scanners do not explicitly account for the effect of diffusion time t because that effect is relatively small for the range of accessible diffusion times; this was especially true in the early days of clinical dMRI (Clark, Hedehus, & Moseley, 2001). With recent improvements in both hardware and postprocessing, reports observing dependence on t in the human brain are increasing, both in normal brains (Baron & Beaulieu, 2014; Fieremans et al., 2016) and in disease states (Baron et al., 2015), yielding several important insights. The lesion contrast of acute ischemic stroke was remarkably smaller with the short diffusion time probed by OGSE than with conventional PGSE (Baron et al., 2015), in agreement with the observations in ischemic brain injury in rats (Does et al., 2003). Novikov et al. (2014) suggested that the data in rats (Does et al., 2003) can be explained by the varicosities along the axons and dendrites, which are present in the normal state and become more prominent in ischemia (Li & Murphy, 2008) (Figure 6.10). Studying neurite morphology by analyzing time-dependent diffusion could be a promising avenue for future research, because alterations of axonal varicosities and dendritic spine density have been observed by microscopy in neurodegenerative diseases like Alzheimer's disease (Ikonomovic et al., 2007) and Parkinson's disease (O'Donnell et al., 2014).

6.6.2 MULTIPLE DIFFUSION ENCODING

Another appealing application of non-conventional dMRI acquisition is double diffusion encoding (Yang et al., 2018) or more generally, multiple diffusion encoding (MDE) (Topgaard, 2017; Westin et al., 2016). In the context of MDE, the b-matrix is often called the b-tensor to emphasize its geometric meaning (Topgaard, 2017; Westin et al., 2016) (Figure 6.15.), which can be understood as analogous to the diffusion tensor. In the conventional PGSE sequence (single diffusion encoding, SDE), the b-tensor (b-matrix) has only one non-zero eigenvalue, and this encoding is therefore called linear tensor encoding. If we add another block of diffusion weighting, in a direction not parallel to the first block, the b-tensor has two non-zero eigenvalues and now it becomes planar tensor encoding. Similarly, having three blocks with orthogonal directions and identical sizes forms spherical tensor

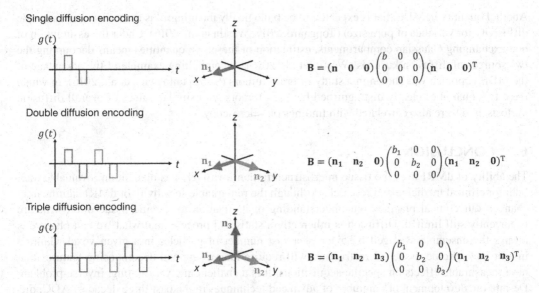

FIGURE 6.15 Illustration of diffusion encoding schemes, where the x-, y-, and z-components are shown in medium, light, and dark gray, respectively. **B** denotes the *b*-tensor.

encoding (Mori and Van Ziji, 1995). One piece of information provided by MDE (and not accessible with SDE alone) is the microscopic diffusion anisotropy (Jespersen et al., 2013; Lasič et al., 2014; Lawrenz & Finsterbusch, 2013; Topgaard, 2017; Westin et al., 2016). A popular normalized parameter that represents this property is microscopic fractional anisotropy (μFA), which is analogous to FA in DTI. In contrast to FA, μFA is not dependent on macroscopic orientation dispersion (Figure 6.16).

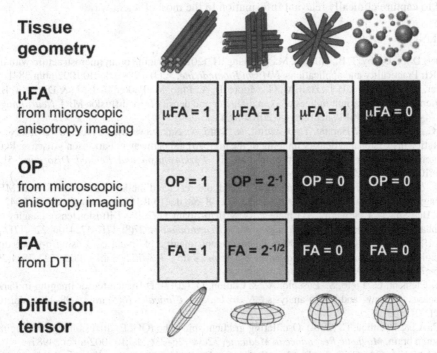

FIGURE 6.16 Idealized tissue geometries with corresponding structure parameters. Consecutive rows show values of the microscopic fractional anisotropy (μFA); orientational order parameter (OP); fractional anisotropy (FA); and diffusion tensors. (Reprinted with permission from Lasič et al., 2014.)

Another quantity in MDE that is expected to be biologically meaningful is the variance of isotropic diffusivity (or variance of pore size) (Topgaard, 2017; Westin et al., 2016). Under the assumption of non-exchanging Gaussian compartments, estimation of these two quantities means decoupling the two sources of diffusion kurtosis (Westin et al., 2016). A compelling example of the advantage of the MDE approach was shown in a study of brain tumors (Szczepankiewicz et al., 2016), in which these two quantities clearly distinguished between tumors with similar values of overall diffusion kurtosis, and were also correlated with findings on microscopy.

6.7   CONCLUSION

The ability of dMRI to probe tissue microstructure non-invasively has made it an invaluable tool, both for clinical medicine and research. Although the remarkable sensitivity of dMRI significantly changed our clinical practice, our understanding of its relation to specific tissue microstructure is currently still limited. Diffusion is inherently a statistical process, and what we can observe is always the ensemble-averaged behavior of a vast number of particles in a given voxel. Gaining insight into specific tissue microstructure will require identifying and decomposing features that have measurable effects on specific acquisitions. It is a challenging, yet exciting, inverse problem. Despite the development of a number of advanced techniques in the past three decades, ADC, diffusion-weighted image, and tractography are still effectively the only applications that have changed our daily clinical practice and gained a major presence in the hospital. However, this does not mean that going beyond conventional DTI has no practical value. Since the first discovery of the ADC reduction in acute ischemia (Moseley et al., 1990) almost 30 years ago, enormous effort has been spent toward understanding its mechanisms and some promising results have been shown very recently (Baron et al., 2015; Budde & Frank, 2010; Novikov et al., 2014). Although the process takes a long time, revealing the microstructural underpinnings behind observations in clinical dMRI, if achieved, will be of major practical benefit, because it will enable us to tailor acquisitions in the hospital to capture clinically relevant information in the most efficient way.

REFERENCES

Alexander, D. C., Dyrby, T. B., Nilsson, M., & Zhang, H. (2019). Imaging brain microstructure with diffusion MRI: Practicality and applications. *NMR in Biomedicine*, *32*(4), e3841. doi:10.1002/nbm.3841

Alexander, D. C., Hubbard, P. L., Hall, M. G., Moore, E. A., Ptito, M., Parker, G. J. M., & Dyrby, T. B. (2010). Orientationally invariant indices of axon diameter and density from diffusion MRI. *NeuroImage*, *52*(4), 1374–1389. doi:10.1016/j.neuroimage.2010.05.043

Andica, C., Kamagata, K., Hatano, T., Okuzumi, A., Saito, A., Nakazawa, M., … Aoki, S. (2018). Neurite orientation dispersion and density imaging of the nigrostriatal pathway in Parkinson's disease: Retrograde degeneration observed by tract-profile analysis. *Parkinsonism and Related Disorders*, 51, 55-60. doi:10.1016/j.parkreldis.2018.02.046

Assaf, Y., & Basser, P. J. (2005). Composite hindered and restricted model of diffusion (CHARMED) MR imaging of the human brain. *NeuroImage*, *27*(1), 48–58. doi:10.1016/j.neuroimage.2005.03.042

Assaf, Y., Ben-Sira, L., Constantini, S., Chang, L. C., & Beni-Adani, L. (2006). Diffusion tensor imaging in hydrocephalus: Initial experience. *American Journal of Neuroradiology*, *27*(8), 1717–1724. doi:27/8/1717 [pii]

Assaf, Y., Mayk, A., & Cohen, Y. (2000). Displacement imaging of spinal cord using q-space diffusion-weighted MRI. *Magnetic Resonance in Medicine*, *44*(5), 713–722. doi:10.1002/1522-2594(200011)4 4:5<713::AID-MRM9>3.0.CO;2-6

Atkinson-Clement, C., Pinto, S., Eusebio, A., & Coulon, O. (2017). Diffusion tensor imaging in Parkinson's disease: Review and meta-analysis. *NeuroImage: Clinical*, *16*(March), 98–110. doi:10.1016/j. nicl.2017.07.011

Baron, C. A., & Beaulieu, C. (2014). Oscillating gradient spin-echo (OGSE) diffusion tensor imaging of the human brain. *Magnetic Resonance in Medicine*, *72*(3), 726–736. doi:10.1002/mrm.24987

Baron, C. A., Kate, M., Gioia, L., Butcher, K., Emery, D., Budde, M., & Beaulieu, C. (2015). Reduction of diffusion-weighted imaging contrast of acute ischemic stroke at short diffusion times. *Stroke*, *46*(8), 2136–2141. doi:10.1161/STROKEAHA.115.008815

Basser, P. J. (1995). Inferring microstructural features and the physiological state of tissues from diffusion-weighted images. *NMR in Biomedicine, 8*(7–8), 333–344. doi:10.1002/nbm.1940080707

Basser, P. J. (1998). *Fiber-Tractography via Diffusion Tensor MRI (DT-MRI)*. In: Proceedings of the 6th Annual Meeting ISMRM, Sydney, Australia. 1998.

Basser, P. J., Mattiello, J., & LeBihan, D. (1994a). Estimation of the effective self-diffusion tensor from the NMR spin echo. *Journal of Magnetic Resonance, Series B, 103*(3), 247–254. doi:10.1006/jmrb.1994.1037

Basser, P. J., Mattiello, J., & LeBihan, D. (1994b). MR diffusion tensor spectroscopy and imaging. *Biophysical Journal, 66*(1), 259–267. doi:10.1016/S0006-3495(94)80775-1

Bassett, D. S., & Bullmore, E. (2006). Small-world brain networks. *Neuroscientist, 12*, 512–523. doi:10.1177/1073858406293182

Bastiani, M., Shah, N. J., Goebel, R., & Roebroeck, A. (2012). Human cortical connectome reconstruction from diffusion weighted MRI: The effect of tractography algorithm. *NeuroImage, 62*(3), 1732–1749. doi:10.1016/j.neuroimage.2012.06.002

Beaulieu, C. (2002). The basis of anisotropic water diffusion in the nervous system - A technical review. *NMR in Biomedicine, 15*(7–8), 435–455. doi:10.1002/nbm.782

Benitez, A., Fieremans, E., Jensen, J. H., Falangola, M. F., Tabesh, A., Ferris, S. H., & Helpern, J. A. (2014). White matter tract integrity metrics reflect the vulnerability of late-myelinating tracts in Alzheimer's disease. *NeuroImage: Clinical, 4*, 64–71. doi:10.1016/j.nicl.2013.11.001

Benitez, A., Jensen, J. H., Falangola, M. F., Nietert, P. J., & Helpern, J. A. (2018). Modeling white matter tract integrity in aging with diffusional kurtosis imaging. *Neurobiology of Aging, 70*, 265–275. doi:10.1016/j.neurobiolaging.2018.07.006

Bonilha, L., Lee, C. Y., Jensen, J. H., Tabesh, A., Spampinato, M. V., Edwards, J. C., ... Helpern, J. A. (2015). Altered microstructure in temporal lobe epilepsy: A diffusional kurtosis imaging study. *American Journal of Neuroradiology, 36*(4), 719–724. doi:10.3174/ajnr.A4185

Budde, M. D., & Frank, J. A. (2010). Neurite beading is sufficient to decrease the apparent diffusion coefficient after ischemic stroke. *Proceedings of the National Academy of Sciences of the United States of America, 107*(32), 14472–14477. doi:10.1073/pnas.1004841107

Bullmore, E., & Sporns, O. (2012). The economy of brain network organization. *Nature Reviews Neuroscience, 13*(5), 336–349. doi:10.1038/nrn3214

Burcaw, L. M., Fieremans, E., & Novikov, D. S. (2015). Mesoscopic structure of neuronal tracts from time-dependent diffusion. *NeuroImage, 114*, 18–37. doi:10.1016/j.neuroimage.2015.03.061

Calabrese, E. (2016). Diffusion tractography in deep brain stimulation surgery: A review. *Frontiers in Neuroanatomy, 10*, 45. doi:10.3389/fnana.2016.00045

Catani, M., & Thiebaut de Schotten, M. (2008). A diffusion tensor imaging tractography atlas for virtual in vivo dissections. *Cortex, 44*(8), 1105–1132. doi:10.1016/j.cortex.2008.05.004

Caverzasi, E., Hervey-Jumper, S. L., Jordan, K. M., Lobach, I. V., Li, J., Panara, V., ... Henry, R. G. (2016). Identifying preoperative language tracts and predicting postoperative functional recovery using HARDI q-ball fiber tractography in patients with gliomas. *Journal of Neurosurgery, 125*(1), 33–45. doi:10.3171/2015.6.JNS142203

Chung, S., Fieremans, E., Wang, X., Kucukboyaci, N. E., Morton, C. J., Babb, J., ... Lui, Y. W. (2018). White matter tract integrity: An indicator of axonal pathology after mild traumatic brain injury. *Journal of Neurotrauma, 35*(8), 1015–1020. doi:10.1089/neu.2017.5320

Citton, V., Burlina, A., Baracchini, C., Gallucci, M., Catalucci, A., Dal Pos, S., ... Manara, R. (2012). Apparent diffusion coefficient restriction in the white matter: Going beyond acute brain territorial ischemia. *Insights into Imaging, 3*(2), 155–164. doi:10.1007/s13244-011-0114-3

Clark, C. A., Hedehus, M., & Moseley, M. E. (2001). Diffusion time dependence of the apparent diffusion tensor in healthy human brain and white matter disease. *Magnetic Resonance in Medicine, 45*(6), 1126–1129. doi.org/10.1002/mrm.1149

Coelho, S., Pozo, J. M., Jespersen, S. N., Jones, D. K., & Frangi, A. F. (2019). Resolving degeneracy in diffusion MRI biophysical model parameter estimation using double diffusion encoding. *Magnetic Resonance in Medicine, 82*, 395–410. https://doi.org/10.1002/mrm.27714

Das, S. K., Wang, J. L., Bing, L., Bhetuwal, A., & Yang, H. F. (2017). Regional values of diffusional kurtosis estimates in the healthy brain during normal aging. *Clinical Neuroradiology, 27*(3), 283–298. doi:10.1007/s00062-015-0490-z

de Kouchkovsky, I., Fieremans, E., Fleysher, L., Herbert, J., Grossman, R. I., & Inglese, M. (2016). Quantification of normal-appearing white matter tract integrity in multiple sclerosis: A diffusion kurtosis imaging study. *Journal of Neurology, 263*(6), 1146–1155. doi:10.1007/s00415-016-8118-z de Swiet, T. M., & Sen, P. N. (1996). Time dependent diffusion coefficient in a disordered medium. *The Journal of Chemical Physics*, *104*(1), 206–209. doi:10.1063/1.470890

Descoteaux, M., Angelino, E., Fitzgibbons, S., & Deriche, R. (2007). Regularized, fast, and robust analytical Q-ball imaging. *Magnetic Resonance in Medicine*, *58*(3), 497–510. doi:10.1002/mrm.21277

Dhital, B., Reisert, M., Kellner, E., & Kiselev, V. G. (2019). Intra-axonal diffusivity in brain white matter. *NeuroImage*, *189*, 543–550. doi:10.1016/j.neuroimage.2019.01.015

Does, M. D., Parsons, E. C., & Gore, J. C. (2003). Oscillating gradient measurements of water diffusion in normal and globally ischemic rat brain. *Magnetic Resonance in Medicine*, *49*(2), 206–215. doi:10.1002/mrm.10385

Douaud, G., Jbabdi, S., Behrens, T. E. J., Menke, R. A., Gass, A., Monsch, A. U., … Smith, S. (2011). DTI measures in crossing-fibre areas: Increased diffusion anisotropy reveals early white matter alteration in MCI and mild Alzheimer's disease. *NeuroImage*, *55*(3), 880–890. doi:10.1016/j.neuroimage.2010.12.008

Einstein, A. (1905). Über die von der molekularkinetischen Theorie der Wärme geforderte Bewegung von in ruhenden Flüssigkeiten suspendierten Teilchen. *Annalen Der Physik*. doi:10.1002/andp.19053220806

Falangola, M. F., Jensen, J. H., Babb, J. S., Hu, C., Castellanos, F. X., Di Martino, A., … Helpern, J. A. (2008). Age-related non-Gaussian diffusion patterns in the prefrontal brain. *Journal of Magnetic Resonance Imaging*, *28*(6), 1345–1350. doi:10.1002/jmri.21604

Falangola, M. F., Jensen, J. H., Tabesh, A., Hu, C., Deardorff, R. L., Babb, J. S., … Helpern, J. A. (2013). Non-Gaussian diffusion MRI assessment of brain microstructure in mild cognitive impairment and Alzheimer's disease. *Magnetic Resonance Imaging*, *31*(6), 840–846. doi:10.1016/j.mri.2013.02.008

Farquharson, S., Tournier, J. D., Calamante, F., Fabinyi, G., Schneider-Kolsky, M., Jackson, G. D., & Connelly, A. (2013). White matter fiber tractography: Why we need to move beyond DTI. *Journal of Neurosurgery*, *118*(6), 1367–1377. doi:10.3171/2013.2.JNS121294

Fick, A. (1855). Ueber diffusion. *Annalen Der Physik*. doi:10.1002/andp.18551700105

Fieremans, E., Burcaw, L. M., Lee, H. H., Lemberskiy, G., Veraart, J., & Novikov, D. S. (2016). In vivo observation and biophysical interpretation of time-dependent diffusion in human white matter. *NeuroImage*, *129*, 414–427. doi:10.1016/j.neuroimage.2016.01.018

Fieremans, E., Jensen, J. H., & Helpern, J. A. (2011). White matter characterization with diffusional kurtosis imaging. *NeuroImage*, *58*(1), 177–188. doi:10.1016/j.neuroimage.2011.06.006

Fornito, A., Zalesky, A., & Breakspear, M. (2015). The connectomics of brain disorders. *Nature Reviews Neuroscience*, *16*(3), 159–172. doi:10.1038/nrn3901

Gauvain, K. M., McKinstry, R. C., Mukherjee, P., Perry, A., Neil, J. J., Kaufman, B. A., & Hayashi, R. J. (2001). Evaluating pediatric brain tumor cellularity with diffusion-tensor imaging. *AJR American Journal of Roentgenology*, *177*(2), 449–454. doi:10.2214/ajr.177.2.1770449

Genc, S., Malpas, C. B., Holland, S. K., Beare, R., & Silk, T. J. (2017). Neurite density index is sensitive to age related differences in the developing brain. *NeuroImage*, *148*, 373–380. doi:10.1016/j.neuroimage.2017.01.023

Giacci, M. K., Bartlett, C. A., Huynh, M., Kilburn, M. R., Dunlop, S. A., & Fitzgerald, M. (2018). Three dimensional electron microscopy reveals changing axonal and myelin morphology along normal and partially injured optic nerves. *Scientific Reports*, *8*(1), 3979. doi:10.1038/s41598-018-22361-2

Glasser, M. F., Coalson, T. S., Robinson, E. C., Hacker, C. D., Harwell, J., Yacoub, E., … Van Essen, D. C. (2016). A multi-modal parcellation of human cerebral cortex. *Nature*, *536*(7615), 171–178. doi:10.1038/nature18933

Glenn, G. R., Helpern, J. A., Tabesh, A., & Jensen, J. H. (2015). Quantitative assessment of diffusional kurtosis anisotropy. *NMR in Biomedicine*, *28*(4), 448–459. doi:10.1002/nbm.3271

Goveas, J., O'Dwyer, L., Mascalchi, M., Cosottini, M., Diciotti, S., De Santis, S., … Giannelli, M. (2015). Diffusion-MRI in neurodegenerative disorders. *Magnetic Resonance Imaging*, *33*(7), 853–876. doi:10.1016/j.mri.2015.04.006

Granberg, T., Fan, Q., Treaba, C. A., Ouellette, R., Herranz, E., Mangeat, G., … Mainero, C. (2017). In vivo characterization of cortical and white matter neuroaxonal pathology in early multiple sclerosis. *Brain*, *140*(11), 2912–2926. doi:10.1093/brain/awx247

Grussu, F., Schneider, T., Tur, C., Yates, R. L., Tachrount, M., Ianuş, A., … Gandini Wheeler-Kingshott, C. A. M. (2017). Neurite dispersion: A new marker of multiple sclerosis spinal cord pathology? *Annals of Clinical and Translational Neurology*, *4*(9), 663–679. doi:10.1002/acn3.445

Guglielmetti, C., Veraart, J., Roelant, E., Mai, Z., Daans, J., Van Audekerke, J., … Verhoye, M. (2016). Diffusion kurtosis imaging probes cortical alterations and white matter pathology following cuprizone induced demyelination and spontaneous remyelination. *NeuroImage*, *125*, 363–377. doi:10.1016/j.neuroimage.2015.10.052

Hagmann, P., Cammoun, L., Gigandet, X., Meuli, R., Honey, C. J., Wedeen, V. J., & Sporns, O. (2008). Mapping the structural core of human cerebral cortex. *PLoS Biology*, *6*(7), e159. doi:10.1371/journal. pbio.0060159

Hansen, B., Lund, T. E., Sangill, R., & Jespersen, S. N. (2013). Experimentally and computationally fast method for estimation of a mean kurtosis. *Magnetic Resonance in Medicine*, *69*(6), 1754–1760. doi:10.1002/mrm.24743

Harper, P. A. W., Healy, P. J., & Dennis, J. A. (1990). Animal model of human disease. Maple syrup urine disease (branched chain ketoaciduria). *American Journal of Pathology*, *136*(6), 1445–1447.

Honey, C. J., Thivierge, J. P., & Sporns, O. (2010). Can structure predict function in the human brain? *NeuroImage*, *52*(3), 766–776. doi:10.1016/j.neuroimage.2010.01.071

Horsfield, M. A., & Jones, D. K. (2002). Applications of diffusion-weighted and diffusion tensor MRI to white matter diseases - A review. *NMR in Biomedicine*, *15*(7–8), 570–577. doi:10.1002/nbm.787

Hui, E. S., Fieremans, E., Jensen, J. H., Tabesh, A., Feng, W., Bonilha, L., ... Helpern, J. A. (2012). Stroke assessment with diffusional kurtosis imaging. *Stroke*, *43*(11), 2968–2973. doi:10.1161/ STROKEAHA.112.657742

Hutchinson, E. B., Avram, A. V., Irfanoglu, M. O., Koay, C. G., Barnett, A. S., Komlosh, M. E., ... Pierpaoli, C. (2017). Analysis of the effects of noise, DWI sampling, and value of assumed parameters in diffusion MRI models. *Magnetic Resonance in Medicine*, *78*(5), 1767–1780. doi:10.1002/mrm.26575

Ikonomovic, M. D., Abrahamson, E. E., Isanski, B. A., Wuu, J., Mufson, E. J., & DeKosky, S. T. (2007). Superior frontal cortex cholinergic axon density in mild cognitive impairment and early Alzheimer disease. *Archives of Neurology*, *64*(9), 1312–1317. doi:10.1001/archneur.64.9.1312

Innocenti, G. M., Caminiti, R., & Aboitiz, F. (2015). Comments on the paper by Horowitz et al. (2014). *Brain Structure and Function*, *220*(3), 1789–1790. doi:10.1007/s00429-014-0974-7

Jbabdi, S., & Johansen-Berg, H. (2011). Tractography: Where do we go from here? *Brain Connectivity*, *1*(3), 169–183. doi:10.1089/brain.2011.0033

Jelescu, I. O., & Budde, M. D. (2017). Design and validation of diffusion MRI models of white matter. *Frontiers in Physics*, *5*. doi:10.3389/fphy.2017.00061

Jelescu, I. O., Veraart, J., Adisetiyo, V., Milla, S. S., Novikov, D. S., & Fieremans, E. (2015). One diffusion acquisition and different white matter models: How does microstructure change in human early development based on WMTI and NODDI? *NeuroImage*, *107*, 242–256. doi:10.1016/j. neuroimage.2014.12.009

Jelescu, I. O., Veraart, J., Fieremans, E., & Novikov, D. S. (2016). Degeneracy in model parameter estimation for multi-compartmental diffusion in neuronal tissue. *NMR in Biomedicine*, *29*(1), 33–47. doi:10.1002/ nbm.3450

Jelescu, I. O., Zurek, M., Winters, K. V., Veraart, J., Rajaratnam, A., Kim, N. S., ... Fieremans, E. (2016). In vivo quantification of demyelination and recovery using compartment-specific diffusion MRI metrics validated by electron microscopy. *NeuroImage*, *132*, 104–114. doi:10.1016/j.neuroimage.2016.02.004

Jensen, J. H., & Helpern, J. A. (2010). MRI quantification of non-Gaussian water diffusion by kurtosis analysis. *NMR in Biomedicine*, *23*(7), 698–710. doi:10.1002/nbm.1518

Jensen, J. H., Helpern, J. A., Ramani, A., Lu, H., & Kaczynski, K. (2005). Diffusional kurtosis imaging: The quantification of non-Gaussian water diffusion by means of magnetic resonance imaging. *Magnetic Resonance in Medicine*, *53*(6), 1432–1440. doi:10.1002/mrm.20508

Jensen, J. H., Russell Glenn, G., & Helpern, J. A. (2016). Fiber ball imaging. *NeuroImage*, *124*(A), 824–833. doi:10.1016/j.neuroimage.2015.09.049

Jespersen, S.N., Kroenke, C. D., Østergaard, L., Ackerman, J. J. H., & Yablonskiy, D. A. (2007). Modeling dendrite density from magnetic resonance diffusion measurements. *NeuroImage*, *34*(4), 1473–1486. doi:10.1016/j.neuroimage.2006.10.037

Jespersen, S.N, Lundell, H., Sønderby, C. K., & Dyrby, T. B. (2013). Orientationally invariant metrics of apparent compartment eccentricity from double pulsed field gradient diffusion experiments. *NMR in Biomedicine*, *26*(12), 1647–1662. doi:10.1002/nbm.2999

Jeurissen, B., Descoteaux, M., Mori, S., & Leemans, A. (2019). Diffusion MRI fiber tractography of the brain. *NMR in Biomedicine*, *32*(4), e3785. doi:10.1002/nbm.3785

Jeurissen, B., Leemans, A., Tournier, J. D., Jones, D. K., & Sijbers, J. (2013). Investigating the prevalence of complex fiber configurations in white matter tissue with diffusion magnetic resonance imaging. *Human Brain Mapping*, *34*(11), 2747–2766. doi:10.1002/hbm.22099

Jeurissen, B., Tournier, J. D., Dhollander, T., Connelly, A., & Sijbers, J. (2014). Multi-tissue constrained spherical deconvolution for improved analysis of multi-shell diffusion MRI data. *NeuroImage*, *103*, 411–426. doi:10.1016/j.neuroimage.2014.07.061

Jones, D. K. (2010). Challenges and limitations of quantifying brain connectivity in vivo with diffusion MRI. *Imaging in Medicine*, *2*(3), 341–355.

Jones, D. K., Knösche, T. R., & Turner, R. (2013). White matter integrity, fiber count, and other fallacies: The do's and don'ts of diffusion MRI. *NeuroImage*, *73*, 239–254. doi:10.1016/j.neuroimage.2012.06.081

Kaden, E., Kelm, N. D., Carson, R. P., Does, M. D., & Alexander, D. C. (2016). Multi-compartment microscopic diffusion imaging. *NeuroImage*, *139*, 346–359. doi:10.1016/j.neuroimage.2016.06.002

Kamagata, K., Tomiyama, H., Hatano, T., Motoi, Y., Abe, O., Shimoji, K., … Aoki, S. (2014). A preliminary diffusional kurtosis imaging study of Parkinson disease: Comparison with conventional diffusion tensor imaging. *Neuroradiology*, *56*(3), 251–258. doi:10.1007/s00234-014-1327-1

Kiselev, V. G. (2013). The cumulant expansion: An overarching mathematical framework for understanding diffusion NMR. *Diffusion MRI*. doi:10.1093/med/9780195369779.003.0010

Kiselev, V. G. (2017). Fundamentals of diffusion MRI physics. *NMR in Biomedicine*, *30*(3), e3602. doi:10.1002/nbm.3602

Koay, C. G., Chang, L.-C., Carew, J. D., Pierpaoli, C., & Basser, P. J. (2006). A unifying theoretical and algorithmic framework for least squares methods of estimation in diffusion tensor imaging. *Journal of Magnetic Resonance*, *182*(1), 115–125. doi:10.1016/j.jmr.2006.06.020

Konno, T., Kasanuki, K., Ikeuchi, T., Dickson, D. W., & Wszolek, Z. K. (2018). CSF1R-related leukoencephalopathy: A major player in primary microgliopathies. *Neurology*, *91*(24), 1092–1104. doi:10.1212/WNL.0000000000006642

Kunz, N., da Silva, A. R., & Jelescu, I. O. (2018). Intra- and extra-axonal axial diffusivities in the white matter: Which one is faster? *NeuroImage*, *181*, 314–322. doi:10.1016/j.neuroimage.2018.07.020

Lampinen, B., Szczepankiewicz, F., Mårtensson, J., van Westen, D., Sundgren, P. C., & Nilsson, M. (2017). Neurite density imaging versus imaging of microscopic anisotropy in diffusion MRI: A model comparison using spherical tensor encoding. *NeuroImage*, *147*, 517–531. doi:10.1016/j.neuroimage.2016.11.053

Lasič, S., Szczepankiewicz, F., Eriksson, S., Nilsson, M., & Topgaard, D. (2014). Microanisotropy imaging: Quantification of microscopic diffusion anisotropy and orientational order parameter by diffusion MRI with magic-angle spinning of the q-vector. *Frontiers in Physics*, *2*, 1–14. doi:10.3389/fphy.2014.00011

Lawrenz, M., & Finsterbusch, J. (2013). Double-wave-vector diffusion-weighted imaging reveals microscopic diffusion anisotropy in the living human brain. *Magnetic Resonance in Medicine*, *69*(4), 1072–1082. doi:10.1002/mrm.24347

Le Bihan, D., Breton, E., Lallemand, D., Grenier, P., Cabanis, E., & Laval-Jeantet, M. (1986). MR imaging of intravoxel incoherent motions: Application to diffusion and perfusion in neurologic disorders. *Radiology*, *161*(2), 401–407. doi:10.1148/radiology.161.2.3763909

Lee, H.-H., Yaros, K., Veraart, J., Pathan, J. L., Liang, F.-X., Kim, S. G., … Fieremans, E. (2019). Along-axon diameter variation and axonal orientation dispersion revealed with 3D electron microscopy: Implications for quantifying brain white matter microstructure with histology and diffusion MRI. *Brain Structure and Function*, *224*(4), 1469–1488. doi:10.1007/s00429-019-01844-6

Li, J. P., Pan, P. L., Song, W., Huang, R., Chen, K., & Shang, H. F. (2012). A meta-analysis of diffusion tensor imaging studies in amyotrophic lateral sclerosis. *Neurobiology of Aging*, *33*(8), 1833–1838. doi:10.1016/j.neurobiolaging.2011.04.007

Li, P., & Murphy, T. H. (2008). Two-photon imaging during prolonged middle cerebral artery occlusion in mice reveals recovery of dendritic structure after reperfusion. *Journal of Neuroscience*, *28*(46), 11970–11979. doi:10.1523/JNEUROSCI.3724-08.2008

McKinnon, E. T., Jensen, J. H., Glenn, G. R., & Helpern, J. A. (2017). Dependence on b-value of the direction-averaged diffusion-weighted imaging signal in brain. *Magnetic Resonance Imaging*, *36*, 121–127. doi:10.1016/j.mri.2016.10.026

Mitra, P. P., Sen, P. N., Schwartz, L. M., & Le Doussal, P. (1992). Diffusion propagator as a probe of the structure of porous media. *Physical Review Letters*, *68*(24), 3555–3558. doi:10.1103/PhysRevLett.68.3555

Mori, S., & Van Zijl, P. C. M. (1995). Diffusion weighting by the trace of the diffusion tensor within a single scan. *Magnetic Resonance in Medicine*, *33*(1), 41–52. doi:10.1002/mrm.1910330107

Moseley, M. E., Cohen, Y., Kucharczyk, J., Mintorovitch, J., Asgari, H. S., Wendland, M. F., … Norman, D. (1990). Diffusion-weighted MR imaging of anisotropic water diffusion in cat central nervous system. *Radiology*, *176*(2), 439–445. doi:10.1148/radiology.176.2.2367658

Moseley, M. E., Cohen, Y., Mintorovitch, J., Chileuitt, L., Shimizu, H., Kucharczyk, J., … Weinstein, P. R. (1990). Early detection of regional cerebral ischemia in cats: Comparison of diffusion- and T2-weighted MRI and spectroscopy. *Magnetic Resonance in Medicine*, *14*(2), 330–346. doi:10.1002/mrm.1910140218

Nilsson, M., Lasič, S., Drobnjak, I., Topgaard, D., & Westin, C. F. (2017). Resolution limit of cylinder diameter estimation by diffusion MRI: The impact of gradient waveform and orientation dispersion. *NMR in Biomedicine, 30*(7), 1–13. doi:10.1002/nbm.3711

Novikov, D. S., Fieremans, E., Jespersen, S. N., & Kiselev, V. G. (2019). Quantifying brain microstructure with diffusion MRI: Theory and parameter estimation. *NMR in Biomedicine, 32*(4), e3998. doi:10.1002/nbm.3998

Novikov, D. S., Jensen, J. H., Helpern, J. A., & Fieremans, E. (2014). Revealing mesoscopic structural universality with diffusion. *Proceedings of the National Academy of Sciences of the United States of America, 111*(14), 5088–5093. doi:10.1073/pnas.1316944111

Novikov, D. S., & Kiselev, V. G. (2011). Surface-to-volume ratio with oscillating gradients. *Journal of Magnetic Resonance, 210*(1), 141–145. doi:10.1016/j.jmr.2011.02.011

Novikov, D. S., Kiselev, V. G., & Jespersen, S. N. (2018). On modeling. *Magnetic Resonance in Medicine, 79*(6), 3172–3193. doi:10.1002/mrm.27101

Novikov, D. S., Veraart, J., Jelescu, I. O., & Fieremans, E. (2018). Rotationally-invariant mapping of scalar and orientational metrics of neuronal microstructure with diffusion MRI. *NeuroImage, 174*(1), 518–538. doi:10.1016/j.neuroimage.2018.03.006

O'Connor, K. M., Barest, G., Moritani, T., Sakai, O., & Mian, A. (2013). Dazed and diffused: Making sense of diffusion abnormalities in neurologic pathologies. *British Journal of Radiology, 86*(1032). doi:10.1259/bjr.20130599

O'Donnell, K. C., Lulla, A., Stahl, M. C., Wheat, N. D., Bronstein, J. M., & Sagasti, A. (2014). Axon degeneration and PGC-1α-mediated protection in a zebrafish model of α-synuclein toxicity. *DMM Disease Models and Mechanisms, 7*(5), 571–582. doi:10.1242/dmm.013185

Panagiotaki, E., Schneider, T., Siow, B., Hall, M. G., Lythgoe, M. F., & Alexander, D. C. (2012). Compartment models of the diffusion MR signal in brain white matter: A taxonomy and comparison. *NeuroImage, 59*(3), 2241–2254. doi:10.1016/j.neuroimage.2011.09.081

Pasternak, O., Sochen, N., Gur, Y., Intrator, N., & Assaf, Y. (2009). Free water elimination and mapping from diffusion MRI. *Magnetic Resonance in Medicine, 62*(3), 717–730. doi:10.1002/mrm.22055

Price, W. S. (2009). *NMR Studies of Translational Motion. NMR Studies of Translational Motion.* Cambridge: Cambridge University Press. doi:10.1017/CBO9780511770487

Qi, S., Meesters, S., Nicolay, K., ter Haar Romeny, B. M., & Ossenblok, P. (2015). The influence of construction methodology on structural brain network measures: A review. *Journal of Neuroscience Methods, 253*, 170–182. doi:10.1016/j.jneumeth.2015.06.016

Raab, P., Hattingen, E., Franz, K., Zanella, F. E., & Lanfermann, H. (2010). Cerebral gliomas: Diffusional kurtosis imaging analysis of microstructural differences. *Radiology, 254*(3), 876–881. doi:10.1148/radiol.09090819

Reisert, M., Kellner, E., Dhital, B., Hennig, J., & Kiselev, V. G. (2017). Disentangling micro from mesostructure by diffusion MRI: A Bayesian approach. *NeuroImage, 147*, 964–975. doi:10.1016/j.neuroimage.2016.09.058

Reisert, M., Kiselev, V. G., & Dhital, B. (2019). A unique analytical solution of the white matter standard model using linear and planar encodings. *Magnetic Resonance in Medicine, 81*(6), 3819–3825. doi:10.1002/mrm.27685

Reynaud, O. (2017). Time-dependent diffusion MRI in cancer: Tissue modeling and applications. *Frontiers in Physics, 5*, 1–16. doi:10.3389/fphy.2017.00058

Rovaris, M., Gass, A., Bammer, R., Hickman, S. J., Ciccarelli, O., Miller, D. H., & Filippi, M. (2005). Diffusion MRI in multiple sclerosis. *Neurology, 65*(10), 1526–1532. doi:10.1212/01.wnl.0000184471.83948.e0

Rubinov, M., & Sporns, O. (2010). Complex network measures of brain connectivity: Uses and interpretations. *NeuroImage, 52*(3), 1059–1069. doi:10.1016/j.neuroimage.2009.10.003

Sakai, M., Inoue, Y., Oba, H., Ishiguro, A., Sekiguchi, K., Tsukune, Y., … Nakamura, H. (2005). Age dependence of diffusion-weighted magnetic resonance imaging findings in maple syrup urine disease encephalopathy. *Journal of Computer Assisted Tomography, 29*(4), 524–527. doi:10.1097/01.rct.0000164667.65648.72

Salvador, R., Peña, A., Menon, D. K., Carpenter, T. A., Pickard, J. D., & Bullmore, E. T. (2005). Formal characterization and extension of the linearized diffusion tensor model. *Human Brain Mapping, 24*(2), 144–155. doi:10.1002/hbm.20076

Schilling, K. G., Gao, Y., Stepniewska, I., Janve, V., Landman, B. A., & Anderson, A. W. (2019). Histologically derived fiber response functions for diffusion MRI vary across white matter fibers—An ex vivo validation study in the squirrel monkey brain. *NMR in Biomedicine, 32*(6), e4090. doi:10.1002/nbm.4090

Schilling, K. G., Janve, V., Gao, Y., Stepniewska, I., Landman, B. A., & Anderson, A. W. (2018). Histological validation of diffusion MRI fiber orientation distributions and dispersion. *NeuroImage*, *165*, 200–221. doi:10.1016/j.neuroimage.2017.10.046

Sen, P. N. (2004). Time-dependent diffusion coefficient as a probe of geometry. *Concepts in Magnetic Resonance Part A: Bridging Education and Research*, *23*(1), 1–21. doi:10.1002/cmr.a.20017

Sexton, C. E., Kalu, U. G., Filippini, N., Mackay, C. E., & Ebmeier, K. P. (2011). A meta-analysis of diffusion tensor imaging in mild cognitive impairment and Alzheimer's disease. *Neurobiology of Aging*, *32*(12), 2322.e5–2322.e18. doi:10.1016/j.neurobiolaging.2010.05.019

Slattery, C. F., Zhang, J., Paterson, R. W., Foulkes, A. J. M., Carton, A., Macpherson, K., ... Schott, J. M. (2017). ApoE influences regional white-matter axonal density loss in Alzheimer's disease. *Neurobiology of Aging*, *57*, 8–17. doi:10.1016/j.neurobiolaging.2017.04.021

Sone, J., Mori, K., Inagaki, T., Katsumata, R., Takagi, S., Yokoi, S., ... Sobue, G. (2016). Clinicopathological features of adult-onset neuronal intranuclear inclusion disease. *Brain*, *139*(Pt 12), 3170–3186. doi:10.1093/brain/aww249

Sotiropoulos, S. N., & Zalesky, A. (2019). Building connectomes using diffusion MRI: Why, how and but. *NMR in Biomedicine*, *32*(4), e3752. doi:10.1002/nbm.3752

Sporns, O., Tononi, G., & Kötter, R. (2005). The human connectome: A structural description of the human brain. *PLoS Computational Biology*, *1*(4), e42. doi:10.1371/journal.pcbi.0010042

Stanisz, G. J., Szafer, A., Wright, G. A., & Henkelman, R. M. (1997). An analytical model of restricted diffusion in bovine optic nerve. *Magnetic Resonance in Medicine*, *37*(1), 103–111. doi:10.1002/mrm.1910370115

Stejskal, E. O., & Tanner, J. E. (1965). Spin diffusion measurements: Spin echoes in the presence of a time-dependent field gradient. *The Journal of Chemical Physics*, *42*(1), 288–292. doi:10.1063/1.1695690

Szczepankiewicz, F., van Westen, D., Englund, E., Westin, C. F., Ståhlberg, F., Lätt, J., ... Nilsson, M. (2016). The link between diffusion MRI and tumor heterogeneity: Mapping cell eccentricity and density by diffusional variance decomposition (DIVIDE). *NeuroImage*, *142*, 522–532. doi:10.1016/j.neuroimage.2016.07.038

Takanashi, J. I. (2009). Two newly proposed infectious encephalitis/encephalopathy syndromes. *Brain and Development*, *31*(7), 521–528. doi:10.1016/j.braindev.2009.02.012

Tang-Schomer, M. D., Johnson, V. E., Baas, P. W., Stewart, W., & Smith, D. H. (2012). Partial interruption of axonal transport due to microtubule breakage accounts for the formation of periodic varicosities after traumatic axonal injury. *Experimental Neurology*, *233*(1), 364–372. doi:10.1016/j.expneurol.2011.10.030

Thiebaut de Schotten, M., Dell'Acqua, F., Ratiu, P., Leslie, A., Howells, H., Cabanis, E., ... Catani, M. (2015). From Phineas gage and monsieur leborgne to H.M.: Revisiting disconnection syndromes. *Cerebral Cortex*, *25*(12), 4812–4827. doi:10.1093/cercor/bhv173

Thomas, C., Ye, F. Q., Irfanoglu, M. O., Modi, P., Saleem, K. S., Leopold, D. A., & Pierpaoli, C. (2014). Anatomical accuracy of brain connections derived from diffusion MRI tractography is inherently limited. *Proceedings of the National Academy of Sciences of the United States of America*, *111*(46), 16574–16579. doi:10.1073/pnas.1405672111

Topgaard, D. (2017). Multidimensional diffusion MRI. *Journal of Magnetic Resonance*, *275*, 98–113. doi:10.1016/j.jmr.2016.12.007

Torrey, H. C. (1956). Bloch equations with diffusion terms. *Physical Review*, *104*(3), 563. doi:10.1103/PhysRev.104.563

Tournier, J.-D., Calamante, F., & Connelly, A. (2012). MRtrix: Diffusion tractography in crossing fiber regions. *International Journal of Imaging Systems and Technology*, *22*(1), 53–66. doi:10.1002/ima.22005

Tournier, J. D., Calamante, F., & Connelly, A. (2007). Robust determination of the fibre orientation distribution in diffusion MRI: Non-negativity constrained super-resolved spherical deconvolution. *NeuroImage*, *35*(4), 1459–1472. doi:10.1016/j.neuroimage.2007.02.016

Tournier, J. D., Calamante, F., Gadian, D. G., & Connelly, A. (2004). Direct estimation of the fiber orientation density function from diffusion-weighted MRI data using spherical deconvolution. *NeuroImage*, *23*(3), 1176–1185. doi:10.1016/j.neuroimage.2004.07.037

Tournier, J. D., Yeh, C. H., Calamante, F., Cho, K. H., Connelly, A., & Lin, C. P. (2008). Resolving crossing fibres using constrained spherical deconvolution: Validation using diffusion-weighted imaging phantom data. *NeuroImage*, *42*(2), 617–625. doi:10.1016/j.neuroimage.2008.05.002

Tuch, D. S. (2004). Q-ball imaging. *Magnetic Resonance in Medicine*, *52*(6), 1358–1372. doi:10.1002/mrm.20279

Van Cauter, S., Veraart, J., Sijbers, J., Peeters, R. R., Himmelreich, U., De Keyzer, F., ... Sunaert, S. (2012). Gliomas: Diffusion kurtosis MR imaging in grading. *Radiology*, *263*(2), 492–501. doi:10.1148/radiol.12110927 van den Heuvel, M. P., & Sporns, O. (2011). Rich-club organization of the human connectome. *Journal of Neuroscience, 31*(44), 15775–15786. doi:10.1523/JNEUROSCI.3539-11.2011

van den Heuvel, M. P., & Sporns, O. (2013). An anatomical substrate for integration among functional networks in human cortex. *Journal of Neuroscience , 33*(36), 14489–14500. doi:10.1523/JNEUROSCI.2128-13.2013

Veraart, J., Fieremans, E., & Novikov, D. S. (2019). On the scaling behavior of water diffusion in human brain white matter. *NeuroImage, 185*, 379–387. doi:10.1016/j.neuroimage.2018.09.075

Veraart, J., Novikov, D. S., & Fieremans, E. (2018). TE dependent Diffusion Imaging (TEdDI) distinguishes between compartmental T2 relaxation times. *NeuroImage, 182*, 360–369. doi:10.1016/j.neuroimage.2017.09.030

Veraart, J., Poot, D. H. J., Van Hecke, W., Blockx, I., Van der Linden, A., Verhoye, M., & Sijbers, J. (2011). More accurate estimation of diffusion tensor parameters using diffusion kurtosis imaging. *Magnetic Resonance in Medicine, 65*(1), 138–145. doi:10.1002/mrm.22603

Veraart, J., Sijbers, J., Sunaert, S., Leemans, A., & Jeurissen, B. (2013). Weighted linear least squares estimation of diffusion MRI parameters: Strengths, limitations, and pitfalls. *NeuroImage, 81*, 335–346. doi:10.1016/j.neuroimage.2013.05.028

Voets, N. L., Bartsch, A., & Plaha, P. (2017). Brain white matter fibre tracts: A review of functional neuro-oncological relevance. *Journal of Neurology, Neurosurgery and Psychiatry, 88*(12), 1017–1025. doi:10.1136/jnnp-2017-316170

Wakana, S., Caprihan, A., Panzenboeck, M. M., Fallon, J. H., Perry, M., Gollub, R. L., … Mori, S. (2007). Reproducibility of quantitative tractography methods applied to cerebral white matter. *NeuroImage, 36*(3), 630–644. doi:10.1016/j.neuroimage.2007.02.049

Wedeen, V. J., Hagmann, P., Tseng, W. Y. I., Reese, T. G., & Weisskoff, R. M. (2005). Mapping complex tissue architecture with diffusion spectrum magnetic resonance imaging. *Magnetic Resonance in Medicine, 54*(6), 1377–1386. doi:10.1002/mrm.20642

Wesbey, G. E., Moseley, M. E., & Ehman, R. L. (1984). Translational molecular self-diffusion in magnetic resonance imaging: II. Measurement of the self-diffusion coefficient. *Investigative Radiology, 19*(6), 491–498. doi:10.1097/00004424-198411000-00005

Westin, C. F., Knutsson, H., Pasternak, O., Szczepankiewicz, F., Özarslan, E., van Westen, D., … Nilsson, M. (2016). Q-space trajectory imaging for multidimensional diffusion MRI of the human brain. *NeuroImage, 135*, 345–362. doi:10.1016/j.neuroimage.2016.02.039

Yang, G., Tian, Q., Leuze, C., Wintermark, M., & McNab, J. A. (2018). Double diffusion encoding MRI for the clinic. *Magnetic Resonance in Medicine, 80*(2), 507–520. doi:10.1002/mrm.27043

Zhang, H., Schneider, T., Wheeler-Kingshott, C. A., & Alexander, D. C. (2012). NODDI: Practical in vivo neurite orientation dispersion and density imaging of the human brain. *NeuroImage, 61*(4), 1000–1016. doi:10.1016/j.neuroimage.2012.03.072

7 Magnetic Particle Imaging

Keiji Enpuku and Takashi Yoshida

CONTENTS

7.1 Operating Principle of MPI 155
    7.1.1 Magnetic Nanoparticles (MNPs) 156
    7.1.2 DC Magnetization of MNPs 158
    7.1.3 Frequency Response of MNP 159
    7.1.4 MNP Response Under AC and DC Gradient Field 160
    7.1.5 Imaging Technique 162
7.2 Point Spread Function 163
    7.2.1 Harmonic Signals 163
    7.2.2 Effect of the Magnetic Moment 166
    7.2.3 Effect of the Amplitude of the AC Field 166
    7.2.4 Relative Direction Between the AC and DC Fields 167
    7.2.5 Hysteresis of Magnetization 169
7.3 MPI Hardware 170
    7.3.1 Excitation Coil 171
    7.3.2 Pickup (Receiver) Coil 172
    7.3.3 Generation and Scanning of the FFP 172
    7.3.4 Generation and Scanning of the FFL 173
    7.3.5 Drive Coil 175
7.4 Imaging Technique 176
    7.4.1 Signal Voltage and Inversion Problem 176
    7.4.2 System Function Matrix 178
    7.4.3 Imaging System 179
References 181

Magnetic particle imaging (MPI) is an imaging technology to visualize and quantify magnetic nanoparticles (MNPs) for biomedical application [1–10]. In this method, MNPs are accumulated in a human (or animal) body and their signal field is measured. The three-dimensional positions and number of MNPs can then be reconstructed from the measured data by solving an inverse problem. MPI allows for the direct quantitative mapping of the spatial distribution of MNPs with high temporal and spatial resolution, and has several merits compared with other medical imaging modalities. MPI is expected to offer a new biomedical tool based on MNPs.

MPI exploits the unique characteristics of MNPs when AC and DC gradient fields are applied. The DC gradient field produces the so-called field-free point (FFP) or field-free line (FFL) where the field becomes zero. The FFP or FFL makes it possible to detect MNPs with high spatial resolution because the magnetic signal is selectively generated from the MNPs located near the FFP (or FFL).

This chapter discusses the basics of MPI, including operating principle, dynamic magnetization of MNPs under AC and DC gradient fields, MPI hardware, and imaging technology.

7.1 OPERATING PRINCIPLE OF MPI

The operating principle of MPI is schematically depicted in Figure 7.1. MNPs are accumulated in a human (or animal) body and are magnetized using an excitation field consisting of an AC and a DC

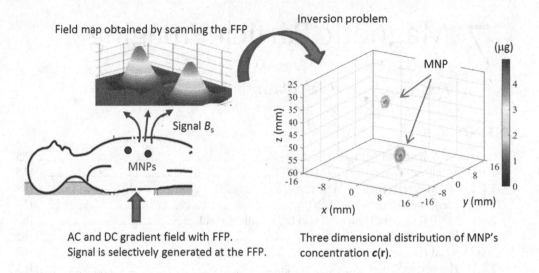

FIGURE 7.1 Schematic description of magnetic particle imaging (MPI).

gradient field. The AC field, H_{AC}, is spatially uniform, while the DC gradient field, H_{DC}, changes with position. The DC gradient field is designed to produce the so-called field-free point. The DC field becomes $H_{DC}=0$ at the FFP, and changes linearly with position with a field gradient G. Because of the magnetic properties of MNPs, the AC signal field, B_s, is generated only from the MNPs that are located near the FFP. Therefore, we can selectively detect MNPs near the FFP, significantly improving the spatial resolution in MNP detection. The field gradient is typically $G=1$ T/m, and a spatial resolution on the order of millimeters can be realized.

In MPI, the position of the FFP, r_p, is scanned three dimensionally, and the signal field, B_s, is measured at each FFP position. As a result, we can obtain the field map of B_s as a function of the FFP position r_p, as schematically shown in Figure 7.1. Then, the spatial distribution of the MNP concentration $c(r)$ can be obtained from $B_s(r_p)$ by solving the inversion problem, as shown in Figure 7.1. This imaging technique can further improve the spatial resolution of MNP detection.

MPI can sensitively detect a very small amount of MNPs, down to less than 1 ng of MNPs located at a distance of 10 cm. This sensitivity means that a few µm cubes of an MNP sample can be detected at a distance of 10 cm. Therefore, MPI enables highly sensitive detection of MNPs with high spatial resolution.

The DC gradient field with the FFL is also used in MPI, in which the gradient field is designed to become $H_{DC}=0$ along a line.

The operating principle of MPI will be discussed in more detail in the following section.

7.1.1 MAGNETIC NANOPARTICLES (MNPs)

Biocompatible MNPs have been widely studied for biomedical applications, including purification of biological targets, the detection of targets, an MRI contrast agent, drug delivery, and hyperthermia therapy [11–18]. Figure 7.2(a) shows a schematic of a multiparticle-based MNPs. The magnetic core consists of an agglomerate of elementary particles. The elementary particle is usually Fe_3O_4 or Fe_2O_3 and its size is less than 10 nm. The agglomerate is approximated using an effective magnetic core with diameter d_c, which typically ranges from 20 to 40 nm.

The magnetic core is covered with a coating material. For use in biological-target detection, detection antibodies are fixed on the surface of the coating material. This biofunctionalized MNP is called a magnetic marker. The hydrodynamic diameter of the marker is denoted by d_H, and is given by $d_H=d_c + 2t$, where t is the sum of the coating material thickness and the effective length of the detecting antibody.

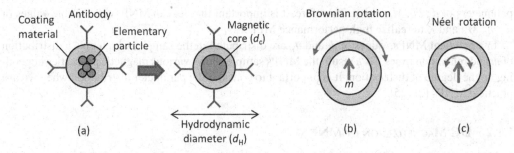

FIGURE 7.2 (a) Schematic figures of the marker made from multicore particles, (b) Brownian relaxation, and (c) Néel relaxation.

In MPI applications, dynamic magnetization of the marker under AC and DC excitation fields is used. In this case, important marker parameters are the magnetic moment m and relaxation time. A high value of m is desirable to generate a strong magnetic signal. Alternatively, an appropriate relaxation time is required for the MNP to respond to the AC excitation field.

The value of m is given by

$$m = M_s V_c \text{ with } V_c = \frac{\pi}{6} d_c^3 \qquad (7.1)$$

where M_s is the saturation magnetization and V_c is the volume of the magnetic core.

There are two types of relaxation times for the MNP, the Brownian relaxation time, τ_B, and the Néel relaxation time τ_N [19, 20]. The Brownian relaxation is caused by the physical rotation of the MNP when suspended in water, as shown in Figure 7.2(b). Although the magnetic moment m is fixed inside the magnetic core, m is rotated by the physical rotation of the MNP. This rotation is called Brownian rotation and is caused by the thermal noise. The relaxation time, τ_B, is given by

$$\tau_B = \frac{3\eta V_H}{k_B T} \text{ with } V_H = \frac{\pi}{6} d_H^3 \qquad (7.2)$$

where η is the viscosity of water, k_B is the Boltzmann constant, T is the absolute temperature, and V_H is a hydrodynamic volume of the marker.

The Néel relaxation is caused by the rotation of m inside the magnetic core, as shown in Figure 7.2(c). This situation occurs when the MNP is immobilized and physical rotation of the MNP is prevented. The Néel rotation is also caused by thermal noise and the relaxation time, τ_N, is given by

$$\tau_N = \tau_0 \frac{\sqrt{\pi}}{2} \frac{1}{\sqrt{\sigma}} \exp(\sigma) \text{ with } \sigma = \frac{K V_c}{k_B T} \qquad (7.3)$$

where $\tau_0 = 10^{-9}$ s is the characteristic time and K is the anisotropy energy density.

When the MNPs are suspended in water, both τ_B and τ_N affect the dynamic magnetization properties of the MNPs. In this case, the effective relaxation time, τ, is given by

$$\frac{1}{\tau} = \frac{1}{\tau_B} + \frac{1}{\tau_N} \qquad (7.4)$$

Alternatively, when the MNPs are immobilized, the magnetic properties are then determined solely by the Néel relaxation.

Notably, MNPs with high m value and appropriate relaxation time are necessary for MPI application. As shown in Eqs. (7.1) to (7.3), the values of m, τ_B and τ_N are determined by the MNP parameters of d_c, d_H, M_s, and K. Therefore, it is important to select an MNP with suitable values of d_c, d_H, M_s, and K to realize high-performance MPI.

In a practical MNPs, values of d_c and d_H are distributed in the sample, which give a distribution of m and relaxation times. As a result, the MNPs sample shows various magnetic properties depending on the degree of distribution. It is important to consider the parameter distribution when we use practical MNPs [21–25].

7.1.2 DC MAGNETIZATION OF MNPS

In MPI applications, nonlinear magnetization of magnetic nanoparticles is used. DC magnetization of the MNP has been described using the Langevin function as [19, 20]:

$$\frac{M}{M_s} = L(\xi) = \coth(\xi) - \frac{1}{\xi} \tag{7.5}$$

with

$$\xi = \frac{\mu_0 m H}{k_B T} \tag{7.6}$$

where H is an applied field and μ_0 is the permeability of vacuum. As shown in Eq. (7.5), the Langevin function is determined by parameter ξ, i.e., nonlinearity of the magnetization is determined by ξ. Therefore, it is important to select the appropriate value of ξ in practical application.

Figure 7.3(a) shows the H vs. M_{dc} curves calculated from Eqs. (7.1) and (7.5) for different values of the magnetic core diameter d_c. We used the saturation magnetization $M_s = 300$ kA/m and the d_c value was chosen as 15, 20, and 25 nm. The H vs. M_{dc} curve becomes linear when the H value is small. In this linear region, the Langevin function given in Eq. (7.5) can be approximated as $L(\xi) = \xi/3$. The H vs. M_{dc} curve becomes nonlinear when the H value becomes large. Using Eq. (7.5), the nonlinearity occurs when H satisfies the condition $\xi \geq 2$. When the H value becomes sufficiently large satisfying the condition $\xi \geq 10$, the M_{dc} value saturates to M_s.

As shown in Figure 7.3(a), the H vs. M_{dc} curve strongly depends on the d_c value. Saturation of M_{dc} occurs at lower H values when the d_c value becomes larger. In the case of $d_c = 25$ nm, saturation occurs at $\mu_0 H \approx 10$ mT. Alternatively, we need a field as large as $\mu_0 H = 50$ mT when $d_c = 15$ nm, which indicates that the nonlinearity of the magnetization curve becomes larger for larger d_c values. Therefore, an MNP with larger d_c is suitable for use in MPI applications.

In MPI applications, the derivative dM/dH also becomes important as it affects the performance of MPI, such as the signal strength and spatial resolution. The derivative is given from Eq. (7.5) as

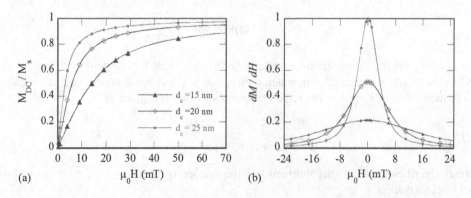

FIGURE 7.3 (a) DC magnetization of MNPs and (b) derivative dM/dH. The results calculated for $d_c = 15$, 20 and 25 nm are shown.

$$\frac{dM}{dH} = \frac{\mu_0 m M_s}{k_B T}\left(\frac{1}{\xi^2} - \frac{1}{\sinh^2(\xi)}\right) \quad \xi \neq 0 \tag{7.7}$$

$$= \frac{1}{3}\frac{\mu_0 m M_s}{k_B T} \quad \xi = 0$$

Figure 7.3(b) shows the H vs. dM/dH curves calculated for different values of d_c. The vertical axis is normalized by the value at $H=0$ and $d_c=25$ nm. The H vs. dM/dH curve becomes much sharper with the increase in d_c. Specifically, the peak of dM/dH increases with d_c, while the full width at half maximum (FWHM) in the H vs. dM/dH curve decreases with d_c. Notably, a sharper H vs. dM/dH curve is desired to improve MPI performance.

7.1.3 FREQUENCY RESPONSE OF MNP

In the MPI application, the magnetic signal from MNPs is generated by applying an AC excitation field given by

$$H_{AC} = H_0 \sin(2\pi f t), \tag{7.8}$$

where H_0 and f is the amplitude and frequency of the excitation field, respectively.

In order for the MNPs to respond to the AC field, the Brownian relaxation time, τ_B, and the Néel relaxation time, τ_N, must be shorter than the period of the AC field, i.e., $1/f > \tau_B$ and $1/f > \tau_N$.

Figure 7.4(a) shows the dependences of τ_B and τ_N on d_c that are calculated from Eqs. (7.2) and (7.3), respectively. In the calculation of τ_B, we used $\eta = 1$ mPa·s and the relation that $d_H = d_c + 2t$ with $t = 15$ nm for the thickness of the coating material. Since τ_N also depends on the anisotropy energy density, K, the values of τ_{N1} and τ_{N2} are calculated for $K = 5$ and 13 kJ/m³, respectively.

As shown, τ_B becomes longer as d_c increases; τ_B ranges from 24 µs to 82 µs when d_c varies between 10 and 30 nm. Alternatively, the value of τ_N has a significant dependence on d_c; τ_N varies more than four orders of magnitude with a small increase of d_c. We can also see that $\tau_{N2} \gg \tau_{N1}$, which indicates that the Néel relaxation time becomes much shorter for the MNPs with smaller K value.

We consider the case when the frequency of the AC field is set as $f = 10$ kHz, i.e., $1/f = 10^{-4}$ s. As shown in Figure 7.4(a), the condition that $1/f > \tau_B$ and $1/f > \tau_{N1}$ can be satisfied when $d_c = 25$ nm and $K = 5$ kJ/m³. In this case, MNPs can respond to the AC field. Therefore, the MNP with $d_c = 25$ nm and $K = 5$ kJ/m³ will be suitable for MPI application because it gives high nonlinearity in magnetization

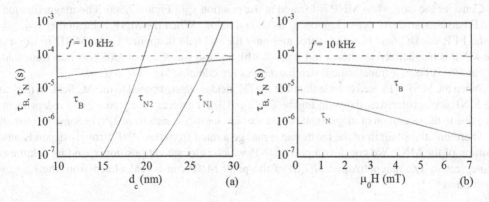

FIGURE 7.4 (a) Dependences of Brownian relaxation time, τ_B, and Néel relaxation time, τ_N, on the core diameter d_c. The values of τ_{N1} and τ_{N2} are calculated for $K_1 = 5$ and $K_2 = 13$ kJ/m³, respectively. The d_H value is assumed to be given by $d_H = d_c + 30$. (b) Field-dependent relaxation times for the case of $d_c = 25$ nm, $d_H = 55$ nm.

and can respond to the high frequency. We note, however, the effect of the relaxation time cannot be completely ignored. The phase of the magnetic signal is delayed compared with the excitation field due to the relaxation time. This phase lag must be considered in practical cases.

τ_B and τ_N given in Eqs. (7.2) and (7.3) represent the relaxation time when the excitation field is small. When the amplitude of the excitation field, H_0, becomes large, the values of τ_B and τ_N become dependent of H_0 [26, 27]. This field-dependent relaxation time is given by

$$\tau_B(H_0) = \frac{\tau_B(0)}{\sqrt{1+0.21\xi^2}} \tag{7.9}$$

$$\tau_N(H_0) = \frac{\tau_0\sqrt{\pi}}{\sqrt{\sigma}(1-h^2)}\left\{(1+h)e^{-\sigma(1+h)^2} + (1-h)e^{-\sigma(1-h)^2}\right\}^{-1} \tag{7.10}$$

with

$$h = \frac{\mu_0 M_s H_0}{2K} \tag{7.11}$$

where $\tau_B(0)$ is the relaxation time given in Eq. (7.2).

In Figure 7.4(b), the field-dependent relaxation time is shown for the case of $d_c = 25$ nm, $d_H = 55$ nm, $M_s = 300$ kA/m, $K = 5$ kJ/m^3, and $\tau_0 = 10^{-9}$ s. τ_B and τ_N decrease when H_0 is increased. The τ_N value becomes considerably smaller with the increase of H_0. Therefore, the condition that $1/f > \tau_B$ and $1/f > \tau_N$ can be satisfied more easily when the excitation field with large amplitude is used. This property is helpful when we use MNPs with large d_c value.

7.1.4 MNP Response Under AC and DC Gradient Field

In MPI, AC and DC gradient fields are applied to MNPs to obtain spatial information of MNPs. In the following, we consider the case of a one-dimensional MPI. In this case, the field is given by

$$H(x,t) = H_0\sin(2\pi ft) + G(x - x_p) \tag{7.12}$$

The first term represents the AC field, which is spatially uniform. The second term represents the DC gradient field, which changes linearly with position x with a field gradient G, as shown in Figure 7.5(a). The gradient field becomes zero at $x = x_p$. This position is called the FFP.

Consider the case when MNP is located at the position x_s in Figure 7.5(a). The magnetization of MNP becomes strongly dependent on x_s, as shown below. When the MNP is located at $x_s = x_p$, i.e., at the FFP, the DC field becomes zero, and only the AC field is applied to the MNP. In this case, the MNP is magnetized as shown in Figure 7.5(b), and the waveform of $M(t)$ becomes trapezoidal. When $M(t)$ is Fourier transformed, rich harmonics are contained in its spectrum.

When the MNP is located outside the FFP, the DC field is superimposed to the AC field. In this case, the MNP is magnetized as shown in Figure 7.5(c), and $M(t)$ becomes nearly constantly independent of time due to the saturation of magnetization. As a result, harmonic signals of $M(t)$ become very small.

Therefore, the strength of the harmonic signal generated from the MNP strongly depends on the position of the MNP. We consider a point MNP, which is located at position x_s and its volume can be neglected. The magnetization $M(x_s,t)$ of the point MNP can be calculated from the Langevin function as

$$M(x_s,t)/M_s = L\{\xi(x_s,t)\} = \coth\xi(x_s,t) - \frac{1}{\xi(x_s,t)} \tag{7.13}$$

with

$$\xi\left(x_s, t\right) = \frac{\mu_0 m}{k_B T}\left\{H_0 \sin\left(2\pi ft\right) + G\left(x_s - x_p\right)\right\} \tag{7.14}$$

Performing the Fourier transform of $M(t)$, we can obtain the n-th harmonic signal, M_n. For simplicity, we study the third harmonic signal, M_3, when the MNP position x_s is changed. The calculated x_s vs. M_3 curve is shown in Figure 7.5(d). In the calculation, we assume an MNP with $m = 3 \times 10^{-18}$ Am^2, corresponding to $d_c = 27$ nm and $M_s = 300$ kA/m. The AC and DC gradient field is chosen as $\mu_0 H_0 = 4$ mT and $G = 1$ T/m, respectively. The M_3 value is normalized by the value at $x_s = x_p$.

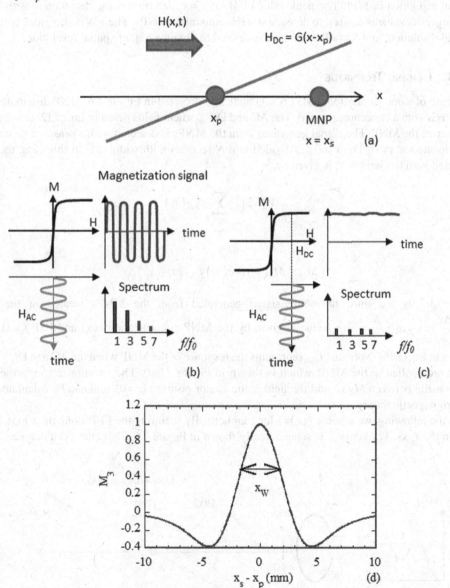

FIGURE 7.5 (a) DC gradient field. (b) Magnetization of the MNP at FFP, i.e., $x = x_p$. (c) Magnetization of the MNP outside the FFP. (d) Amplitude of the third harmonic signal when the MNP position, x_s, is changed.

As shown in Figure 7.5(d), M_3 is maximized when MNP is located at $x_s = x_p$. The M_3 value decreases with increasing or decreasing x_s from $x_s = x_p$. The FWHM of the x_s vs. M_3 curve is defined by x_w and becomes $x_w = 3.3$ mm in this case. The x_s vs. M_3 curve shown in Figure 7.5(d) was called the point spread function (PSF) and is expressed by f_{PSF} $(x_s$-$x_p)$.

Figure 7.5(d) indicates that the third harmonic signal can be obtained only from the MNPs that are located around the FFP. This property is used to obtain the MNP position, x_s, as follows. In MPI, the FFP, x_p, is spatially scanned. In this case, we obtain the signal from the MNP when the FFP coincides with the MNP position, i.e., $x_s = x_p$. As a result, the MNP position x_s can be obtained from x_p that generated the signal from the MNP.

Therefore, spatial resolution in MNP detection can be significantly improved using the FFP. We note that the PSF (x_s vs. M_3 curve) shown in Figure 7.5(d) determines the detection sensitivity and spatial resolution in MPI. The peak value of M_3 at $x_s = x_p$ determines the detection sensitivity, and the larger M_3 value is desired to detect a smaller amount of MNPs. The FWHM, x_w, determines the spatial resolution, and a smaller x_w value is desired to obtain a higher spatial resolution.

7.1.5 IMAGING TECHNIQUE

The case of a one-dimensional MPI is schematically depicted in Figure 7.6. MNP distributes along the x axis with a concentration $c(x)$. The AC and DC gradient fields given in Eq. (7.12) are applied to magnetize the MNP. The signal generated from the MNPs is detected with a sensor, e.g., induction coil, located at $x = 0$. The x axis is divided into N sections with width Δx. In this case, the signal detected with the sensor, V, is given by

$$V\left(x_p\right) = \sum_i M_{3i} g\left(x_i\right) \tag{7.15}$$

with

$$M_{3i} = M_3\left(x_i\right) = c\left(x_i\right) f_{PSF}\left(x_i - x_p\right) \Delta x \tag{7.16}$$

where M_{3i} is the third harmonic signal generated from the MNPs located in the region $x_i - \dfrac{\Delta x}{2} < x < x_i + \dfrac{\Delta x}{2}$. This value is given by the MNP concentration $c(x_i)$ and PSF $f_{PSF}(x$-$x_p)$, as shown in Eq. (7.16). Note that f_{PSF} represents the response of the MNP when the AC and DC gradient fields are applied to the MNP, which is shown in Figure 7.5(d). The function $g(x_i)$ represents the relationship between $M_3(x_i)$ and the field at the sensor position ($x = 0$), and can be calculated using electromagnetic theory.

In the following, we assume $g(x_i) = 1$ for simplicity. By scanning the FFP along the x axis, we can obtain the x_p vs. $V(x_p)$ curve, as schematically shown in Figure 7.6. This curve is a magnetic image

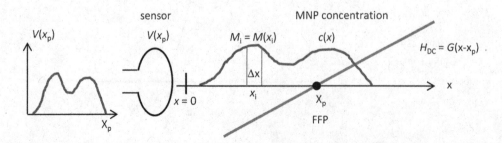

FIGURE 7.6   One-dimensional MPI using the scanning of FFP.

used in MPI. We first consider the limiting case when f_{PSF} can be approximated by a delta function $\delta(x\text{-}x_p)$. This situation occurs when the field gradient G is extremely high. In this case, we obtain

$$V\left(x_p\right) = \sum_i M_i = \sum_i c\left(x_i\right)\delta\left(x_i - x_p\right)\Delta x = c\left(x_p\right)\Delta x \qquad (7.17)$$

Therefore, $V(x_p)$ is directly given by the $c(x_p)$ value, and the measured x_p vs. $V(x_p)$ curve directly produces the x_p vs. $c(x_p)$ curve, i.e., the spatial distribution of the MNP.

Practically, however, f_{PSF} has a finite width, x_w, as shown in Figure 7.5(d). In this case, we have to reconstruct the x_p vs. $c(x_p)$ curve from the measured x_p vs. $V(x_p)$ curve. This can be done by solving an inversion problem, as shown below. The x_p axis in the measured x_p vs. $V(x_p)$ curve is divided into M sections. We then obtain a signal vector, V, whose component is given by the value of $V(x_p)$ at $x_p = x_{pk}$. Similarly, the x axis is divided into N sections, and we obtain a concentration vector, c, whose component is given by the value of $c(x)$ at $x = x_i$. The relationship between V and c is given by the following equation.

$$V = Ac \qquad (7.18)$$

where A is called the system function matrix with $N \times M$ components [1]. Note that the component A_{ik} is given by PSF as $A_{ik} = f_{PSF}(x_i \text{-} x_{pk})$.

The concentration vector, c, can be obtained by solving the inversion problem given by Eq. (7.18). For this purpose, several mathematical techniques have been developed. For example, Eq. (7.18) can be solved using the nonlinear nonnegative least square (NNLS) method, as shown:

$$c = \operatorname{argmin}_c \left\| Ac - v^2 \right\| + \lambda \left\| c^2 \right\| \ \left(c \geq 0\right) \qquad (7.19)$$

where λ is a regularization parameter.

The spatial resolution in MPI is primary determined by the FWHM x_w of the point spread function shown in Figure 7.5(d). Further improvements in the spatial resolution are possible using the imaging technique described above.

An example of MPI is shown in Figure 7.7. A letter "K" was constructed using 11 cylindrical containers as shown in Figure 7.7(a), where MNPs were enclosed in each container with a diameter of 1.2 mm and a height of 2 mm. The AC field with $\mu_0 H_0 = 3.5$ mT and $f = 3$ kHz and the DC gradient field with $G = 1$ T/m (x axis) and 2 T/m (y axis) were applied to magnetize the sample. The FFP, r_p, was scanned two dimensionally, and field map generated from the letter "K" was obtained. The field map $V(r_p)$ was then analyzed using the imaging procedure described above, and the concentration image of MNP, $c(r_p)$, was reconstructed as shown in Figure 7.7(b).

7.2 POINT SPREAD FUNCTION

As shown in Section 7.1, the signal strength and spatial resolution of MPI are dominated by the point spread function f_{PSF}. This point spread function is determined by the response of MNP when the AC and DC gradient fields are applied to the point MNP sample. In the following, we study the properties of PSF, and discuss factors that affect the signal strength and spatial resolution of MPI.

7.2.1 HARMONIC SIGNALS

We consider the case when the AC and DC fields are applied in the same direction. In this case, the applied field is given by

$$H(t) = H_0 \sin\left(2\pi ft\right) + H_{DC} \qquad (7.20)$$

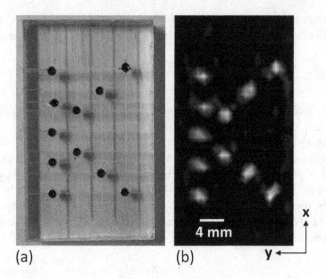

FIGURE 7.7 (a) Photo of letter "K" made of 12 MNPs. (b) MPI of the letter "K". Concentration image of MNP is shown.

The DC field is given by $\mu_0 H_{\text{DC}} = G x_s$, when the gradient field with field gradient G is applied and the sample is located at $x = x_s$.

Magnetization of the MNP sample caused by the field H is given by the Langevin function as

$$M(t) = M_s L(\xi(t)) \tag{7.21}$$

with

$$\xi(t) = \frac{\mu_0 m}{k_B T} \left\{ H_0 \sin\left(2\pi f t\right) + H_{\text{DC}} \right\} \tag{7.22}$$

When the signal generated from the MNPs sample is detected with a pickup coil, the voltage signal V is given by $V = dM/dt$. The voltage signal V can be obtained as

$$V(t) = \frac{dM}{dt} = \frac{dM}{dH} \times \frac{dH}{dt} \tag{7.23}$$

with

$$\frac{dM}{dH} = \frac{\mu_0 m M_s}{k_B T} \left(\frac{1}{\xi^2} - \frac{1}{\sinh^2(\xi)} \right) \quad \xi \neq 0$$

$$= \frac{1}{3} \frac{\mu_0 m M_s}{k_B T} \quad \xi = 0 \tag{7.24}$$

First, we study harmonic signals of the magnetization, $M(t)$. Performing the Fourier transform of $M(t)$, we can obtain the n-th harmonic signal as follows:

$$M_n\left(H_{\text{DC}}\right) = \frac{2}{T} \int_0^T M(t) \sin\left(2n\pi f t\right) dt, \quad \text{with} \quad T = 1/f \tag{7.25}$$

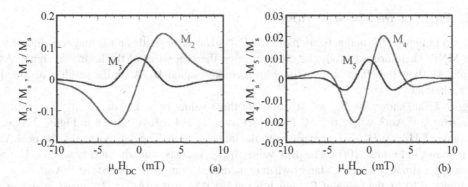

FIGURE 7.8 Dependence of the harmonic signal on the DC field H_{DC}. (a) M_2 and M_3, (b) M_4 and M_5.

Harmonics of the voltage signal are given by $V_n = 2n\pi f M_n$.

Figure 7.8 shows the dependence of M_n on the DC field H_{DC} when we choose the parameters of $\mu_0 H_{AC} = 4$ mT and $m = 3 \times 10^{-18}$ Am^2. Figure 7.8(a) represents M_2 and M_3, while M_4 and M_5 are shown in Figure 7.8(b). The fundamental component M_1 is usually not used in MPI because M_1 is considerably affected by the interference of the excitation field.

As shown in Figure 7.8, the H_{DC} vs. M_n curves of odd number harmonics (M_3 and M_5) are symmetric with respect to H_{DC}. M_3 and M_5 have peak values at $H_{DC} = 0$ and decrease with increasing or decreasing H_{DC} from $H_{DC} = 0$. The peak value of M_5 becomes smaller than that of M_3, but M_5 decreases with H_{DC} faster than M_3. Alternatively, the H_{DC} vs. M_n curves of even number harmonics (M_2 and M_4) becomes asymmetric with respect to H_{DC}. M_2 and M_4 become zero at $H_{DC} = 0$ and have a peak value at a specific value of H_{DC}.

The H_{DC} vs. M_n curves shown in Figure 7.8 can be represented as the x_s vs. M_n curves using the relation $\mu_0 H_{DC} = G x_s$. Note that the x_s vs. M_n curves represent the response of the MNP sample when the position x_s of the MNP sample moves away from the FFP ($x = 0$). Therefore, these curves give the PSF. In a practical MPI, combinations of the harmonic signals are used to construct the PSF.

The PSF can also be constructed using the waveform of the signal voltage $V(t)$. Figure 7.9(a) shows the waveforms of V when $\mu_0 H_{DC} = 0$ and 3 mT. The waveform of V is changed when the DC field is applied. The time, t_p, at which V becomes a peak value, is changed, and the peak value of V also decreases when H_{DC} is applied. This change of the waveform of V can be used to construct the PSF. For example, Figure 7.9(b) shows the voltage V at $t = 0$ when the DC field H_{DC} is changed. The voltage is maximized when $H_{DC} = 0$ and decreases with increasing or decreasing H_{DC} from $H_{DC} = 0$. This H_{DC} vs. $V(t=0)$ curve can be used to construct the PSF using the relation $\mu_0 H_{DC} = G x_s$. The time, t_p, can also be used to construct the PSF, as will be shown in Section 7.3.

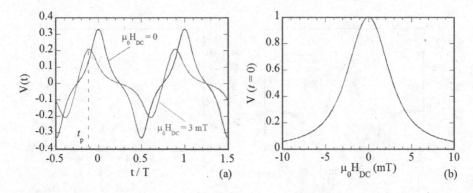

FIGURE 7.9 (a) Waveforms of the signal voltage $V(t)$ when H_{DC} is changed. (b) Dependence of the signal voltage $V(t=0)$ on H_{DC}.

7.2.2 EFFECT OF THE MAGNETIC MOMENT

As shown in Figure 7.3, magnetization of the MNP strongly depends on the magnetic moment, m, of the MNP. Therefore, we study the effect of m on the properties of the harmonic signal. As an example, we show the effect of m on the third harmonic signal M_3. A similar result is obtained for other harmonics.

Figure 7.10(a) shows the H_{DC} vs. M_3 curve for three values of m, i.e., $m_1 = 2 \times 10^{-18}$, $m_2 = 4 \times 10^{-18}$, and $m_3 = 6 \times 10^{-18}$ Am^2, where the AC field is set as $\mu_0 H_{AC} = 4$ mT. As shown in Figure 7.10(a), the peak value of M_3 at $\mu_0 H_{DC} = 0$ increases with the increase in m. The dependence of the peak value on m is shown in Figure 7.10(b). The peak value, M_{3peak}, increases nearly linearly with m. The peak value is approximately six times larger when m increased from 2×10^{-18} to 6×10^{-18} Am^2.

In Figure 7.10(b), the value of B_w, which is the FWHM of the H_{DC} vs. M_3 curve, is shown as a function of m. B_w decreases with the increase in m. The value of B_w changes from 4 mT to 2.8 mT when m increased from 2×10^{-18} to 6×10^{-18} Am^2.

The H_{DC} vs. M_3 curve can be expressed as the x vs. M_3 curve when the DC field is a gradient field given by $\mu_0 H_{DC} = Gx$, as already shown in Figure 7.5. The FWHM, x_w, of the x vs. M_3 curve is given by $x_w = B_w/G$. Therefore, B_w determines the spatial resolution in MPI. Alternatively, the peak value, M_{3peak}, determines the signal strength in MPI. Therefore, a large M_{3peak} and small B_w are desired to improve the performance of MPI. Figure 7.10(b) indicates that MNPs with large m value should be used for MPI.

7.2.3 EFFECT OF THE AMPLITUDE OF THE AC FIELD

The H_{DC} vs. M_3 curve is also affected by the amplitude H_0 of the AC excitation field. Figure 7.11(a) shows the H_{DC} vs. M_3 curve for three values of H_0, i.e., $\mu_0 H_{01} = 3$ mT, $\mu_0 H_{02} = 5$ mT, and $\mu_0 H_{03} = 7$ mT, where m is fixed to $m = 3 \times 10^{-18}$ Am^2. The vertical axis is normalized by the value of M_3 at $H_{DC} = 0$ and $\mu_0 H_0 = 2$ mT. The H_{DC} vs. M_3 curve significantly depends on the amplitude H_0 of the AC excitation field. The peak value of M_3 increases, but the H_{DC} vs. M_3 curve becomes broader with the increase in H_0.

The dependences of the M_{3peak} and B_w on H_0 are shown in Figure 7.11(b). The M_{3peak} increases with H_0. The peak value becomes about 12 times larger when $\mu_0 H_0$ increased from 2 mT to 8 mT, which indicates that the signal strength from the MNPs can be increased by increasing the amplitude of the AC excitation field. However, B_w also increases with H_0, as shown in Figure 7.11(b). The value of B_w increases from 2.5 mT to 5.5 mT when $\mu_0 H_0$ increased from 2 mT to 8 mT. This increase in B_w means that the spatial resolution of MPI is degraded when we use a large amplitude of the AC excitation field.

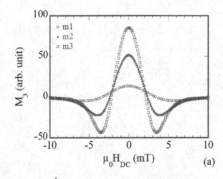
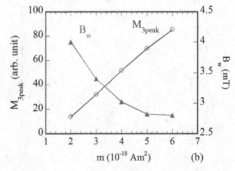

FIGURE 7.10 Effect of the magnetic moment m of the MNP on the H_{DC} vs. M_3 curve. (a) The H_{DC} vs. M_3 curves for different values of m. (b) Dependences of M_{3peak} and B_w on m.

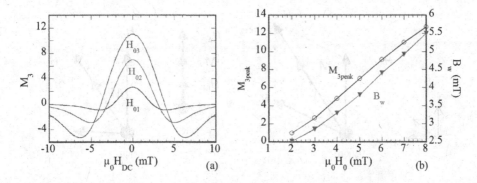

FIGURE 7.11 (a) H_{DC} vs. M_3 curve when the amplitude, H_0, of the AC excitation field is changed. (b) Dependences of M_{3peak} and B_w on H_0.

The above result indicates that the signal strength becomes large, but the spatial resolution is degraded when we use large H_0. Alternatively, when we use small H_0, the signal strength becomes small, but the spatial resolution is improved. Therefore, the amplitude of the AC field should be determined by the tradeoff between the signal strength and spatial resolution [28, 29].

For example, we consider the case when MNPs with $m = 3 \times 10^{-18}$ Am^2 are used and an AC field with amplitude $\mu_0 H_0 = 5$ mT is applied. The DC gradient field is applied in parallel to the AC field with a field gradient of $G = 1$ T/m, i.e., $\mu_0 H_{DC} = Gx$. In this case, we obtain $B_w = 3.8$ mT from Figure 7.11(b). Therefore, the spatial resolution is given by $x_w = B_w/G = 3.8$ mm. The spatial resolution can be further improved using the imaging technique shown in Section 7.1.5.

The sensitivity of MNP detection is given as follows. It is shown that the amplitude of M_3 becomes $M_3 = 0.47\ M_s$ when $\mu_0 H_0 = 5$ mT and $m = 3 \times 10^{-18}$ Am^2. We consider the case when the MNP with small volume V is located at a distance r from the sensor. In this case, the magnetic field B at the sensor position is given by

$$B = \frac{\mu_0 M_3 V}{2\pi r^3} = \frac{0.47 \mu_0 M_s V}{2\pi r^3} \tag{7.26}$$

The magnetic field is usually detected with a pickup coil. When we use a large pickup coil, it is possible to detect the magnetic field of $B = 1$ fT $= 10^{-15}$ T when the signal frequency is high. When we substitute the values of $M_s = 300$ kA/m, $r = 10$ cm, and $B = 1$ fT, we can estimate that the minimum detectable volume of the MNP becomes $V = 35$ μm^3, corresponding to a 3.3 μm cube. The minimum detectable weight of the MNP can be calculated as $2 \times 10^{-10} = 0.2$ ng using the MNP's relative gravity value of 5.2. Therefore, highly sensitive detection of MNPs located at a distance of 10 cm is possible.

7.2.4 RELATIVE DIRECTION BETWEEN THE AC AND DC FIELDS

As shown in Section 7.1, the signal strength and spatial resolution of MPI is dominated by the point spread function f_{PSF}. This point spread function is determined by the response of the MNP when the AC and DC fields are applied to the MNP. In the previous section, we studied the PSF for the simple case where the AC and DC fields are parallel with each other. In a practical case, however, this condition is not always satisfied. In MPI, the DC field is given by the gradient field, and the gradient field is a vector quantity that has x, y, and z components in the xyz coordinate. This means that the relative direction between the AC and DC fields becomes arbitrary in practical MPI. Therefore, to obtain the PSF for MPI, we have to study the magnetization of the MNP when the AC and DC fields have arbitrary direction with each other.

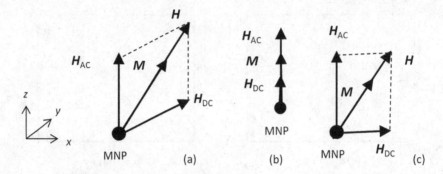

FIGURE 7.12 Relative direction between the AC and DC fields. (a) Arbitrary case, (b) parallel case ($H_{AC}//H_{DC}$), and (c) perpendicular case ($H_{AC} \perp H_{DC}$).

Figure 7.12(a) depicts the case where the AC and DC fields are applied to a point MNP. For simplicity, we assume that the AC field has only the z component. In this case, we obtain

$$H_{AC} = \left(0, 0, H_0 \sin\left(2\pi ft\right)\right), \tag{7.27}$$

$$H_{DC} = \left(H_{DC,x}, H_{DC,y}, H_{DC,z}\right) \tag{7.28}$$

Therefore, the external field becomes a vector quantity given by

$$H\left(r, t\right) = H_{AC} + H_{DC} = \left(H_{DC,x}, H_{DC,y}, H_0 \sin\left(2\pi ft\right) + H_{DC,z}\right) \tag{7.29}$$

When the magnetization of the MNP is determined by the Langevin function, the direction of M of the MNP is aligned to that of H, as shown in Figure 7.12. We note that the direction of H rotates with time as shown in Eq. (7.29). Therefore, the direction of M also rotates with time.

The amplitude of M is given by the Langevin function as

$$\left|M\right|(t) = M_s L\left\{\xi\left(t\right)\right\} \tag{7.30}$$

with

$$\xi\left(t\right) = \frac{\mu_0 m}{k_B T} \left|H\right|(t) = \frac{\mu_0 m}{k_B T} \sqrt{H_{DC,x}^2 + H_{DC,y}^2 + \left(H_0 \sin\left(2\pi ft\right) + H_{DC,z}\right)^2} \tag{7.31}$$

The magnetization vector M of MNP can be given by

$$M\left(r, t\right) = \left|M\right|(t) \frac{H\left(r, t\right)}{\left|H\right|(t)} = \left(M_x, M_y, M_z\right) \tag{7.32}$$

Figure 7.12(b) and 7.12(c) depict two typical cases. In Figure 7.12(b), the DC field is applied in parallel to the AC field. In this case, the M of the MNP has only the z component, M_z, and M_z vary with time. In Figure 7.12(c), the DC field is applied perpendicular to the AC field. In this case, M has x and z components (M_x and M_z) and these components vary with time.

Equation (7.32) provides the point spread function when the AC and DC fields are applied in an arbitrary direction. We now discuss how the point spread function is affected by the relative direction between the AC and DC fields using the third harmonic signal, M_3, as an example. The third

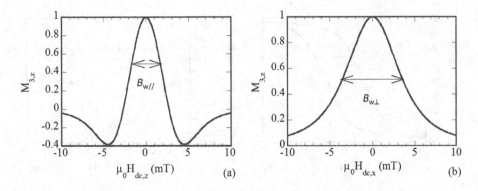

FIGURE 7.13 (a) H_{DC} vs. M_3 curve when $H_{AC}//H_{DC}$. (b) H_{DC} vs. M_3 curve when $H_{AC} \perp H_{DC}$.

harmonic signal can be obtained from the Fourier transform of $M(t)$ given in Eq. (7.32). In Figure 7.13(a), the result is shown for the case when the DC field is parallel to the AC field. As shown in Figure 7.12(b), both the DC and AC fields are applied in the z direction, and the magnetization of MNP has the z component, M_z, in this case. In the calculation, we assume the MNP with $m = 3 \times 10^{-18}$ Am^2, and the amplitude of the AC field as $\mu_0 H_0 = 4$ mT. The horizontal axis is the DC field, $H_{DC,z}$, while the vertical axis represents $M_{3,z}$ normalized by the value at $H_{DC,z} = 0$. As shown in Figure 7.13(a), $M_{3,z}$ is maximized when $H_{DC,z} = 0$, and decreases with increasing or decreasing $H_{dc,z}$ from $H_{dc,z} = 0$. The FWHM of the $H_{DC,z}$ vs. $M_{3,z}$ curve is defined by $B_{w//}$, and becomes $B_{w//} = 3.3$ mT in this case.

Figure 7.13(b) shows the result when the DC field is perpendicular to the AC field. As shown in Figure 7.12(c), the AC field is applied to the z direction, while the DC field is applied to the x direction. In this case, the M of the MNP has x and z components (M_x and M_z) with M_x having even number harmonics and M_z having odd number harmonics. The horizontal axis in Figure 7.13(b) represents the DC field, $H_{DC,x}$, while the vertical axis represents $M_{3,z}$ normalized by the value at $H_{DC,x} = 0$. The $H_{DC,x}$ vs. $M_{3,z}$ curve is different from the parallel case shown in Figure 7.13(a). Particularly, the value of $B_{w\perp}$ becomes 7.2 mT, which is approximately two times greater than the parallel case.

Different H_{DC} vs. M_3 curves shown in Figure 7.13(a) and 7.13(b) indicate that the spatial resolution of MPI becomes different between the case of $H_{AC}//H_{DC}$ and $H_{AC} \perp H_{DC}$ [30]. Since the spatial resolution is determined by the value of B_w, the spatial resolution for the case of $H_{AC} \perp H_{DC}$ is two times worse than the case of $H_{AC}//H_{DC}$.

7.2.5 HYSTERESIS OF MAGNETIZATION

In the previous sections, we assume the ideal case where the magnetization M of the MNP is determined by the Langevin function. However, magnetization of the practical MNP becomes different from the ideal case. Particularly, we have to consider the finite relaxation time of the MNP. Owing to the relaxation time, a time delay occurs in M compared with the AC field. As a result, hysteresis appears in the AC magnetization curve, as shown in Figure 7.14(a).

Figure 7.14(b) represents the H vs. dM/dt curve. Since dM/dt = dM/d$H \times$dH/dt, Figure 7.14(b) corresponds to the H vs. dM/dH curve. We can compare the curve with that obtained from the Langevin function shown in Figure 7.3(b). Hysteresis occurs in the H vs. dM/dt curve, and dM/dt has peak values at $\mu_0 H = \pm 3$ mT. This behavior is different from that shown in Figure 7.3(b), where the value of dM/dH is maximized at $H = 0$. This hysteresis behavior must be considered when we construct the point spread function in a practical MPI.

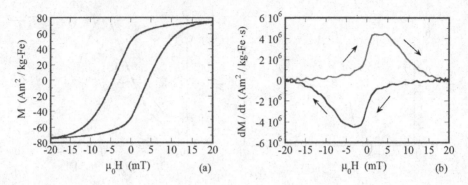

FIGURE 7.14   (a) AC magnetization curve of a practical MNP. (b) H vs. dM/dt curve obtained from (a).

Notably, the time delay in M causes the phase lag of harmonic signals. In this case, the harmonic signal is expressed as

$$M_n(t) = M_n\sin(2n\pi ft - \theta_n) = M_n\cos(\theta_n)\sin(2n\pi ft) - M_n\sin(\theta_n)\cos(2n\pi ft) \qquad (7.33)$$

where θ_n is the phase lag of the n-th harmonic signal. The first and second term in the right side of Eq. (7.33) represents the in-phase and out-of-phase component, respectively. Therefore, both of the components have to be used to construct the point spread function when hysteresis appears in the magnetization.

7.3   MPI HARDWARE

MPI hardware consists of several coils and magnets. Figure 7.15 depicts a schematic figure of a one-dimensional MPI system. The excitation coil, which usually consists of a solenoid coil, is used to apply the AC excitation field to the MNP sample. A pair of permanent magnets is used to produce a DC gradient field with the FFP. The MNP sample is magnetized with the AC and DC fields, and the signal field is generated from the MNP sample that is located around the FFP. This signal field is detected with the pickup coil.

A shift coil is used to electrically scan the FFP. The shift coil consists of a Helmholtz coil and produces a spatially uniform DC field. The FFP can be moved by adding a uniform DC field to the gradient field. The position of the FFP can be scanned by changing the current supplied to the shift coil.

We discuss the properties of these coils and magnets in the following section.

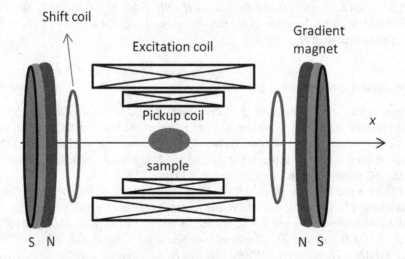

FIGURE 7.15   Schematic figure of a one-dimensional MPI system.

7.3.1 EXCITATION COIL

An AC excitation field is usually applied to the MNPs sample using a solenoid coil. The amplitude and frequency of the AC field is typically $\mu_0 H_0 = 3\text{–}10$ mT and $f_e = 5\text{–}25$ kHz, respectively. Figure 7.16 represents a circuit diagram to supply the AC current to the excitation coil with inductance, L_{coil}, and resistance, R_{coil}. The AC power source V_e is connected to the excitation coil through two filters. One is a bandpass filter that consists of L_{coil} and a resonant capacitance C_1. The resonant frequency of the L_{coil} - C_1 circuit is set to be equal to the frequency of the AC field, i.e., $f_r = 1/(2\pi(L_{coil}C_1)^{1/2}) = f_e$. In this case, the impedance seen from the AC source becomes $Z = R_{coil}$ at $f = f_e$, and a large AC current $I_e = V_e/R_{coil}$ can be supplied to the coil.

The other is a band-stop (or notch) filter that consists of L_2 and C_2, as shown in Figure 7.16. This notch filter is used to prevent the third harmonic current, I_{e3}, from flowing into the excitation coil. Notably, the AC power source also generates the third harmonic voltage, V_{e3}, due to its nonlinearity, and the nonlinearity becomes large with the increase of the supplied power. Since the harmonic signals from the MNP sample are used in MPI, it is necessary to suppress I_{e3} for precise measurement of the harmonic signals.

The L_2 - C_2 circuit in Figure 7.16 is a parallel resonant circuit, whose resonant frequency is set as $3f_e$, i.e., $1/(2\pi(L_2C_2)^{1/2}) = 3f_e$. Therefore, the impedance of the L_2 - C_2 circuit is approximately given by $Z = L_2/(C_2R_2)$ at $f = 3f_e$. If we use the Q factor of the resonant circuit, $Q = (L_2/C_2)^{1/2}/R_2$, we obtain $Z = Q^2R_2$ and $I_{e3} = V_{e3}/Z = V_{e3}/(Q^2R_2)$. Therefore, we can sufficiently suppress I_{e3} when the Q factor of the resonant circuit is large. The L_2 - C_2 circuit has a small reactance component at $f = f_e$. The capacitance, C_3, is used to compensate this reactance component. We note that a notch filter for the fifth harmonic current can also be added if necessary.

The frequency and amplitude of the AC field are determined considering the following items. First, the signal from the MNP sample is detected with the pickup coil. In this case, the signal strength becomes proportional to the signal frequency. Therefore, a higher frequency is desired to improve the sensitivity in MNP detection. A higher frequency is also necessary to ensure sufficient temporal resolution in MPI. When we measure the time trace of the MNPs movement, the measurement time at each MNP position must be short enough compared with the movement of the MNPs. To reduce the measurement time, a higher frequency of the excitation field is necessary.

The amplitude of the excitation field, H_0, is determined by the tradeoff between the signal strength and spatial resolution in MPI because the signal strength from the MNP sample increases with H_0, but the spatial resolution degrades with H_0, as shown in Section 7.2.3. Power consumption in the excitation coil, $P = I_e^2 R_{coil}/2$, is also an important factor from an engineering point of view.

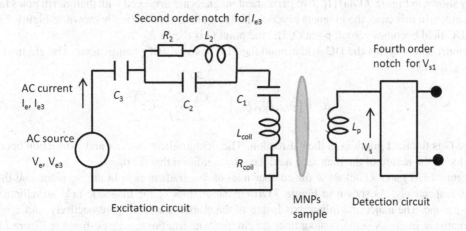

FIGURE 7.16 Diagram of the excitation and detection circuits. The bandpass filter at $f = f_e$ and notch filter at $f = 3f_e$ are used in the excitation circuit. The notch filter at $f = f_e$ is used in the detection circuit.

The excitation coil is usually water-cooled to protect the heating of the coil. Since H_0 is proportional to I_e, P is proportional to $H_0{}^2$, and we need a more powerful cooling system when H_0 becomes larger.

Finally, we note that the biological side effects of the AC field must be considered for biomedical application [4, 31]. The AC field is known to generate magneto-stimulation when it is applied to human or animal subjects. The AC field also causes heating of tissues, whose safety limit is expressed as the specific absorption rate (SAR). These safety limits are determined by both the amplitude and frequency of the AC field. Fortunately, these safety conditions are satisfied when $\mu_0 H_0 = 3\text{--}10$ mT and $f_e = 5\text{--}25$ kHz.

7.3.2   PICKUP (RECEIVER) COIL

The magnetic signal generated from the MNP sample is usually detected with the pickup coil. In this case, the output voltage across the pickup coil, V_s, is given by $V_s = dM/dt$, where M is the magnetization of the sample. We note that the interference of the excitation field becomes a significant problem in the measurement. As previously shown, an excitation field on the order of mT is applied. Alternatively, the signal field is on the order of pT when a very small amount of the MNP sample is detected, which means that the ratio between the excitation and signal field becomes as large as 10^9. Therefore, the performance of the detection circuit is significantly degraded even when a very small amount of the excitation field interlinks the pickup coil.

In a practical case, the configuration of the pickup coil is used so that the interlinkage of the excitation field becomes as small as possible, e.g., a gradiometer-type pickup coil. However, it is impossible to completely eliminate the interference of the excitation field. Therefore, a fundamental component of the signal field, V_{s1}, is excluded in the MPI using a notch filter for V_{s1}, as schematically shown in Figure 7.16. We note that the L_2 - C_2 circuit used in the excitation circuit is a second order notch filter. In the receiver circuit, a fourth order notch filter can be used to suppress V_{s1} more.

The signal field from the MNP sample has x, y, and z components in the xyz coordinate. Therefore, we have to use three pickup coils if we need to measure all three components of the signal field. We note that an array of the small pickup coils can also be used instead of one large pickup coil. When the array of coils is used, we can obtain more information on the spatial distribution of the MNPs sample, as will be discussed in Section 7.4.

7.3.3   GENERATION AND SCANNING OF THE FFP

The gradient field with the FFP is usually constructed with a pair of permanent magnets, as schematically shown in Figure 7.17(a) [1]. Two permanent magnets are arranged with their north poles facing each other. In this case, the magnets produce the DC field as schematically shown in Figure 7.17(a). The DC field becomes zero at point O, i.e., the point O is the FFP.

It must be noted that the DC field around the FFP has x, y, and z components. The gradient field $\boldsymbol{B}_G$ is given by

$$\boldsymbol{B}_G = \left(B_{Gx}, B_{Gy}, B_{Gz} \right) = \left(\frac{Gx}{2}, \frac{Gy}{2}, -Gz \right) \tag{7.34}$$

where G is the field gradient in the z direction. The field gradient in the x and y direction becomes $G/2$ due to the nature of the magnetic field, i.e., the condition that $\mathrm{div}\boldsymbol{B}_G = 0$.

Figure 7.17(b) and 7.17(c) show the contour map of the gradient field in the x-z plane and the x-y plane, respectively. As shown in Figure 7.17(b), contour lines of the magnetic field are elliptical in the x-z plane. The major and minor axis length of the ellipse is x_w and z_w, respectively, and $z_w = x_w/2$. Alternatively, in the x-y plane, contour lines are circular with a radius $x_w = y_w$, as shown in Figure 7.17(c).

We note that a magnetic field from MNPs is generated when they are located inside the ellipse with axis length x_w and z_w in the x-z plane or inside the circular with radius $x_w = y_w$ in the x-y plane.

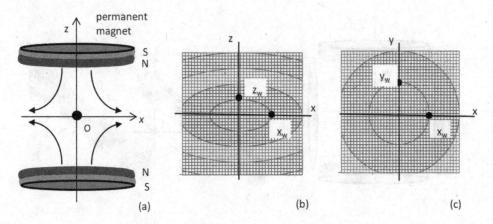

FIGURE 7.17 (a) Gradient magnet to produce the FFP. (b) Contour map of the gradient field in the x-z plane. (c) Contour map of the gradient field in the x-y plane.

Using the B_w value shown in Section 7.2 and Eq. (7.34), $x_w = y_w = 2B_w/G$ and $z_w = B_w/G$. As shown in Figure 7.13, we should choose either $B_{w//}$or $B_{w\perp}$ depending on the relative direction between the AC field and gradient field.

The FFP can be electrically scanned using the shift coil, as schematically shown in Figure 7.15. The shift coil consists of a Helmholtz coil, and produces a spatially uniform magnetic field when a current I is supplied to the shift coil. Three sets of the shift coil are used to generate the x, y, and z component of the magnetic field. When current I_x, I_y, and I_z is supplied to each shift coil, the magnetic field is given by

$$B_{sh} = \left(B_{shx}, B_{shy}, B_{shz}\right) = \left(k_x I_x, k_y I_y, k_z I_z\right) \tag{7.35}$$

where k_x, k_y, and k_z are constants that give the relation between the current and generated field.

Therefore, the summation of the gradient field and field generated by the shift coil is given by

$$B_G + B_{sh} = \left(\frac{Gx}{2} + k_x I_x, \frac{Gy}{2} + k_y I_y, -Gz + k_z I_z\right) \tag{7.36}$$

The FFP, at which the field becomes zero, is given by

$$r_{FFP} = \left(x_p, y_p, z_p\right) = \left(-2k_x I_x / G, -2k_y I_y / G, k_z I_z / G\right) \tag{7.37}$$

As shown in Eq. (7.37), the FFP can be scanned three dimensionally by controlling the current I_x, I_y, and I_z supplied to the shift coils. For this purpose, a sinusoidal current with different phase is supplied to each shift coil [32, 33]. The frequency of the current is much lower than that of the excitation field.

7.3.4 Generation and Scanning of the FFL

The FFL, along which the DC field becomes zero, is also used in MPI [34, 35]. Figure 7.18 shows typical configurations to produce the FFL. In Figure 7.18(a), two pairs of permanent magnets are used. One pair of the magnets is arranged along the z axis, and another pair is arranged along the x axis. Each pair of magnets produces a gradient field, as shown in Figure 7.17(a).

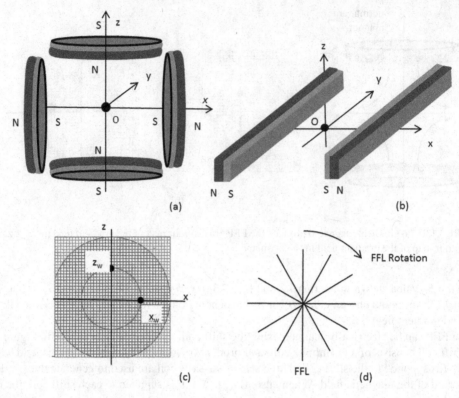

FIGURE 7.18 (a) Four permanent magnets to produce the FFL. (b) Rectangular magnet to produce the FLL. (c) Contour map of the gradient field in the x-z plane. (d) Rotation of FLL in the xyz coordinates.

Using Eq. (7.34), we can obtain the superposition of the gradient field generated from two pairs of magnets as

$$\boldsymbol{B}_G = \left(B_{Gx}, B_{Gy}, B_{Gz}\right) = \left(\frac{Gx}{2}, \frac{Gy}{2}, -Gz\right) + \left(Gx, \frac{-Gy}{2}, \frac{-Gz}{2}\right) = \left(\frac{3Gx}{2}, 0, \frac{-3Gz}{2}\right) \quad (7.38)$$

Equation (7.38) means that the gradient field becomes zero along the y axis. Therefore, the FFL is produced along the y axis.

Figure 7.18(b) depicts a pair of rectangular permanent magnets, where the length l of the magnet is much longer than the width w, i.e., $l \gg w$. In this case, the FLL is formed along the y axis, and the gradient field is given by

$$\boldsymbol{B}_G = \left(B_{Gx}, B_{Gy}, B_{Gz}\right) = \left(Gx, 0, -Gz\right) \quad (7.39)$$

Figure 7.18(c) shows the contour map of the gradient field in the x-z plane. The contour lines of the magnetic field are circular with radius $x_w = z_w$. Therefore, the magnetic field is generated when the MNPs are located inside the cylinder installed in the y axis direction of radius along $x_w = z_w$.

In this case, the magnetic signal from the MNPs is given by

$$M \propto \pi x_w^2 \int c(y) dy \quad (7.40)$$

where $c(y)$ is the concentration of MNPs along the y axis. This means that the information on $c(y)$ along the y axis cannot be obtained when the FFL is used. To obtain three-dimensional information on the MNP concentration, the FFL is rotated in the xyz coordinate, as schematically shown in Figure 7.18(d). The coil system for the FFL rotation has been developed. Mechanical rotation of the subject can be also used to simplify the coil system.

7.3.5 Drive Coil

As mentioned previously, excitation and shift coils are used to magnetize the MNP sample and scan the FFP, respectively. These two coils can be combined into one coil, which is called a drive coil. When we use the drive coil, magnetization of the MNPs sample and scanning of the FFP are done simultaneously. The PSF can be obtained by analyzing the waveform of the detected signal, as shown below.

We consider the one-dimensional case, where the AC field and DC gradient field is given by

$$H(x,t) = H_0 \sin(2\pi ft) + (G / \mu_0)x \tag{7.41}$$

A point MNP sample is located at $x=x_s$. In this case, MNP is magnetized by the field given in Eq. (7.41), and its magnetization becomes

$$M(x_s,t) = M_s L\{\xi(x_s,t)\} \tag{7.42}$$

with

$$\xi(x_s,t) = \frac{\mu_0 m H_0}{k_B T}\{\sin(2\pi ft) + (G / \mu_0)x_s\} \tag{7.43}$$

When the magnetic signal is detected with the pickup coil, the signal voltage is given by

$$V(x_s,t) = \frac{dM}{dt} = \frac{dM}{dH} \times \frac{dH}{dt} \tag{7.44}$$

where dM/dH is given by Eq. (7.24) in Section 7.2.

Figure 7.19(a) represents the waveform of the AC field H_{AC} and signal voltage V calculated from Eq. (7.44). In the calculation, we used the following parameters: $\mu_0 H_0 = 6$ mT, $G = 1$ T/m, $m = 4 \times 10^{-18}$ Am^2, and $x_s = 2$ mm. The voltage, V, has sharp peaks at $t=t_1$ and t_2. The voltage becomes a positive peak at t_1, while it becomes a negative peak at t_2.

As shown in Eq. (7.41), the field $H(x,t)$ becomes zero at the position, x_p, given by

$$x_p(t) = \frac{-\mu_0}{G} H_0 \sin(2\pi ft) \tag{7.45}$$

Therefore, the position $x=x_p$ gives the FFP, and the FFP is moved with time according to Eq. (7.45). Figure 7.19(b) depicts $x_p(t)$ and $V(t)$. The FFP becomes $x_p=x_s=2$ mm at $t=t_1$ and t_2. This means that the sharp signal voltage is generated when the FFP passes the sample position, x_s.

Figure 7.19(c) represents the x_p-x_s vs. V curve, which is obtained from the result shown in Figure 7.19(b); for simplicity, only a half cycle $(-0.25 < t/T < 0.25)$ result of Figure 7.19(b) was used. This curve shows the response of the point MNP as a function of x_p-x_s, and gives the PSF. Therefore, we can obtain the PSF by analyzing the waveform of the signal voltage. The PSF obtained in this case is a little different from that shown in Figure 7.5(d).

For three-dimensional scanning of the FFP, three drive coils are used. Each of the drive coils generates the x, y, and z component of the AC field. The three-dimensional scanning of the FFP is performed by controlling the current supplied to each drive coil.

We note that the FFP can be scanned in the area $-\mu_0 H_0 / G < x_p < \mu_0 H_0 / G$, as shown in Eq. (7.45). In the present case, we use $\mu_0 H_0 = 6$ mT and $G = 1$ T/m. In this case, we obtain -6 mm $< x_p < 6$

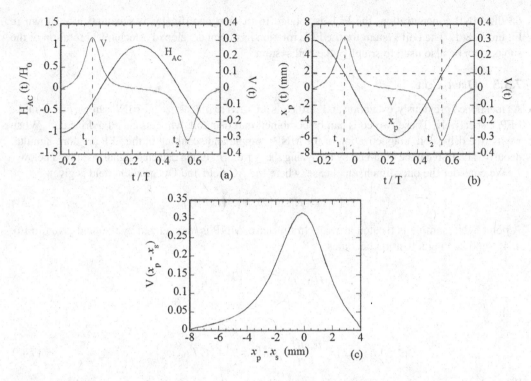

FIGURE 7.19 (a) Waveform of the signal voltage, V, generated with the drive coil. (b) Signal voltage, V, and FFP movement generated with the drive coil. (c) PSF obtained with the drive coil.

mm, i.e., a scanning area of the FFP is not large. This means that the high-speed scanning of the FFP is possible only over a small region using the drive coil. However, it is impossible to scan the FFP across all regions of the subject. Therefore, an additional coil similar to the shift coil is necessary for a wide but low speed scan of the FFP. This coil is called a selection- or focused-field coil.

7.4 IMAGING TECHNIQUE

7.4.1 SIGNAL VOLTAGE AND INVERSION PROBLEM

First, we obtain the expression for the signal voltage when the magnetic signal generated from the object is detected with the pickup coils. Figure 7.20 depicts a case when the gradient field with the FFP is used, and the signal is detected with an array of pickup coils.

We divide the object into N sections with volume δV. The position vector of the i-th section is expressed with $r_i = (x_i, y_i, z_i)$. At the i-th section, the external field and magnetization are expressed as H_i and M_i, respectively. The external field consists of the AC field $H_{AC,i}$ and DC field $H_{DC,i}$. For simplicity, we consider the case when a spatially uniform AC field is applied in the z direction. In this case, we obtain

$$H_i = H_{AC,i} + H_{DC,i} \tag{7.46}$$

with

$$H_{AC,i} = \begin{pmatrix} 0 \\ 0 \\ H_{AC,i} \end{pmatrix} = \begin{pmatrix} 0 \\ 0 \\ H_0 \sin(2\pi ft) \end{pmatrix} \tag{7.47}$$

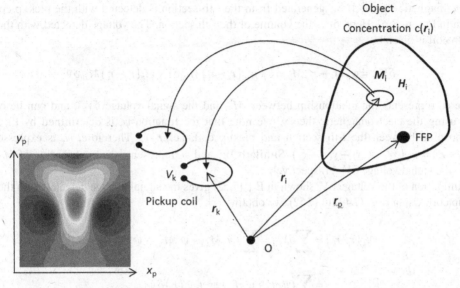

Map of V_k when the FFP is scanned

FIGURE 7.20 Signal voltage obtained in MPI.

$$\mu_0 \boldsymbol{H}_{\text{DC},i} = \begin{pmatrix} \mu_0 H_{\text{DC}x,i} \\ \mu_0 H_{\text{DC}y,i} \\ \mu_0 H_{\text{DC}z,i} \end{pmatrix} = \boldsymbol{G}(\boldsymbol{r}_p - \boldsymbol{r}_i) = \begin{pmatrix} G/2 & 0 & 0 \\ 0 & G/2 & 0 \\ 0 & 0 & -G \end{pmatrix} \begin{pmatrix} x_p - x_i \\ y_p - y_i \\ z_p - z_i \end{pmatrix} \quad (7.48)$$

where the position vector of the FFP is expressed as $\boldsymbol{r}_p = (x_p, y_p, z_p)$. As shown in Eq. (7.48), the DC field is the gradient field and is given by a matrix $\boldsymbol{G}$. When the gradient field is generated using a pair of permanent magnets as shown in Figure 7.17, the component of $\boldsymbol{G}$ is given by Eq. (7.48).

The magnetization of $\boldsymbol{M}_i$ is determined by the dynamic magnetization of the MNPs when the AC and DC fields are applied. The magnetization has x, y, z components and each is given by

$$M_{x,i} = c(\boldsymbol{r}_i) f_x (H_{\text{DC}x,i}, H_{\text{DC}y,i}, H_{\text{DC}z,i}, H_{\text{AC}z,i})$$
$$= c(\boldsymbol{r}_i) f_x (x_p - x_i, y_p - y_i, z_p - z_i; G, H_0) \quad (7.49)$$

$$M_{y,i} = c(\boldsymbol{r}_i) f_y (x_p - x_i, y_p - y_i, z_p - z_i; G, H_0) \quad (7.50)$$

$$M_{z,i} = c(\boldsymbol{r}_i) f_z (x_p - x_i, y_p - y_i, z_p - z_i; G, H_0) \quad (7.51)$$

where $c(\boldsymbol{r}_i)$ is the concentration of MNPs at position $\boldsymbol{r}_i$.

We note that the functions f_x, f_y, and f_z are determined only by the relative position between the FFP and the i-th section, i.e., $\boldsymbol{r}_p - \boldsymbol{r}_i$ when G and H_0 are fixed. These functions are the point spread function. As shown in Figure 7.5, PSF is maximized when $\boldsymbol{r}_i = \boldsymbol{r}_p$, i.e., when the i-th section is at the FFP. The values of the PSF rapidly decrease with the increase of the distance between the FFP and the i-th section. In the ideal case where the G value is extremely large, PSF can be approximated by the delta function, i.e., $f_x = \delta(x_p - x_i)\delta(y_p - y_i)\delta(z_p - z_i)$.

The magnetic signal, $M_i \delta v$, generated from the i-th section is detected with the pickup coil, as shown in Figure 7.20. Here, δv is the volume of the i-th section. The voltage detected with the k-th pickup coil is given by

$$\delta V_{k,i} = \psi_x (r_k - r_i) M_{x,i} \delta v + \psi_y (r_k - r_i) M_{y,i} \delta v + \psi_z (r_k - r_i) M_{z,i} \delta v \qquad (7.52)$$

where ψ_x represents the relationship between $M_{x,i}$ and the signal voltage $\delta V_{k,i}$, and can be calculated using the electromagnetic theory. We note that the function ψ_x is determined by the relative position between the i-th section and pickup coil, i.e., r_k-r_i. Therefore, ψ_x is expressed as $\psi_x (r_k - r_i) = \psi_x (x_k - x_i, y_k - y, z_k - z_i)$. Similarly, ψ_y and ψ_z represents the contribution of $M_{y,i}$ and $M_{z,i}$ to the signal voltage $\delta V_{k,i}$, respectively.

Summation of the voltage δV_k shown in Eq. (7.52) gives the signal voltage detected with the k-th pickup coil. Using Eqs. (7.49) to (7.52), we obtain

$$V_k (r_p, r_k) = \sum_i \delta V_{k,i} = \sum_i (\psi_x M_{x,i} + \psi_y M_{y,i} + \psi_z M_{z,i}) \delta v$$

$$= \sum_i (\psi_x f_x + \psi_y f_y + \psi_z f_z) c(r_i) \delta v \qquad (7.53)$$

$$= \sum_i a(r_p, r_k, r_i) c(r_i) \delta v$$

with

$$a(r_p, r_k, r_i) = \psi_x f_x + \psi_y f_y + \psi_z f_z \qquad (7.54)$$

Note that the value a given in Eq. (7.54) is a function of $\mathbf{r}_p - \mathbf{r}_i$ and $\mathbf{r}_k - \mathbf{r}_i$, and the signal voltage, V_k, detected with the pickup coil is determined by the position r_p of the FFP. Therefore, by scanning the FFP over the object volume, we can obtain the signal voltage $V_k (r_p)$ as a function of the FFP position r_p, as schematically shown in Figure 7.20. We define the signal vector, V_k, whose component is given by $V_k (r_p)$, and a concentration vector, c, whose component is given by the value of $c(r_i)$. In this case, the relationship between V_k and c is given by

$$V_k = A_k c \qquad (7.55)$$

where A_k is called the system function matrix, whose element is given by $a(r_p, r_k, r_i) \delta v$, as shown in Eq. (7.54).

The concentration vector, c, can be obtained by solving the inversion problem given by Eq. (7.55). For this purpose, several mathematical techniques have been developed [2, 3].

7.4.2 System Function Matrix

To solve Eq. (7.55), we have to know the system function, A_k, in advance. Since the system function greatly affects the accuracy of the solution method, it is very important to obtain accurate A_k. The matrix element of A_k, i.e., $a(\mathbf{r}_p, \mathbf{r}_k, \mathbf{r}_i)$ given in Eq. (7.54), can be determined theoretically or experimentally. The former is called a model-based system function, while the latter is called measurement-based system function [36–38].

In the model-based system function, functions ψ_x, ψ_y, and ψ_z in Eq. (7.54), which give the relationship between magnetization M_i of the MNP at r_i and the voltage signal V_k at r_k, can be exactly calculated with electromagnetic theory. The functions f_x, f_y, and f_z in Eq. (7.54), which give the magnetization M_i of the MNP at r_i, are determined by the field H_i and dynamic magnetization of the MNPs. The field H_i at r_i consists of the AC excitation field and DC gradient field and can be exactly calculated with electromagnetic theory. Alternatively, the dynamic magnetization of MNP can be calculated based on the Langevin function as shown previously. However, dynamic magnetization of the practical MNP sample cannot be exactly modeled with the Langevin function. Therefore, functions f_x, f_y, and f_z obtained with the Langevin function include errors. To obtain an accurate model-based system function, accurate modeling of the dynamic magnetization of practical MNPs sample is necessary.

The measurement-based system function can be obtained as shown below. We rewrite Eq. (7.55) as follows,

$$\begin{pmatrix} V_1 \\ V_2 \\ \vdots \\ V_N \end{pmatrix} = \begin{pmatrix} a_{11} & a_{12} & \cdots & a_{1N} \\ a_{21} & a_{22} & \cdots & a_{2N} \\ \vdots & \vdots & \vdots & \vdots \\ a_{N1} & a_{N2} & \cdots & a_{NN} \end{pmatrix} \begin{pmatrix} c_1 \\ c_2 \\ \vdots \\ c_N \end{pmatrix} \tag{7.56}$$

We consider the case when point MNP is placed at position $i=1$. The concentration vector in the right-hand side of Eq. (7.56) becomes $c_1=1$ and $c_i=0$ for $i \geq 2$. When the FFP is scanned, we can obtain the signal vector in the left-hand side of Eq. (7.56). In this case, we obtain $V_1=a_{11}$, $V_2=a_{21}$, ..., and $V_N=a_{N1}$. Therefore, the first column of the system function can be obtained experimentally. Next, the point MNP is placed at position $i=2$, and the signal vector is measured by scanning the FFP to obtain the second column of the system function. In this way, we can experimentally determine the system function. The measurement-based system function is very accurate; however, it takes a lot of time for the measurement.

7.4.3 IMAGING SYSTEM

The system function, A_k, gives spatial information of the position of MNPs when the concentration vector c is reconstructed from the measured the signal vector V_k. Several types of imaging systems have been developed depending on the construction method of the system function. Figure 7.21 schematically depicts four types of imaging systems.

Figure 7.21(a) is an imaging system using the FFP and a large pickup coil [32–36]. In this case, we can approximate $\psi_x = \psi_y = \psi_z = 1$ in Eq. (7.54), and obtain the element of the system function matrix as

$$a(r_p, r_i) \approx f_x + f_y + f_z \tag{7.57}$$

The above equation means that the position information of the MNP is given by the relative distance $r_p - r_i$ between the FFP and MNP positions. The signal vector, V_k, is obtained by scanning the FFP over the object volume, and the concentration vector, c, is reconstructed from Eq. (7.55).

Figure 7.21(b) is an imaging system using the FFL and a large pickup coil. The position information of the MNP is also given by the relative distance between FFL and the MNP position. The signal vector, V_k, is obtained by rotating the FFL, and the concentration vector, c, is reconstructed through the solution of the inversion problem [37–42].

The systems shown in Figure 7.21(a) and (b) are the main MPI systems. Imaging of MNP distribution with high spatial resolution can be performed using a strong gradient field. However, a system for the generation and scanning of the FFP or FFL becomes bigger and more complicated when the object size becomes larger. We have to develop a sophisticated scanner system for human-size MPI.

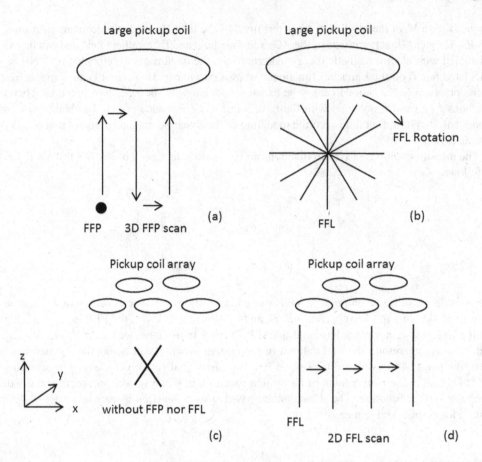

FIGURE 7.21 Imaging systems using (a) FFP and a large pickup coil, (b) FFL and a large pickup coil, (c) a pickup coil array without FFP or FFL, and (d) a pickup coil array and FFL.

When the requirement for the spatial resolution is not so severe, another imaging approach is possible. Figure 7.21(c) shows a limiting case where the DC gradient field is not used. In this case, we can eliminate a system for the generation and scanning of the FFP or FFL, and the imaging system becomes very simple. We use only the excitation field and a pickup coil array, as shown in Figure 7.21(c). In this case, we can approximate $f_x = f_y = f_z = 1$ in Eq. (7.54). When the AC field is applied in the z direction, the magnetization M_i of the MNPs sample has only z components. Therefore, the element of the system function matrix is given by

$$a(r_k, r_i) \approx \psi_z \tag{7.58}$$

The above equation means that the position information of the MNP is given by the relative distance, $r_k - r_i$, between the pickup coil and the MNP position. The signal vector, V_k, obtained with a pickup coil array is used to reconstruct the concentration vector, c, from the solution of Eq. (7.55). This procedure is the same as that used in magnetoencephalography (MEG) and magnetocardiography (MCG). We note that inversion problem becomes simpler compared with MEG and MCG because magnetization has only the z component. The system is useful when MNPs are located in a shallow position from the body surface, e.g., for application in breast cancer and sentinel lymph node detection [43–46].

Figure 7.21(d) is the imaging system that combines the FFL and the pickup coil array. The element of the system function matrix is given by

$$a(r_p, r_k, r_i) = \psi_x f_x + \psi_y f_y + \psi_z f_z \tag{7.59}$$

In this case, FFL is mainly used to obtain position information of the MNP in the xy plane. The functions f_x, f_y and f_z in Eq. (7.59) give the position information in the xy plane. Alternatively, the coil array is mainly used to obtain the position information of the MNP in the z direction. The functions $\psi_x, \psi_y,$ and ψ_z give the position information in the z direction.

Therefore, two-dimensional scanning of the FFL in the xy plane is sufficient to obtain the three-dimensional position information of the MNP [47–49]. As a result, the scanner system becomes simpler than the case of Figure 7.21(b). The system can be used to detect MNPs located much deeper from the body surface than the system shown in Figure 7.21(c).

REFERENCES

1. T. Knopp and T.M. Buzug, *Magnetic Particle Imaging*, Springer-Verlag, Berlin, 2012.
2. B. Gleich and J. Weizenecker, "Tomographic imaging using the nonlinear response of magnetic particles", *Nature* **435**(7046), 1214 (2005).
3. J. Weizenecker, B. Gleich, J. Rahmer, H. Dahnke, and J. Borgert, "Three-dimensional real-time in vivo magnetic particle imaging", *Phys. Med. Biol.* **54**(5), L1–L10 (2009).
4. E.U. Saritas, P.W. Goodwill, L.R. Croft, J.J. Konklw, K. Lu, B. Zheng, and S.M. Conolly, "Magnetic particle imaging (MPI) for NMR and MRI researchers", *J. Magn. Reson.* **229**, 116–126 (2013).
5. M.H. Pablico-Lansigan, S.F. Situ, and A.C.S. Samia, "Magnetic particle imaging: Advancements and perspectives for real-time in vivo monitoring and image-guided therapy", *Nanoscale* **5**(10), 4040 (2013).
6. N. Panagiotopoulos, R.L. Duschka, M. Ahlborg, G. Bringout, C. Debbeler, M. Graeser, C. Kaethner, K. Lüdtke-Buzug, H. Medimagh, J. Stelzner, T.M. Buzug, J. Barkhausen, F.M. Vogt, and J. Haegele, "Magnetic particle imaging: Current developments and future directions", *Int. J. Nanomed.* **10**, 3097–3114 (2015).
7. J.W.M. Bulte, P. Walczak, M. Janowski, K.M. Krishnan, H. Arami, A. Halkola, B.Gleich, and J. Rahmer, "Quantitative "Hot-Spot" imaging of transplanted stem cells using superparamagnetic tracers and magnetic particle imaging", *Tomography* **1**(2), 91–97 (2015).
8. B. Zheng, M.P. See, E. Yu, B. Gunel, K. Lu, T. Vazin, D.V. Schaffer, P.W. Goodwill, and S.M. Conolly, "Quantitative magnetic particle imaging monitors the transplantation, biodistribution, and clearance of stem cells in vivo", *Theranostics* **6**(3), 291–301 (2016).
9. E.Y. Yu, M. Bishop, B. Zheng, R.M. Ferguson, A.P. Khandhar, S.J. Kemp, K.M. Krishnan, P.W. Goodwill, and S.M. Conolly, "Magnetic particle imaging: A novel in vivo imaging platform for cancer detection", *Nano Lett.* **17**(3), 1648–1654 (2017).
10. P. Ludewig, N. Gdaniec, J. Sedlacik, N.D. Forkert, P. Szwargulski, M. Graeser, G. Adam, M.G. Kaul, K.M. Krishnan, R.M. Ferguson, A.P. Khandhar, P. Walczak, J. Fiehler, G. Thomalla, C. Gerloff, T. Knopp, and T. Magnus, "Magnetic particle imaging for real-time perfusion imaging in acute stroke", *ACS Nano* **11**(10), 10480–10488 (2017).
11. N.T.K. Thanh ed., *Magnetic Nanoparticles from Fabrication to Clinical Applications*, CRC Press, 2012.
12. C. Sun, Jerry S.H. Lee, and M. Zhang, "Magnetic nanoparticles in MR imaging and drug delivery", *Adv. Drug Deliv. Rev.* **60**(11), 1252–1265 (2008).
13. R.S. Gaster, L. Xu, S.J. Han, R.J. Wilson, D.A. Hall, S.J. Osterfeld, H. Yu, and S.X. Wang, "Quantification of protein interactions and solution transport using high-density GMR sensor arrays", *Nat. Nanotech.* **6**(5), 314–320 (2011).
14. A.B. Salunkhe, V.M. Khot, and S.H. Pawar, "Magnetic hyperthermia with magnetic nanoparticles: A status review", *Curr. Top. Med. Chem.* **14**(5), 572–594 (2014).
15. T.H. Shin, Y. Choi, S. Kim, and J. Cheon, "Recent advances in magnetic nanoparticle-based multimodal imaging", *Chem. Soc. Rev.* **44**(14), 4501–4516 (2015).
16. S. Schrittwieser, B. Pelaz, W.J. Parak, S. Lentijo-Mozo, K. Soulantica, J. Dieckhoff, F. Ludwig, A. Guenther, A. Tschöpe, and J. Schotter, "Homogeneous biosensing based on magnetic particle labels", *Sensors* **16**(6), 828 (2016).
17. K. Enpuku, Y. Tsujita, K. Nakamura, T. Sasayama, and T. Yoshida, "Biosensing utilizing magnetic markers and superconducting quantum interference devices", *Supercond. Sci. Technol.* **30**(5), 053002 (2017).
18. Y.T. Chen, A.G. Kolhatkar, O. Zenasni, S. Xu, and T.R. Lee, "Biosensing using magnetic particle detection techniques", *Sensors* **17**(10), 2300 (2017).

19. J.L. Dormann, D. Fiorani, and E. Tronc, "Magnetic relaxation in finite-particle systems", *Adv. Chem. Phys.* **98**, I. Prigogine and S.A. Rice, Eds., Wiley, New York (1997).

20. W.T. Coffey, P.J. Cregg, and Y.P. Kalmykov, "On the theory of debye and Néel relaxation of single domain ferromagnetic particles", *Adv. Chem. Phys.* **83**, I. Prigogine and S.A. Rice, Eds., Wiley, New York (1993).

21. R.M. Ferguson, K.R. Minard, and K.M. Krishnan, "Optimization of nanoparticle core size for magnetic particle imaging", *J. Magn. Magn. Mater.* **321**(10), 1548 (2009).

22. S. Dutz, J.H. Clement, D. Eberbeck, T. Gelbrich, R. Hergt, R. Müller, J. Wotschadlo, and M. Zeisberger, "Ferrofluids of magnetic multicore nanoparticles for biomedical applications", *J. Magn. Magn. Mater.* **321**(10), 1501–1504 (2009).

23. D. Eberbeck, F. Wiekhorst, S. Wagner, and L. Trahms, "How the size distribution of magnetic nanoparticles determines their magnetic particle imaging performance", *Appl. Phys. Lett.* **98**(18), 182502 (2011).

24. F. Ludwig, T. Wawrzik, T. Yoshida, N. Gehrke, A. Briel, D. Eberbeck, and M. Schilling, "Optimization of magnetic nanoparticles for magnetic particle imaging", *IEEE Trans. Magn.* **48**(11), 3780–3783 (2012).

25. T. Yoshida, N.B. Othman, and K. Enpuku, "Characterization of magnetically fractionated magnetic nanoparticles for magnetic particle imaging", *J. Appl. Phys.* **114**(17), 173908 (2013).

26. T. Yoshida and K. Enpuku, "Simulation and quantitative clarification of AC susceptibility of magnetic fluid in nonlinear Brownian relaxation region", *Jpn. J. Appl. Phys.* **48**(12), 127002 (2009).

27. J. Dieckhoff, D. Eberbeck, M. Schilling, and F. Ludwig, "Magnetic-field dependence of Brownian and Neel relaxation times", *J. Appl. Phys.* **119**(4), 043903 (2016).

28. K. Enpuku, S. Bai, A. Hirokawa, K. Tanabe, T. Sasayama, and T. Yoshida, "The effect of Neel relaxation on the properties of the third harmonic signal of magnetic nanoparticles for use in narrow-band magnetic nanoparticle imaging", *Jpn. J. Appl. Phys.* **53**(10), 103002 (2014).

29. L.R. Croft, P.W. Goodwill, J.J. Konkle, H. Arami, D.A. Price, A.X. Li, E.U. Saritas, and S.M. Conoly, "Low drive field amplitude for improved image resolution in magnetic particle imaging", *Med. Phys.* **43**(1), 424 (2016).

30. S. Bai, A. Hirokawa, K. Tanabe, T. Yoshida, and K. Enpuku, "Magnetic particle imaging utilizing orthogonal gradient field and third-harmonic signal detection", *IEEE Trans. Magn.* **50**(11), 5101304 (2014).

31. E.U. Saritas, P.W. Goodwill, G.Z. Zhang, and S.M. Conolly, "Magnetostimulation limits in magnetic particle imaging", *IEEE Trans. Med.* **32**(9), 1600 (2013).

32. P.W. Goodwill, K. Lu, B. Zheng, and S.M. Conolly, "An *x*-space magnetic particle imaging scanner", *Rev. Sci. Instr.* **83**(3), 033708 (2012).

33. P. Vogel, M.A. Rückert, P. Klauer, W.H. Kullmann, P.M. Jakob, and V.C. Behr, "Traveling wave magnetic particle imaging", *IEEE Trans. Med. Imaging* **33**(2), 400–407 (2014).

34. K. Gräfe, A. Gladiss, G.Bringout, M.Ahlborg, and T.M. Buzug, "2D images recorded with a single-sided magnetic particle imaging scanner", *IEEE Trans. Med. Imaging* **35**(4), 1056–1065 (2016).

35. J. Franke, U. Heinen, H. Lehr, A. Weber, F. Jaspard, W. Ruhm, M. Heidenreich, and V. Schulz, "System characterization of a highly integrated preclinical hybrid MPI-MRI scanner", *IEEE Trans. Med. Imaging* **35**(9), 1993–2004 (2016).

36. J.Salamon, M. Hofmann ,C. Jung, M.G. Kaul,F. Werner, K. Them, R. Reimer, P. Nielsen, A. Vom Scheidt, G. Adam, T. Knopp, and H. Ittrich, "Magnetic particle / magnetic resonance imaging: In-vitro MPI-guided real time catheter tracking and 4D angioplasty using a road map and blood pool tracer approach", *PLoS One*. doi:10.1371/journal.pone.0156899 June 1, 2016

37. P.W. Goodwill, J.J. Konkle, B. Zheng, E.U. Saritas, and S.M. Conolly, "Projection X-space magnetic particle imaging", *IEEE Trans. Med. Imaging* **31**(5), 1076–1085 (2012).

38. P.W. Goodwill, E.U. Saritas, L.R. Croft, T.N. Kim, K.M. Krishnan, D.V. Schaffer, and S.M. Conolly, "X-space MPI: Magnetic nanoparticles for safe medical imaging", *Adv. Mater.* **24**(28), 3870 (2012).

39. E.E. Mason, C.Z. Cooley, S.F. Cauley, M.A. Griswo, S.M. Conolly, and L.L. Walda, "Design analysis of an MPI human functional brain scanner", *Int. J. Magn. Part. Imaging* **3**, 1703008 (2017).

40. T. Knopp, T.F. Sattel, S. Biederer, J. Rahmer, J. Weizenecker, B. Gleich, J. Borgert, and T.M. Buzug, "Model-based reconstruction for magnetic particle imaging", *IEEE Trans. Med. Imaging* **29**(1), 12–18 (2010).

41. J. Rahmer, J. Weizenecker, B. Gleich, and J. Borgert, "Analysis of a 3-D system function measured for magnetic particle imaging", *IEEE Trans. Med. Imaging* **31**(6), 1289–1299 (2012).

42. J. Rahmer, A. Halkola, B. Gleich, I. Schmale, and J. Borgert, "First experimental evidence of the feasibility of multi-color magnetic particle imaging", *Phys. Med. Biol.* **60**(5), 1775 (2015).

43. H.J. Hathaway, K.S. Butler, N.L. Adolphi, D.M. Lovato, R. Belfon, D. Fegan, T.C. Monson, J.E. Trujillo, T.E. Tessier, H.C. Bryant, D.L. Huber, R.S. Larson, and E.R. Flynn, "Detection of breast cancer cells using targeted magnetic nanoparticles and ultra-sensitive magnetic field sensors", *Breast Cancer Res.* **13**(5), R108 (2011).

44. L.P. Haroa, T. Karaulanova, E.C. Vreelanda, B. Anderson, H.J. Hathaway, D.L. Huber, A.N. Matlashov, C.P. Nettles, A.D. Price, T.C. Monson, and E.R. Flynn, "Magnetic relaxometry as applied to sensitive cancer detection and localization", *Biomed. Eng. Biomed. Tech.* **60**, 445–455 (2015).

45. D. Baumgarten, F. Braune, E. Supriyanto, and J. Haueisen, "Plane-wise sensitivity based inhomogeneous excitation fields for magnetorelaxometry imaging of magnetic nanoparticles", *J. Magn. Magn. Mater.* **380**, 255–260 (2015).

46. B.W. Ficko, P. Giacometti, and S.G. Diamond, "Nonlinear susceptibility magnitude imaging of magnetic nanoparticles", *J. Magn. Magn. Mater.* **378**, 267–277 (2015).

47. T. Sasayama, Y. Tsujita, M. Morishita, M. Muta, T. Yoshida, and K. Enpuku, "Three-dimensional magnetic nanoparticle imaging using small field gradient and multiple pickup coils", *J. Magn. Magn. Mater.* **427**, 143 (2017).

48. M. Muta, S. Hamanaga, N. Tanaka, T. Sasayama, T. Yoshida, and K. Enpuku, "Three-dimensional imaging of magnetic nanoparticles using multiple pickup coils and field-free line", *Jpn. J. Appl. Phys.* **57**(2), 023002 (2018).

49. S. Hamanaga, T. Yoshida, T. Sasayama, A.L. Elrefai, and K. Enpuku, "Three-dimensional detection of magnetic nanoparticles using a field-free line with weak field gradient and multiple pickup coils", *Jpn. J. Appl. Phys.* **58**(6), 061001 (2019).

8 Sensing of Magnetic Nanoparticles for Sentinel Lymph Nodes Biopsy

Masaki Sekino and Moriaki Kusakabe

CONTENTS

8.1 Introduction 185
8.2 Current State and Challenges of Sentinel Lymph Node Biopsy 186
8.3 Sentinel Lymph Node Biopsy Using Magnetic Techniques 188
8.4 Magnetic Probe 189
8.5 Clinical Trials 193
8.6 Magnetic Immunostaining for Intraoperative Diagnosis of Cancer Metastasis 194
8.7 Design of a Permanent Magnet for Immunostaining Based on Dynamics
of Magnetic Beads 194
8.8 Automatic Magnetic Immunostaining System 195
8.9 Future Prospects 196
References 197

8.1 INTRODUCTION

Cancer is prone to metastasis which may complicate its treatment. Breast cancer is known to spread to the axillary lymph nodes, and in order to determine the stage of the disease and select the appropriate treatment, it must be confirmed whether there are metastases to the axillary lymph nodes. Diagnosis of lymph node metastases requires a biopsy, a technique to resect lymph nodes for pathological examination. This is a very invasive procedure for patients. Resection of all the axillary lymph nodes for examination (radical dissection) blocks the flow of lymph fluid and causes side effects such as severe swelling. Therefore, in early-stage breast cancer surgery, the patient's quality of life (QoL) may be increased by using minimally invasive breast-conserving surgery as well as avoiding radical dissection of axillary lymph nodes. In other words, examination techniques that do not involve unnecessary lymph node resection in patients without lymph node metastases are important. The lymph nodes that are reached first by cancer cells traveling from the primary lesion through the lymphatic vessels are termed sentinel lymph nodes (SLNs). Once these are identified, tissue is obtained selectively and subjected to pathological examination [1–3]. A common method to identify SLNs is to inject a tracer into or in close proximity to the primary lesion and observe, from the outside, the agent passing through the lymphatic vessels and accumulating in the SLNs. The SLN is then excised and examined for cancer cells, and if metastases to the SLNs are found, radical dissection of the axillary lymph nodes in addition to the SLNs is necessary. However, if none are found, most of the axillary lymph nodes are preserved [4]. Under this principle, side effects can be limited by avoiding unnecessary lymph node dissection [5–7]. The important factor here is the method to identify SLNs.

With the current radioisotope (RI) method that uses radioisotopes, patients are given a subcutaneous subareola injection of technetium (^{99m}Tc)-labeled phytic acid one day prior to surgery. The next day, SLNs where the radioactive tracer has accumulated are identified using a radiation detection device and the SLNs are excised. However, in addition to patients' direct exposure to radioactive isotopes, this method requires specific controlled areas for radiation work and is therefore difficult to implement in small to medium-sized hospitals.

To solve these issues, other methods are deployed where SLNs are identified visually using non-RI tracers such as dye (e.g., patent blue, indigo carmine) or fluorescence (e.g., indocyanine green, ICG) methods. However, these methods are subjective and have relatively low identification rates.

This chapter introduces the recently developed mammary sentinel lymph node biopsy (SLNB) system that uses a magnetic probe and magnetic nanoparticles (ferucarbotran).

8.2   CURRENT STATE AND CHALLENGES OF SENTINEL LYMPH NODE BIOPSY

Medical technologies have undergone terrific advancements in recent years and promising drug developments include molecular targeted drugs that are based on gene data and gene editing technologies. On the other hand, in medical device development, new devices are being developed by merging robot technology and artificial intelligence (AI).

However, cancer is still the leading cause of death, representing 30% of all-cause mortality in Japan. The estimated number of cancer patients in 2013 was 862,452 (498,720 males and 363,732 females) and the number of cancer deaths in 2016 was 372,986 (219,785 males and 153,201 females) [8]. Assessing the 2013 cancer data, one in two people are predicted to be diagnosed with a cancer. For breast cancer, this represents one out of every 11 people. The reason that cancer death risk is higher for women up to the age of 50 is the high risk of breast cancer in middle-aged women. Although five-year relative survival rates for breast cancer between 2006 and 2008 were high at 98.8% for local (intraductal) cancers and 88.4% for regional (spread to extraductal areas) cancers, survival rates dropped substantially for distant (metastatic) cancer at 33.7%. In other cancers, cases with distant metastases had markedly lower survival rates.

This clearly indicates that cancer infiltration and distant metastasis associated with the progression of breast cancer greatly affects death rates, and this is also the case for other cancers. Early detection and treatment have been recognized to have contributed to reduced cancer death rates since the early 2000s.

As can be seen from the 2013 survey, detection through breast self-examination or mammography increased and although incidence for relatively early detected cancer in women ranks first, due to the introduction of metastasis diagnosis through sentinel lymph node biopsy (SLNB), the number of deaths ranks fifth. This is believed to be due to the effect of the selection of surgical procedure and subsequent determination of treatment strategy due to the SLNB being performed for the purpose of diagnosing metastases in the diagnostic process after detection. In fact, breast-conserving surgery rates rank high in recent breast surgery procedures at around 60% and in an increasing number of cases, mastectomy and lymph node dissection are not applied. When determining if early detected cancer has metastasized, initial metastatic foci are SLN and establishing the presence or absence of cancerous cells in SLN are very important in subsequent selection of surgical procedures and determination of the treatment strategy. Currently available SLNB techniques include the RI method which uses a radioactive tracer and a Geiger counter, dye techniques where SLNs are identified with the naked eye, and the fluorescence method where accumulations of fluorescent dye are identified through their fluorescence.

To prevent systemic metastasis, breast cancer surgery traditionally involved complete dissection of axillary lymph nodes. This significantly damaged lymph flow in the upper arm, causing swelling of the entire upper arm and substantially reducing patients' QoL.

During metastasis of breast cancer cells, first, cancer cells that have broken away from the primary tumor enter into the lymphatic vessels, then travel on the lymph fluid, and from the venous angles that the lymph fluid drains into (junction of the internal jugular and subclavian veins), they enter the bloodstream and are carried throughout the body by the circulatory system forming new tumors after reaching other organs (Figure 8.1) [9]. Cancer cells that do not form tumors but are in

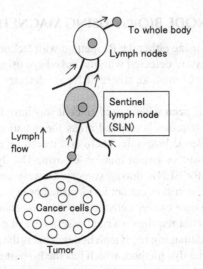

FIGURE 8.1 Cancer metastasis and the sentinel lymph node [9].

the lymph fluid and bloodstream are single cells in the initial metastatic phase, but because they form a number of small cell clusters, they cannot be easily detected. However, lymph fluid, which is present throughout the body, carries water from the local tissue back into the bloodstream and, at the same time, also ends up carrying foreign bodies away. It is for this reason that lymph nodes exist, acting as filtering devices throughout the lymphatic vessel network (macrophages, B lymphocytes, tissue made up from reticuloendothelial system cells) to trap foreign bodies. Also during metastasis of cancer cells, cancer cells that have broken away from the tumor and have entered the lymphatic vessels are trapped in lymph nodes. The lymph nodes that cancer cells that have broken away from the primary tumor first enter into are the SLNs. The surgical technique adopted is as follows. SLNs are accurately identified and excised during surgery and are promptly diagnosed by a pathologist. That is, a frozen section is produced from the excised SLN which is stained with hematoxylin and eosin. The pathologist then identifies cancer cells under the microscope. Consequently, no lymphadenectomy is performed unless metastases are seen, and lymphadenectomy is only performed when the sample is positive for metastasis (Figure 8.2) [10]. Through the introduction of these SLNB, there has been a reduction in unnecessary lymphadenectomies resulting in improved patients' QoL.

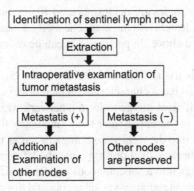

FIGURE 8.2 Flow of sentinel lymph node biopsy (SLNB) [10].

8.3   SENTINEL LYMPH NODE BIOPSY USING MAGNETIC TECHNIQUES

Currently, SLNs are identified using either the RI method with technetium phytate and an RI detector, the dye method, where they are detected with the naked eye using dyes such as indigo carmine or patent blue, or the fluorescence method, where SLNs are detected using ICG and a fluorescence camera.

Although the RI method has been widely used since it was introduced for breast cancer SLNB, the use of radioactive tracers requires controlled areas for radiation work and can only be done in large hospitals such as academic hospitals. There are, therefore, also hospitals that use the dye method involving indigo carmine or patent blue on its own. The dye method does not require a controlled area and it allows for SLNB during surgery. However, since SLN location cannot be identified from the outside, incision sites cannot be identified prior to surgery. Moreover, these dyes easily travel from the SLNs where cancer cells end up first to secondary lymph nodes over time and excising them from the axilla requires a visual assessment of color tone, so delicate surgical technique is required to avoid damaging the lymph nodes. This is the reason that current SLNB that combines the RI method and the dye method, which has the highest detection rate, has become the mainstream method. The reported detection rate is up to 97% [11, 12]. Furthermore, some facilities are using the fluorescence method with ICG on its own. With this method, ICG is injected *in vivo*, and since fluorescence of 780–950 nm is visible when excited at wavelength 700–850 nm, the SLN's location can be identified as it is lit up by ICG reaching SLN. However, as with the dye method, observing fluorescence from the outside is difficult and requires the surgeon to make an incision in the approximate SLN location and proceed from there. Furthermore, this solution also easily travels to secondary lymph nodes, and if a lymphatic vessel is damaged, the entire surgical area becomes fluorescent due to the leaking ICG, making identification difficult. Another drawback is that SLN detection rates are not that high either.

We believe that SLNB through the magnetic method using magnetic nanoparticles as a tracer and a detector device meets the following requirements [13–19].

1. The tracer can be safely administered to the human body.
2. Allows for detection from the body surface as with isotopes.
3. No need to restrict facilities using it.
4. Comparable or higher (not inferior) identification rate as other methods.
5. Detection is simple.
6. The detector is inexpensive.
7. Excellent accumulation and stability.

We use ferucarbotran (Resovist), which was developed as a liver magnetic resonance imaging (MRI) contrast agent as a magnetic tracer. This one vial contains 864 mg ferucarbotran (44.6 mg/1.6 mL as iron). These magnetic nanoparticles are dextran-coated 10 nm diameter gamma-ferric oxide particles forming 60 nm diameter clusters and the magnetic method using these magnetic nanoparticles meets the requirements outlined above. In particular, it can be offered both in a small size and at a low price.

The lymph flow that returns tissue fluid in the body to the bloodstream has organ-dependent characteristics. Breast tissue also has a characteristic lymph flow.

Blind ending lymphatic capillaries merge to form a network which then merges further to form a lymphatic trunk. The lymph in the upper right body (including the right side of the head) is the right lymphatic duct and drains into the right venous angle whereas the lymph flow of the lower half and the upper left body is the thoracic duct draining into the left venous angle. Lymphatic capillaries in the mammary tissue merge into a subcutaneous mammary network structure and the flow is toward the papilla, collected subcutaneously via subareola and drains in the direction of the axilla. Lymph flow then drains into the axillary lymph nodes.

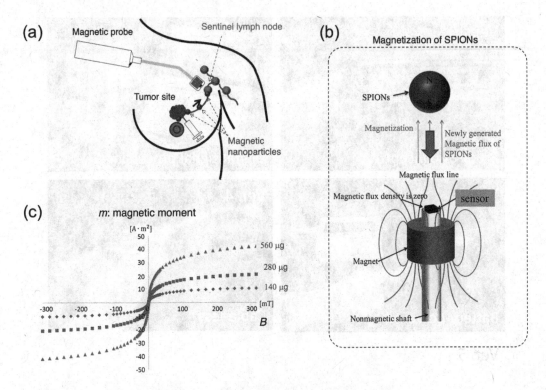

FIGURE 8.3 Sentinel lymph node biopsy using magnetic nanoparticles [20].

This means that individual cancer cells that have broken away from the primary cancer lesion enter the lymphatic capillaries, first travel to the subcutaneous subareola area and from there enter the axillary lymph nodes through one lymphatic vessel directed toward the axilla. Taking this into consideration, most tracer injection sites are established as subareola intradermally or subcutaneously. Naturally, subcutaneous injection right above the primary breast cancer lesion or injections into the lesion have been attempted too, but a multicenter study with a large number of cases concluded that SLN can be identified through subareole subcutaneous administration. In the following studies, magnetic nanoparticles were injected subcutaneously subareole (Figure 8.3(a)) [20].

8.4 MAGNETIC PROBE

Figure 8.3(b) outlines the principle of the magnetic probe we developed [20–23]. For this device, a Hall effect sensor was attached to the tip of a brass pipe, and the sensor was combined with a ring-shaped neodymium magnet. The magnetic field is 0 at a point on the axis of the ring-shaped neodymium magnet so the Hall effect sensor is aligned to that position. Unless magnetic bodies enter in the area ahead of the sensor, it does normally not detect any magnetic field. In actual measurements, magnetic nanoparticles are searched using the magnetic probe and if the magnetic nanoparticles are present in the detection area ahead of the probe head, these are magnetized by the magnetic field of the neodymium magnet (Figure 8.3(c)). Then, a magnetic field is formed from the nanoparticles and it is detected by the Hall effect sensor.

Furthermore, to eliminate the geomagnetic effect, the probe is equipped with a second Hall effect sensor in addition to the one on the probe head, with the difference between the readouts for both sensors being displayed. This magnetic probe has undergone many improvements (Figure 8.4), most recently CE mark approval in July 2019 [24]. Figure 8.5 shows a diagram of identifying the sentinel lymph nodes using the magnetic probe.

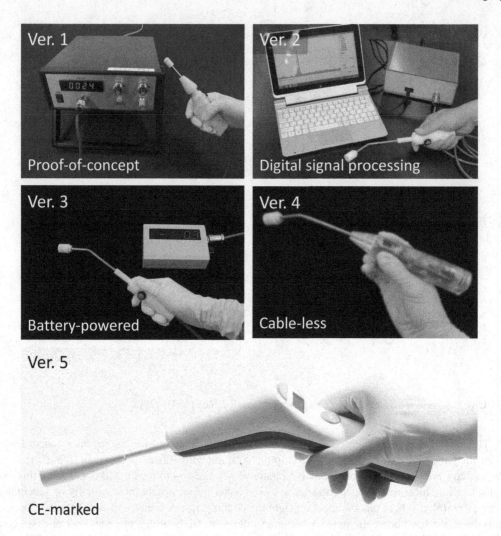

FIGURE 8.4 Development and improvement of the magnetic probe. The latest version is CE-marked.

We will initially outline the magnetic detection principles of the first developed device [25]. The magnetic field was restricted by inserting a (permanent) bar magnet into a high permeability permalloy (iron and nickel) tube. This combination of magnet and permalloy was attached to both sides, N-pole facing forward of the detection area with a Hall effect sensor attached to a non-magnetic pipe. To ensure that the magnetic field of the neodymium magnet is not directly sensed by the Hall effect sensor, the permalloy and bar magnet positions were aligned. This way, an area without any magnetic field was formed in close proximity to the magnet, and by aligning the Hall effect sensor to this point where magnetic force is 0, the sensor does not detect anything unless something is present in front of the sensor. Thus, if magnetic nanoparticles enter into the magnetic field ahead of the detection part, the nanoparticles are magnetized and the magnetic field in front of the sensor is disturbed and a new magnetic field is formed. This results in the Hall effect sensor detecting the magnetic field of the nanoparticles. In fact, observation (using the movement of iron sand) of the new magnetic field generated when a magnetic substance enters the magnetic field, shows that, as seen in Figure 8.6(a)(c), if nothing occurs, the iron sand collects in the areas slightly removed from each pole of the permanent magnet, which are the strongest areas of the magnetic field. If a magnetic substance enters this magnetic field, a new magnetic field is formed right opposite the Hall

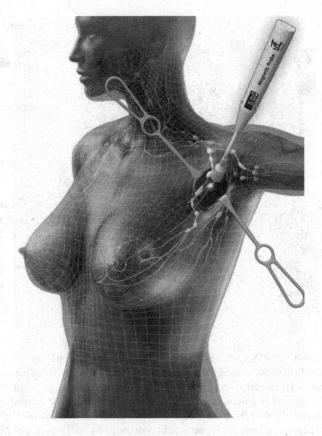

FIGURE 8.5 Identification of axillary sentinel lymph nodes using a magnetic probe.

effect sensor, as seen in (d). However, as seen in Figure 8.6(a)(b), the permanent magnet and the permalloy are insufficiently aligned and, even when a magnetic substance enters the front magnetic field, no formation of a new magnetic field toward the sensor was observed.

This magnetic probe thus consisted of a probe head with a permalloy pipe containing a bar-shaped neodymium magnet placed on both sides of a non-magnetic brass pipe with a Hall element attached to the tip. Detection was performed with a system that varies the luminance of (red and green) LEDs and sound. The sensor is fixed in the location where the magnet's magnetic field is 0 and as well as light and sound alerts, readings were also displayed in numbers. From the second-generation device onward, permalloy was no longer used and a brass pipe with the Hall effect sensor attached was directly inserted into the center of a ring-shaped neodymium magnet, still fixing it in a location where the magnetic force is 0 (Figure 8.3(b)). This version of the device displayed the detected field in μT, making the change from an analog to a digital circuit. The third-generation device had a built-in battery thus removing the power cable. The fourth-generation device was then further improved into a handheld compact device with a built-in power source. The shape was designed to suit requirements established through a survey of breast surgeons. The detection mechanism is formed by a system where the Hall effect sensor is placed on the location where magnetic force is 0 and the magnetic field which is newly formed when magnetic nanoparticles are present in front and when that nanoparticles are magnetized by the neodymium magnet.

For this device, we succeeded in incorporating the circuitry and battery all within the handheld probe. External units connected by cables, as seen for similar devices, are not required. The device can be kept in its entirety in the clean area, and it has no cables to cause an obstruction during surgery. The magnetic probe works on the principle that it magnetizes magnetic nanoparticles in the

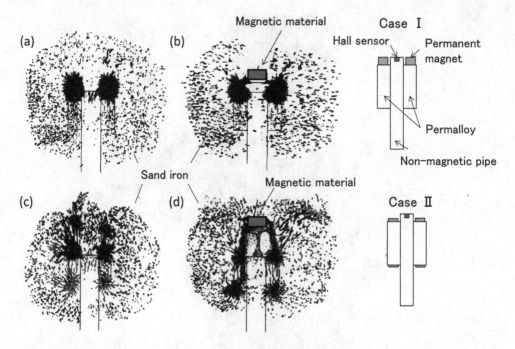

FIGURE 8.6 Formation of the magnetic null field by positioning of the magnet in permalloy. As indicated in Case I, the magnetic field is not formed well if the magnet is fabricated without permalloy. Sand iron sticks directly to magnet (a), and no magnetic null field is formed in front of Hall sensor (b). If positioning between the permalloy and the magnet is well adjusted, the magnetic null field can be formed in front of Hall sensor (c). When magnetic material enters the front area of the Hall sensor, it is magnetized by the magnetic field on both sides. Then, the newly synthesized magnetic field is detected by Hall sensor (d) [25].

body through its permanent magnet and detects the resulting magnetic field through a small magnetic sensor at the probe's tip. The use of a permanent magnet allowed for energy-saving operation, a smaller drive circuit, and a battery-driven configuration. Different equipment with similar functionality uses a cable to connect a handheld probe used within the clear area to a unit that is operated outside. This device's lack of cables does not create any inconveniences. It has a simple operational set-up with three buttons, on/off, reset, and mute, which have been placed in an ergonomic layout making it easy to operate.

While the magnetic probe outputs the magnetic flux density detected by the sensor, quantifying the amount of magnetic nanoparticles in the sentinel lymph node is necessary for optimizing the protocol in terms of the amount of injection and interval between the injection and detection. There are reports of developing dedicated devices of quantifying magnetic nanoparticles [26].

Alongside the magnetic probe, we have also developed non-magnetic titanium retractors. Different from stainless steel, which is commonly used for surgical instruments, titanium is fully non-magnetic and therefore does not interfere with measurements performed by the magnetic probe. An additional benefit is that titanium is lighter than stainless steel. To prevent detection errors by the magnetic probe, we have used non-magnetic titanium to manufacture retractors to keep the surgical field open. The reasons we have not used plastic are the selection of a material that has excellent strength and that biocompatibility is difficult. Titanium instruments are light, and titanium has lately been used in the development of a wide range of instruments for use in MRI-guided surgery, including in the ophthalmology field. However, since few of the relatively large surgical instruments used in common surgeries are available in titanium, we have developed new titanium retractors. These retractors have already been used in clinical research and their usefulness have been confirmed.

8.5 CLINICAL TRIALS

SLNB, commonly used in clinical practice, is currently used in breast cancer diagnosis. Attempts have also been made to diagnose cancer metastases by introducing SLNB, even in operations for cancers prone to metastasis, such as pigment epithelial cancer (melanoma), head and neck cancer (tongue cancer, buccal mucosal cancer), gastric cancer, and lung cancer. Magnetic methods can be applied to all such cases. Usually, the target cases are the ones where cancer is present on the body surface. However, in cases of operations involving internal organs, such as for gastric cancer and endometrial cancer, minimally invasive laparoscopic operations are being promoted, and the use of SLNB together with such procedures is now strongly encouraged. Moreover, while existing RI methods and fluorescence methods have been introduced in various parts of the human body, detections in the abdominal cavity remain difficult. Accordingly, the development of magnetic probes for laparoscopy is also progressing to introduce magnetic methods in these medical situations.

The clinical study is described as follows. A first-generation device using analog circuits was utilized in the first clinical study using magnetic methods [27]. Immediately before the operation, 1% patent blue dye and Resovist were injected to the affected areola subcutaneously, and SLN was identified from the body surface using a magnetic probe during the operation. After skin incision, the magnetized SLN was searched and removed, and backup dissection was performed. A total of 30 cases were divided into three groups named as magnetic nanoparticles group, patent blue dye group, and magnetic nanoparticles and patent blue dye combination group. The detection rate was 77% for the magnetic nanoparticles only group, 80% for the patent blue dye only group, and 90% for the magnetic nanoparticles and patent blue dye combination group. There were no false-negative cases. SLNB using magnetic tracer has been previously reported. However, in this study, the detection of magnetized SLN was attempted by sending magnetic nanoparticles to SLN and applying the magnetic field there. This enabled the identification of lymph nodes in even deeper positions than was previously possible. The detection rate was not as high as that of the RI method, even considering the stability of the equipment and unfamiliarity with the technology. Nevertheless, the results strongly suggested that the usefulness of the magnetic method was promising and that SLNB could likely be performed in facilities without nuclear medical equipment.

The purpose of the next clinical study was to compare and evaluate the identification rate of sentinel lymph nodes. In this study, SLNB was performed using magnetic nanoparticles and magnetic probes, and gamma probes were also used simultaneously [28]. Fourth-generation magnetic probes were used in this clinical trial. On the day before the operation, ferucarbotran was administered to the areola subcutaneously like the technetium-phytic acid tracer for increasing the accumulation rate at the lymph nodes. On day 1 before the operation, Tc-phytic acid together with 1 mL of ferucarbotran was injected to the areola subcutaneously. The following day, the positions of the lymph nodes were detected using a magnetic probe on the body surface and the respective locations were identified. After skin marking and incision, the SLN was detected and excised. Simultaneously, the RI values were measured with a radiation detector. Considering the cases after equipment improvement and procedure changes, the identification rate was 94.1% (80 out of 85 cases). Although the corresponding identification rate using conventional RI was 97.1%, the results of the magnetic method were considered favorable. Moreover, no adverse events due to the administration were observed, suggesting that the method is safe.

Many modern advanced medical devices tend to be more expensive due to their advanced functionality. The introductory price of such equipment in the hospitals is included in the treatment cost of the patients, and this is a factor in the high cost of treatment. Consequently, the expensive treatment costs associated with advanced medical care have a significant impact on the financial pressure of health insurance. Especially in Japan and the developed countries, where the birthrate is declining and the population is aging, it is the most important social problem that needs to be considered and solved as soon as possible, as it creates a huge burden on the younger generation. Thus, the development and deployment of inexpensive devices are very important and produce a great ripple effect on the society. Accordingly, the worldwide spread of SLNB using magnetic nanoparticles will increase the number of hospitals that can perform breast cancer operations, which can currently be performed only in large hospitals having all the required radiation control facilities. This will help ease the congestion and prolonged waiting time caused by the concentration of cancer patients in one facility.

The development of medical devices that are minimally invasive is currently a major challenge. As described above, the intraoperative diagnosis of infiltration and metastasis is important for improving the survival rate. SLNB provides an indication of the need for lymph node dissection and has significant consequences in choosing a suitable treatment. Accordingly, in addition to breast cancer, it is being introduced in other cancer operations where lymph node dissection cannot be avoided, such as head and neck cancer, gastric cancer, endometrial cancer, and cervical cancer.

The transition to endoscopy and laparoscopy is progressing in operations for gastric cancer, endometrial cancer, and cervical cancer, but the introduction of SLNB has been delayed. One factor behind such status is that the identification of SLN is difficult and the procedure does not lead favorably to the avoidance of lymph node dissection. Moreover, guidelines for cancers, such as gastric cancer and endometrial cancer, prescribe lymph node dissection even for early-stage cancers, which significantly degrades the patient's QoL. Accordingly, in such medical situations, the need for introducing magnetic methods is increasing.

8.6   MAGNETIC IMMUNOSTAINING FOR INTRAOPERATIVE DIAGNOSIS OF CANCER METASTASIS

The identified and excised SLNs are sent to the pathological diagnosis department to produce frozen sections. The presence or absence of metastasis is diagnosed by a pathologist using hematoxylin-eosin (HE) staining of these sections. The results determine whether to dissect the downstream lymph nodes [29, 30]. As the HE samples are subjected mainly to morphological observation of cells and nuclei, diagnosis may be difficult depending on the nature of the cancer or if the extent of metastasis is small. In actual clinical practice, an average wait time of approximately one hour is needed, including preparation and diagnosis. Moreover, immunostaining is performed for definitive diagnosis, but it is almost impossible to perform conventional immunostaining during the surgery.

Fluorescent magnetic nanoparticles that can be magnetically accumulated and that emit fluorescent light have been developed for use in immunostaining, with the fluorescent molecules infiltrating into the dispersible polymer-coated ferrite particles, which have been developed as carriers for protein purification [31, 32]. The surface of the particle is coated with a single surface-modifiable organic polymer, and thus various ligands, such as low-molecular-weight compounds, proteins, and nucleic acids can be immobilized. Immobilizing the antibody to fluorescent magnetic nanoparticles enables using it as a detection probe in sandwich immunoassays. Through incorporating a magnetic accumulation process, it was possible to complete the assay in only a few minutes. Thus, it was possible to reduce the assay time dramatically compared with conventional immunoassays and develop a technology enabling a rapid diagnosis of diseases. Rapid immunostaining methods have also been developed to identify cancer cells with high accuracy by producing fluorescent magnetic nanoparticles tagged with antibodies that recognize cancer cells expressing specific antigens. The frequency of reaction with antigens in cancer cells is increased to complete the reaction in less than a minute by increasing the concentration of the antibody through an accumulation of particles on the surface of the lymph node sample by applying magnetic force. It is possible to implement a system that enables intraoperative diagnosis of tumor metastasis using this method. According to this method, even the cancer cells difficult to detect using conventional HE staining can be stained using specific antibodies.

8.7   DESIGN OF A PERMANENT MAGNET FOR IMMUNOSTAINING BASED ON DYNAMICS OF MAGNETIC BEADS

It is important to consider the shape of the magnet used in the staining process to perform a proper and quick diagnosis. Rapid staining and diagnosis can be performed by magnetizing magnetic beads to produce a sufficiently strong magnetic force [33, 34]. Moreover, it is possible to reduce the risk of uneven staining and misdiagnosis by spatially attracting beads uniformly and maintaining the uniformity of beads that reach the tissue during the staining time. There are variations in the size of the sample sections and the position of the sections placed on the slide glass. Accordingly, in magnetic immunostaining, it is necessary to use magnets that generate a uniform and strong magnetic force over the entire slide glass.

A program that can quantitatively evaluate uniformity of nanoparticles in each elapsed time using the motion analysis of the trajectory of beads and a dome-shaped magnet obtained by exploring the best shape of magnet for immunostaining have been developed. Two programs were developed to analyze the motion caused by attracting magnetic beads using permanent magnets and to evaluate their spatial uniformity. One program was intended for setting an arbitrary magnet shape and calculating the magnetic flux density generated from it and the intensity and direction of the magnetic force on the magnetic beads. The other program was intended for analyzing the falling motion of magnetic beads in a staining solution and evaluating the uniformity within the staining time.

First, analysis was performed on a typical cylindrical magnet shape with a diameter of 24 mm. The magnetic force generated in cylindrical magnets was relatively high because the gradient of the magnetic flux density was larger near the edge than at the center of the magnet. It was observed that the direction of magnetic force is created to induce gathering at the edge of the magnet. Moreover, for the beads near the edge of the magnet to reach the sample surface quickly, while the distribution of beads eventually became sufficiently uniform, spatial-non-uniformity did occur during the staining.

A dome-shaped magnet, formed by excluding the corner portions from a cylindrical magnet, has been proposed for attracting beads uniformly without lowering the staining efficiency. It is possible to increase the staining uniformity by preventing the beads from gathering at the edges of the magnet and by creating a gradient of magnetic flux density at the center of the magnet. The aim of our study was to determine the optimal magnet shape by brute-force examination of the curve monotonically reducing along the r-z plane and evaluating the features of dome-shaped magnets of various shapes. Through verifying 3016 pieces of magnets, it was observed that it was optimal to perform staining using a 30-mm-diameter magnet having a dome curve at the top, with a radius of curvature of 50 mm, placed at 1.5 mm from the tissue section. The uniformity was better than that of cylindrical magnets of the same diameter, and a stronger magnetic force could be produced at the center of the magnet. The results demonstrate that, compared with the cylindrical magnet, for the dome-shaped magnet, the bias of the bead distribution during the staining time was small and it was possible to perform staining of the tissue more uniformly. Magnetic forces near the center were stronger than those in cylindrical magnets due to the generation of magnetic flux density gradients. Moreover, the magnetic force was sufficiently large, as all the magnetic beads settled on the tissue surface within 60 s.

8.8 AUTOMATIC MAGNETIC IMMUNOSTAINING SYSTEM

A staining device is also being developed to perform the magnetic immunostaining process automatically in a short time period [35]. To develop the device specifications, immunostaining protocols were carefully investigated, and the elemental technologies to be incorporated into the equipment, such as the shape of the magnet suitable for immunostaining using magnetic accumulation, bead concentration and antibody immobilization amount that enables high-precision immunostaining, immune response conditions, washing conditions, and shape of the equipment, were extracted. Moreover, the specifications of the rapid immunostaining equipment to be developed, such as the number of simultaneously processed samples, reaction liquid volume, cleaning liquid volume, liquid feed method, method of introducing antibody-tagged fluorescent magnetic beads, waste liquid discharge method, shape of magnet, position setting, and range of motion, were decided by considering the opinions of pathologists.

FIGURE 8.7 A prototype of the automatic magnetic immunostaining device.

A tabletop size device that can process at least five samples at a time was conceived, based on the opinion that approximately five sentinel lymph nodes per patient are excised and that examining them simultaneously is desirable. A holder was designed for gripping the slide glass with a sample on it. The extent of placement area for samples on the slide glass was set as 32 mm×22 mm in accordance with the width of the slide glass. For the reaction liquid and the waste liquid, the required minimum amount of formalin for fixing and magnetic beads were used. The process was set such that immunostaining is completed within 10 min of bead injection into the sample.

The unit for injecting beads and chemical solution, the immunostaining unit, and the cleaning unit were fabricated separately. Although all these units were independent, it was possible to automate the rapid magnetic immunostaining scheme established by this method. The prototype of the tabletop type rapid immunostaining device was fabricated by integrating all these units (Figure 8.7).

Biological cancer tissue samples were immunostained using the finished prototype. The samples were obtained using nude mice injected with A431 (epithelioid cells, epidermal growth factor receptor [EGFR] high expression strain) or H69 (lung cancer cells, EGFR non-expression strain). When the size of the tumor was approximately 5 mm in diameter, it was excised to prepare the frozen sections. While beads were bound to cancerous tissues expressed by EGFR, no bead binding was observed in tissues where EGFR was not expressed. The staining process was completed in one minute, and the concept of reducing the reaction time using magnetism was demonstrated. It was also demonstrated that all the processes worked well under the conditions set. Additionally, it was verified that the background fluorescence was maintained at a sufficiently practical low level, that the antibodies had sufficient selectivity, and that the washing conditions were appropriate.

8.9 FUTURE PROSPECTS

If the developed method of combining sentinel lymph node biopsy using magnetic sensors and rapid intraoperative diagnosis is put to practical use, the accuracy of metastasis diagnosis will improve in small- and medium-sized medical institutions that can handle many cases. If intraoperative immunostaining can be realized, it can likely reduce the risk of oversight in conventional rapid pathological diagnosis, and the recurrence rate can be reduced. Moreover, promoting the expansion of facilities where operations can be performed will enable early treatment, improve the QoL of patients, and help reduce the physical, economical, and mental burdens. This method is not limited to breast cancer and it has the potential to be applied to a wide range of cancer cases. Furthermore, theoretically, it is not limited to lymph node metastasis, but can also be applied to lung metastasis and bone metastasis, and it is expected that it will be incorporated in procedures for these cases as well. Socially, it can be promising for reducing medical expenses, extending the survival period of patients, and increasing productive labor time.

Especially in developing countries, where many medical institutions do not have nuclear medicine equipment, establishing the method using magnetic techniques will enable rapid and accurate sentinel lymph node biopsy. Breast cancer is on the rise even in emerging countries, where it is more difficult to maintain nuclear medicine facilities. In China, only approximately 5% of the patients can undergo such minimally invasive operations, and the percentage is even lower in other developing countries. It is believed that this technology, which does not require nuclear medicine facilities, will increase the demand economically in the global market.

REFERENCES

1. Krag D., D. Weaver, J. Alex, and J. Fairbank: Surgical resection and radiolocalization of the sentinel lymph node in breast cancer using a gamma probe. *Surgical Oncology* 2(6): 335–9, 1993.
2. Ahmed M., A. Purushotham, and M. Douek: Novel techniques for sentinel lymph node biopsy in breast cancer: A systematic review. *Lancet Oncology* 15(8): e351–62, 2014.
3. Kim T., A. Giuliano, and G. Lyman: Lymphatic mapping and sentinel lymph node biopsy in early-stage breast carcinoma: A metaanalysis. *Cancer* 106(1): 4–16, 2006.
4. Pesek S., T. Ashikaga, L. Krag, and D. Krag: The false-negative rate of sentinel node biopsy in patients with breast cancer: A meta-analysis. *World Journal of Surgery* 36(9): 2239–51, 2012.
5. Schrenk P., R. Rieger, A. Shamiyeh, and W. Wayand: Morbidity following sentinel lymph node biopsy versus axillary lymph node dissection for patients with breast carcinoma. *Cancer* 88(3): 608–14, 2000.
6. Swenson K., M. Nissen, C. Ceronsky, L. Swenson, M. W. Lee, and T. M. Tuttle: Comparison of side effects between sentinel lymph node and axillary lymph node dissection for breast cancer. *Annals of Surgical Oncology* 9(8): 745–53, 2002.
7. Fleissig A., L. Fallowfield, C. Langridge, L. Johnson, R. G. Newcombe, J. M. Dixon, M. Kissin, and R. E. Mansel: Post-operative arm morbidity and quality of life. Results of the ALMANAC randomised trial comparing sentinel node biopsy with standard axillary treatment in the management of patients with early breast cancer. *Breast Cancer Research & Treatment* 95(3): 279–93, 2006.
8. https://ganjoho.jp/en/professional/statistics/brochure/2017_en.html.
9. Kusakabe M., K. Taruno, T. Kurita, H. Takei, S. Nakamura, and A. Kuwahata: Sentinel lymph node biopsy using magnetic nanoparticles and magnetic probe. *Magnetics Japan* 13(4): 174–80, 2018.
10. Sekino Masaki and Moriaki Kusakabe: Portable magnetic probe for detecting magnetic nanoparticles inside the body. In Sandhu, A. and Handa, H. (eds.) *Magnetic Nanoparticles for Medical Diagnostics*, 1–19. Bristol: IOP Publishing, 2018.
11. Mansel R., L. Fallowfield, M. Kissin, A. Goyal, R. G. Newcombe, J. M. Dixon, C. Yiangou, K. Horgan, N. Bundred, I. Monypenny, and D. England: Randomized multicenter trial of sentinel node biopsy versus standard axillary treatment in operable breast cancer: The ALMANAC Trial. *Journal of the National Cancer Institute* 98(9): 599–609, 2006.
12. Straver M., P. Meijnen, G. van Tienhoven, C. J. van de Velde, R. E. Mansel, J. Bogaerts, N. Duez, L. Cataliotti, J. H. Klinkenbijl, H. A. Westenberg, and H. van der Mijle: Sentinel node identification rate and nodal involvement in the EORTC 10981-22023 AMAROS Trial. *Annals of Surgical Oncology* 17(7): 1854–61, 2010.
13. Karakatsanis A., P. Christiansen, L. Fischer, C. Hedin, L. Pistioli, M. Sund, N. R. Rasmussen, H. Jørnsgård, D. Tegnelius, S. Eriksson, and K. Daskalakis: The Nordic SentiMag trial: A comparison of super paramagnetic iron oxide (SPIO) nanoparticles versus Tc(99) and patent blue in the detection of sentinel node (SN) in patients with breast cancer and a meta-analysis of earlier studies. *Breast Cancer Research & Treatment* 157(2): 281–94, 2016.

14. Thill M., A. Kurylcio, R. Welter, V. van Haasteren, B. Grosse, G. Berclaz, W. Polkowski, and N. Hauser: The Central-European SentiMag study: Sentinel lymph node biopsy with superparamagnetic iron oxide (SPIO) vs. radioisotope. *Breast* 23(2): 175–9, 2014.

15. Houpeau J., M. Chauvet, F. Guillemin, C. Bendavid-Athias, H. Charitansky, A. Kramar, and S. Giard: Sentinel lymph node identification using superparamagnetic iron oxide particles versus radioisotope: The French Sentimag feasibility trial. *Journal of Surgical Oncology* 113(5): 501–7, 2016.

16. Douek M., J. Klaase, I. Monypenny, A. Kothari, K. Zechmeister, D. Brown, L. Wyld, P. Drew, H. Garmo, O. Agbaje, and Q. Pankhurst: Sentinel node biopsy using a magnetic tracer versus standard technique: The SentiMAG Multicentre Trial. *Annals of Surgical Oncology* 21(4): 1237–45, 2014.

17. Zada A., M. Peek, M. Ahmed, B. Anninga, R. Baker, M. Kusakabe, M. Sekino, J. M. Klaase, B. Ten Haken, and M. Douek: Meta-analysis of sentinel lymph node biopsy in breast cancer using the magnetic technique. *The British Journal of Surgery* 103(11): 1409–19, 2016.

18. Ahmed M. and M. Douek: The role of magnetic nanoparticles in the localization and treatment of breast cancer. *BioMed Research International* 2013: 281230, 2013.

19. Kuwahata A., M. Ahmed, K. Saeki, S. Chikaki, M. Kaneko, W. Qiu, Z. Xin, S. Yamaguchi, A. Kaneko, M. Douek, M. kusakabe, and M. Sekino: Combined use of fluorescence with a magnetic tracer and dilution effect upon sentinel node localization in a murine model. *International Journal of Nanomedicine* 13: 2427, 2018.

20. Sekino M., A. Kuwahata, T. Ookubo, M. Shiozawa, K. Ohashi, M. Kaneko, I. Saito, Y. Inoue, H. Ohsaki, H. Takei, and M. Kusakabe: Handheld magnetic probe with permanent magnet and Hall sensor for identifying sentinel lymph nodes in breast cancer patients. *Scientific Reports* 8(1): 1195, 2018.

21. Ookubo T., Y. Inoue, D. Kim, H. Ohsaki, Y. Mashiko, M. Kusakabe, and M. Sekino: Development of a probe for detecting magnetic fluid in lymph nodes. *Electronics & Communications in Japan* 99(3): 13–21, 2016.

22. Kuwahata A., S. Chikaki, A. Ergin, M. Kaneko, M. Kusakabe, and M. Sekino: Three-dimensional sensitivity mapping of a handheld magnetic probe for sentinel lymph node biopsy. *AIP Advances* 7(5): 056720, 2017.

23. Kaneko M., K. Ohashi, S. Chikaki, A. Kuwahata, M. Shiozawa, M. Kusakabe, and M. Sekino: A magnetic probe equipped with small-tip permanent magnet for sentinel lymph node biopsy. *AIP Advances* 7(5): 056713, 2017.

24. http://mag-sense.com/.

25. Hirota, A. and M. Kusakabe: JP2008206578A, Detector and detecting device of magnetic fluid, 2007.

26. Kuwahata A., M. Kaneko, S. Chikaki, M. Kusakabe, and M. Sekino: Development of device for quantifying magnetic nanoparticle tracers accumulating in sentinel lymph nodes. *AIP Advances* 8(5): 056713, 2018.

27. Shiozawa M., A. T. Lefor, Y. Hozumi, K. Kurihara, N. Sata, Y. Yasuda, and M. Kusakabe: Sentinel lymph node biopsy in patients with breast cancer using superparamagnetic iron oxide and a magnetometer. *Breast Cancer (Tokyo, Japan)* 20(3): 223–9, 2013.

28. Taruno Kanae, Tomoko Kurita, Akihiko Kuwahata, Keiko Yanagihara, Katsutoshi Enokido, Yoshihisa Katayose, Seigo Nakamura, Hiroyuki Takei, Masaki Sekino, and Moriaki Kusakabe: Multicenter clinical trial on sentinel lymph node biopsy using superparamagnetic iron oxide nanoparticles (SPIO) and a novel handheld magnetic probe. *Journal of Surgical Oncology* 120(8): 1391–96, 2019.

29. Bonacho T., F. Rodrigues, and J. Liberal: Immunohistochemistry for diagnosis and prognosis of breast cancer: A review. *Biotechnic & Histochemistry* 10: 1, 2019.

30. Osamura R. Y., N. Matsui, M. Okubo, L. Chen, and A. S. Field: Histopathology and cytopathology of neuroendocrine tumors and carcinomas of the breast: A review. *Acta Cytologica* 63(4): 340, 2019.

31. Sakamoto S., K. Omagari, Y. Kita, Y. Mochizuki, Y. Naito, S. Kawata, S. Matsuda, O. Itano, H. Jinno, H. Takeuchi, Y. Yamaguchi, Y. Kitagawa, and H. Handa: Magnetically promoted rapid immunoreactions using functionalized fluorescent magnetic beads: A proof of principle. *Clinical Chemistry* 60(4): 610–20, 2014.

32. Onishi T., S. Matsuda, Y. Nakamura, J. Kuramoto, A. Tsuruma, S. Sakamoto, S. Suzuki, D. Fuchimoto, A. Onishi, S. Chikaki, and M. Kaneko: Magnetically promoted rapid immunofluorescence staining for frozen tissue sections. *Journal of Histochemistry & Cytochemistry* 67(8): 575–87, 2019.

33. Kaneko M., S. Chikaki, S. Matsuda, A. Kuwahata, M. Namita, I. Saito, S. Sakamoto, M. Kusakabe, and M. Sekino: Development of magnet configurations for magnetic immunostaining. *AIP Advances* 8(5): 056732, 2018.

34. Sekino Masaki, Akihiro Kuwahata, Shunsuke Fujita, Sachiko Matsuda, Miki Kaneko, Shinichi Chikaki, Satoshi Sakamoto, Itsuro Saito, Hiroshi Handa, and Moriaki Kusakabe: Development of an optimized dome-shaped magnet for magnetically-promoted rapid immunostaining. *AIP Advances* 10: 025317, 2020.
35. Sekino Masaki, Akihiro Kuwahata, Akinobu Yoshibe, Kiyotaka Imai, Miki Kaneko, Shinichi Chikaki, Itsuro Saito, Akinori Tsuruma, Satoshi Sakamoto, Hiroshi Handa, Sachiko Matsuda, and Moriaki Kusakabe: Development of an automatic magnetic immunostaining system for rapid intraoperative diagnosis of cancer metastasis. *AIP Advances* 10: 015106, 2020.

9 Optimizing Reporter Gene Expression for Molecular Magnetic Resonance Imaging
Lessons from the Magnetosome

Qin Sun, Frank S. Prato, and Donna E. Goldhawk

CONTENTS

9.1 Introduction 201
9.2 Defining Reporter Gene Expression...202
    9.2.1 What Is an Imaging Reporter? 202
    9.2.2 Why Search for an MRI Reporter?..203
    9.2.3 How Do Magnetotactic Bacteria Acquire Their Magnetic Properties? 203
9.3 Understanding the Magnetosome 204
    9.3.1 Stepwise Assembly 204
    9.3.2 Designating the Vesicle 205
    9.3.3 Building the Iron Biomineral 206
9.4 Detecting Reporter Gene Expression with MRI 207
    9.4.1 Imaging the Magnetosome with MRI 207
    9.4.2 Principles of the MRI Signal..207
    9.4.3 Paramagnetic Contrast Agents 208
    9.4.4 Ferromagnetic Contrast Agents..208
    9.4.5 MR Imaging of Magnetosomes...208
9.5 Conclusion 209
References..210

9.1 INTRODUCTION

In magnetic resonance imaging (MRI), methods for the non-invasive detection of cells and their activities often rely on the manipulation of cellular iron. In many aspects, this is ideal. Iron perturbs the magnetic resonance (MR) signal and thus influences measurable MR parameters. All cells, and therefore tissues, require an iron co-factor for various functions, thus iron-related MR measures inherently target integral cellular activities. In addition, differential regulation of iron enhances the distinction between tissue and cell type(s), which enables MRI to non-invasively monitor changes in iron-related organ function. For example, after a hemorrhagic heart attack, cardiac MR can quantitate the iron that often accumulates in infarcted tissue and predisposes a patient to heart failure.[1] In addition, the inflammatory component of heart disease involves immune cell responses that target the iron export activity of monocytes[2] and macrophages.[3,4]

One of the great challenges in MRI is development of sensitive, iron-related measures that complement the exquisite spatial resolution of this modality. There has been short-term success with a variety of superparamagnetic iron oxide (SPIO) nanoparticles that temporarily boost the cellular MR signal.[5] These advancements in exogenously introduced iron contrast have demonstrated the power and feasibility of tracking cells. However, reporting cellular activity throughout the cell's life cycle means tapping into the endogenous workings of the cell. New approaches are required for this type of molecular imaging. The use of molecular cloning techniques, to overexpress iron-handling protein(s) that amplify MR contrast, has opened new avenues in gene-based iron-labeling.[6] Genetically engineered cells clearly possess the ability to express a long-lived signal, suitable for preclinical imaging.[7]

While the search continues for the most efficient expression system(s), the utility of reporter genes for *in vivo* imaging by MRI and other modalities is increasingly recognized. With this type of endogenous contrast, we can examine the regulation of cellular activities that predict disease and offer therapeutic solutions. To realize the full potential of this technology for MRI, we have drawn on the best-known example of efficient, gene-based iron-labeling in single cells: the magnetosome found in magnetotactic bacteria (MTB).[8] The manner of iron compartmentalization in the magnetosome not only reflects fine-tuned regulation of iron, which is a necessary feature of iron metabolism in virtually every cell type[9] but also provides the means by which a cell's magnetic properties may be altered and potentially exploited to provide unique MRI signatures. For example, iron-containing abdominal pathologies are often detected using MRI.[10]

In this chapter, we review concepts in reporter gene expression for molecular MRI. This includes the regulation of iron compartmentalization and metabolism in mammalian cells, and the influence this has on MR detection of iron contrast. In addition, we examine the synthesis of magnetosomes and propose strategies, inspired by this bacterial structure, for improving the detection of reporter gene expression by MRI.

9.2 DEFINING REPORTER GENE EXPRESSION

Genetic encoding of reporter genes enables the tracking of several features of cellular activity, regardless of platform used to detect the encoded proteins. For example, although all somatic cells contain the full complement of an organism's DNA (the genotype), not all of these genes are expressed at once. Each cell type regulates the expression of individual genes to obtain specific cell functionality (the phenotype). The transcription factors that regulate gene expression thus perform a key role in determining what function a cell has, for example, cardiac, hepatic, neural, or other activity. The same transcription factor (TF) regulation applies to reporter genes that are introduced for the purpose of monitoring cellular activity.

Broadly speaking, there are two types of TF regulation: constitutive and selective. On the one hand, genes that are expressed constitutively are generally required for cell functions that operate continuously, like glycolysis or oxidative phosphorylation. The "housekeeping" factors that control these genes are virtually always present and ready to drive the required gene expression to preserve baseline functioning in most cell types. On the other hand, selective TF activity regulates specific gene expression that distinguishes one cell type from another. This category of TF is therefore characteristic of specific cell type(s), microenvironment, stage of development, health, and disease. Every cell contains both types of TF.

Simply stated, a reporter gene encodes protein that can be reliably detected by some method of choice. If the reporter gene is constitutively expressed, it generates a constant label that can theoretically be monitored through all phases of the cell, including its proliferation and differentiation. If the reporter gene is selectively expressed, it imparts a signal that will only be detected when specific TF activity is present, including those that trigger a disease process like inflammation, fibrosis, or metastasis. Traditionally, "reporter gene expression" is a term coined to describe this type of selective gene expression. It is a powerful tool for understanding cellular phenotype and identifying what factors contribute to function and malfunction.

9.2.1 WHAT IS AN IMAGING REPORTER?

Most cells do not have sufficient, inherent contrast to enable *in vivo* molecular imaging by any modality. With little exception (refer to Section 9.4), an imaging reporter is needed to distinguish background signal from cellular target. Reporter gene technology not only serves this purpose but will outperform exogenous contrast with respect to duration of signal and relationship to cellular molecular activity.[11–13] What remains is the challenge of developing imaging reporter genes with high enough contrast to noise ratio (CNR) for satisfactory molecular imaging in living subjects.

This challenge has been met with varying levels of success, depending on the imaging modality.[8,14,15] Toward applications in cell therapy, several methods are available to quantitatively image transplanted cells using reporter genes like thymidine kinase for nuclear medicine or luciferase for optical imaging. However, the use of reporter genes in nuclear medicine requires injection of radioactive substrate and creates a massive background signal, often obscuring the desired signal. In optical techniques, poor depth of penetration requires either the insertion of invasive light detectors and/or surgical exposure of the targeted area.

Although transforming cells with foreign DNA for imaging with reporter genes is not yet readily accepted by regulatory bodies for human use, the preclinical market for this technology is huge and will inform medical practices. For example, gas vesicles (bacterial structures that provide buoyancy) have been expressed in both human cells and *E. coli* to develop ultrasound acoustic reporter gene imaging.[16,17] Similarly, to develop reporter gene imaging for MRI, various MTB transgene expression systems have been developed.[18–22]

9.2.2 Why Search for an MRI Reporter?

Of all the medical imaging platforms, MRI provides the best resolution of soft tissues at any imaging depth using non-ionizing radiation. Already a mainstream clinical imaging platform, what MRI lacks in sensitivity can be compensated for by contrast agents that either amplify the MR signal, like hyperpolarized compounds, or modify the signal, like iron, which amplifies differences in MR signals. In addition, many MRI sequences are available to capture different aspects of the biophysical properties that govern a cell's response to magnetism. Beyond the standard longitudinal (T1) and transverse (T2, T2*) relaxation times of water protons, methods have been developed to monitor changes in sodium, pH, oxygenation of hemoglobin (allowing detection of neuronal activity in the brain[23] and tissue oxygenation in the heart[24]), blood flow in the brain,[25] and diffusion,[26] to name a few. Extensive opportunities for multi-parametric imaging, with or without the addition of contrast agents, mean that scientific breakthroughs facilitated by using MRI afford excellent value for the investment in MR research, including technical improvements in hardware and software by technologists and biomedical personnel. Moreover, the movement toward hybrid MR scanners, like positron emission tomography (PET)/MRI, creates further options for comprehensive imaging with complete registration of signals from multiple organ systems, reducing the negative impact of motion on image interpretation, adding sensitivity of a radiolabel to the MR image, and supporting the acquisition of multiple signals within a single imaging session. In addition, because MR image acquisition is a relatively slow process, when imaging with PET/MRI over a longer time, the concentration of radioisotope can be reduced and the average PET signal collected over that longer time.

In view of these MRI attributes, a reporter gene expression system that enhances T1- or T2-weighted contrast for molecular MRI opens another phase in the evolution of this modality. By drawing on examples in nature of cells that generate their own endogenous magnetism, we have developed a template for further exploration of MRI reporter gene technology. The lesson plan comes from a group of microorganisms called magnetotactic bacteria (MTB) and provides a means to synthesize and regulate iron biominerals with superparamagnetic qualities but no cytotoxicity.

9.2.3 How Do Magnetotactic Bacteria Acquire Their Magnetic Properties?

While iron is a required co-factor in all cells, few if any generate as refined an iron biomineral as found in MTB. These prokaryotes synthesize particles of magnetite (Fe_3O_4) or greigite (Fe_3S_4)

within a membrane-enclosed vesicle called a magnetosome.[27] Protected from iron toxicity by the lipid bilayer, MTB use their magnetosomes to respond to the earth's magnetic field and display magnetotaxis appropriate to Northern and Southern hemispheres.[28,29] A growing body of literature has examined multiple applications of MTB and magnetosomes,[30-32] taking advantage of their navigational properties in response to magnetic fields,[33,34] their ability to deliver thermal therapy,[35-37] and their precise synthesis of crystalline particles,[38-41] which, for example, could be useful as a reporter probe in magnetic particle imaging.[8] In the following section, we describe the application of magnetosomes to MRI reporter gene expression.

9.3   UNDERSTANDING THE MAGNETOSOME

Magnetosome synthesis is a protein-directed process, beginning with expression of structural genes that encode the required magnetosome components. While the nature of these components is still incompletely understood, progress has been made in many areas. The genomes of numerous MTB have now been sequenced, permitting comparison of conserved gene sequences and synteny.[42,43] Despite the breadth of MTB species, there are common genes that specify the main magnetosome structure and approximately two-thirds of these are clustered on a magnetosome genomic island.[44] Removal of this cluster of DNA prevents magnetotaxis in the microorganism but is not lethal, demonstrating that magnetosomes likely confer selective advantage(s) rather than compulsory function(s).

Studies examining specific gene deletions have further categorized which magnetosome gene products are essential to the structure versus auxiliary.[43,45] That is, some proteins regulate critical steps in magnetosome formation while others control optional features like subcellular arrangement and crystal morphology. These findings raise the possibility that certain magnetosome genes may be used as regulatory agents in the design and synthesis of MR contrast for reporter gene expression. As the magnetosome compartment becomes more fully understood, it should also be feasible to genetically program various features of the magnetosome, including variations in the degree and type of biomineralization and strength of magnetism, all of which impinge on the MR signal.[5] Manipulation of these features is thus compatible with the notion that magnetosome-like nanoparticles in any cell type may be tailored for a desired purpose and not restricted to magnetotaxis.

9.3.1   STEPWISE ASSEMBLY

The current body of magnetosome knowledge supports a model of biosynthesis in which there is a hierarchy of protein expression dictating which components create the framework upon which the iron biomineral is compartmentalized, shaped, and localized within the cell. This also implies that no one protein recapitulates the entire structure; rather, its assembly depends on multiple protein–protein interactions.[46,47] Although this design is more complex than previously acknowledged,[18] the elaborate nature of magnetosome nanoparticles provides a wealth of information about how to program a vesicle for any purpose, including iron biomineralization for MRI reporter gene expression. Complexity in design also explains why single gene expression systems have fallen short of the robust MR signals obtained with intact MTB or isolated magnetosomes and their variants.[48] It is the sum of individual steps in magnetosome formation that creates a compartment housing a well-defined crystal. Understanding how these steps work together to produce the magnetosome will enhance how we utilize this structure for molecular MRI among other applications.

Magnetosome biosynthesis begins with vesicle formation, followed by the initiation of iron biomineralization and then maturation of the crystal.[49] The molecular instructions that prescribe vesicle formation must subsequently be linked to biomineralization and candidate proteins involved in rendering these activities are steadily being described.[44] We propose that a subset of these may be sufficient for the formation of a rudimentary magnetosome-like nanoparticle in mammalian cells. For this, we envision that designation of the vesicle membrane involves establishing the point of

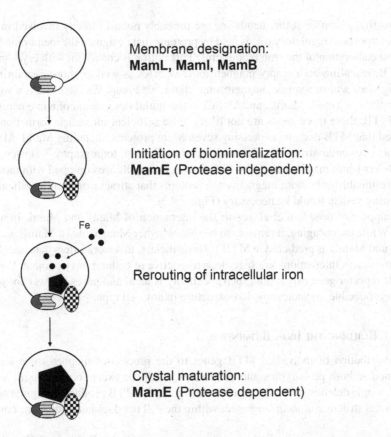

Membrane designation:
MamL, MamI, MamB

Initiation of biomineralization:
MamE (Protease independent)

Rerouting of intracellular iron

Crystal maturation:
MamE (Protease dependent)

FIGURE 9.1 Modeling the roles of essential magnetosome proteins in stepwise formation of the magnetosome. First, the proteins MamL (dark gray), MamI (horizontal dashed), and MamB (white) assemble at the membrane, designating the site for magnetosome formation (solid black circle). Subsequently, MamE (checkered) is recruited to the nascent magnetosome vesicle to initiate iron biomineralization. This step proceeds independent of MamE protease activity.[49] Then, intracellular iron is rerouted to the rudimentary magnetosome compartment, permitting magnetite crystal formation. This enables maturation of the iron biomineral in a step that requires the proteolytic activity of MamE.[49]

contact for initiation of iron biomineralization. This docking site would then constitute the critical protein–protein interactions required for the first step(s) in magnetosome formation. The schematic in Figure 9.1 predicts one potential sequence of interactions based on the essential roles of magnetosome membrane (Mam) proteins I, L, B, and E [45].

9.3.2 DESIGNATING THE VESICLE

Since most magnetosome proteins are membrane-bound,[45] their expression in any cell type involves a lipid bilayer. In prokaryotes, this implicates the cell membrane; however, in eukaryotes, integral membrane proteins are routed through vesicles emanating from the Golgi apparatus and will remain within the cytoplasmic compartment[20] unless directed otherwise by specific membrane localization signals. Regardless of cell type, iron biomineralization is confined to a membrane-enclosed compartment and this is an important protective mechanism, built into the magnetosome blueprint, for avoiding iron cytotoxicity. In species of *Magnetospirillum*, the absence of MamI, MamL, or MamB produces neither magnetosome vesicle nor iron biomineral,[43] establishing the importance of these proteins in key steps of magnetosome synthesis. In contrast, deletion of MamE interferes with biomineralization but still permits vesicle formation. Based on these results, magnetosome proteins that assemble at the membrane are probably not all directly involved in iron-handling. A no less important, regulatory role is held by proteins that designate the membrane location for initiation and elaboration of the iron biomineral. This notion is consistent with cryotomographic images of MTB, revealing both empty magnetosome vesicles as well as those containing biominerals of varying sizes within a single magnetosome chain.[49,50] Figure 9.1 therefore is a working hypothesis that implicates MamI, MamL, and MamB in the initial designation of a magnetosome membrane.

In MTB, these three genes are not likely to be sufficient for vesicle formation. Raschdorf et al. reported that MTB mutants expressing seven Mam proteins, including MamI, MamL, and MamB, will form vesicular structures identified by cryo-electron tomography.[51] However, in mammalian cells, fewer genes may be necessary. Since eukaryotic cells are equipped with intracellular vesicles, then presumably only those magnetosome proteins that attract iron biomineralization function(s) to an existing vesicle would be necessary (Figure 9.2).

In support of this, Sun et al. report the interaction of MamI and MamL in a mammalian cell line.[53] While encouraging, it remains to be seen whether MamB and/or MamE will co-localize with MamI and MamL as predicted in MTB.[44] Nevertheless, this work infers that essential magnetosome protein–protein interactions are specific, irrespective of cellular environment. For the development of MRI reporter gene expression, such specificity is ideal and paves the way for genetically encoding a reproducible magnetosome-like structure in any cell type.[54,55]

9.3.3 BUILDING THE IRON BIOMINERAL

The contribution of individual MTB genes to the process of magnetosome formation has been examined in both prokaryotes and eukaryotes. With the exception of *mamB*, *mamE*, *mamI*, and *mamL*, single deletion of many magnetosome genes in MTB produces irregularities in the biomineral crystal structure or its arrangement within the cell but does not destroy the compartment. There

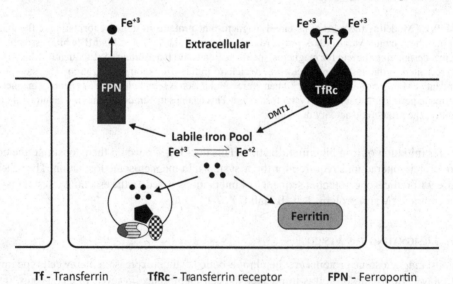

FIGURE 9.2 Modeling key features of mammalian iron regulation in cells producing magnetosome-like nanoparticles. Serum iron in its ferric state (Fe^{+3}) is mainly bound to transferrin (Tf) and taken up by cells through transferrin-receptor (TfRc)- mediated endocytosis. Within the intracellular compartment, ferric iron is reduced to its ferrous state (Fe^{+2}) and delivered to a redox-active, labile iron pool by divalent metal transporter 1 (DMT1).[52] From there, surplus iron is stored in ferritin when not needed for immediate cellular activities. In cells expressing magnetosome-like particles, iron is rerouted to this compartment by magnetosome proteins (Figure 9.1). Only select cells, including monocytes and macrophages, export iron through ferroportin (FPN).

may be redundant functions among magnetosome proteins or subtle changes in magnetosome function that are not readily detected using common microbiology measures, like C_{mag}[56,57] or a simple lab magnet.[58] If instead of responding to the geomagnetic field (25–65 μTesla[59]), the objective is an *in vivo* response to MRI at clinical field strengths on the order of 3 Tesla (3T), then cells bearing the same type of iron biomineral would respond differently to the change in external field strength. As discussed below (in Section 9.4), the interaction of iron biomineral with the magnetic field in turn has an effect on relaxation rates of the proton, which is observed with MRI. This influence of iron will vary depending on the oxidation state, size, and shape of the crystal. Hence, diverse MR signals could be genetically programmed by taking advantage of the remarkable number of magnetosome sizes and shapes evident in diverse species of MTB.

The early report of magnetosomes in *Anisonema* algae indicated that eukaryotic cells are capable of magnetosome synthesis.[60] Since then, expression of select MTB genes in mammalian cells has demonstrated compatibility of such transgene expression systems and enhancement of cellular iron stores, leading to MR contrast improvements. Figure 9.2 depicts key steps in the regulation of iron metabolism to provide some context for the biosynthesis of magnetosome-like nanoparticles in mammalian cells. Just like other iron-requiring cellular activities, the magnetosome-like compartment is shown drawing iron from the labile iron pool. In this way, iron homeostasis is maintained while a new magnetosome-like iron storage compartment is being synthesized. Using single MTB expression systems like *magA*,[61] the response of various features of mammalian iron homeostasis has been examined. In the iron-exporting P19 cell line, MagA expression augments total cellular iron content without altering its iron export function.[20] However, Guan et al. showed that level of transferrin receptor (TfRc) decreases and ferritin increases in MagA-expressing A549 lung cancer cells.[19] Even though further studies are needed to understand the full effect of MTB iron-handling protein on mammalian cell systems, these findings suggest that iron metabolism is compatible with MTB protein expression and any influence it may have on (re)routing intracellular iron. Even the magnetosome-associated magnetic particle membrane specific (Mms) protein, Mms6, has a measurable effect on the MR signal.[22, 62]

9.4 DETECTING REPORTER GENE EXPRESSION WITH MRI

9.4.1 Imaging the Magnetosome with MRI

In Chapter 4, Drs. Sekino and Ueno summarized the principles of [proton] MRI. Here, we provide a very brief summary of the source of the MRI signal and how it can be affected by paramagnetic and ferromagnetic contrast agents.[63] Then, we focus on how the MRI signal could be perturbed by magnetosome particles and estimate, based on the literature, the detection sensitivity of an MRI reporter that produces the same concentration of iron particles in mammalian cells as produced within MTB.

9.4.2 Principles of the MRI Signal

The MRI signal comes from protons (hydrogen nuclei) in the subject being imaged. As there are many protons in biological samples, the signal is strong and allows resolution of structures in small animals (e.g., rodents) at fractions of a mm and in larger animals (e.g., dogs, pigs) and humans at approximately 1 mm[3]. However, discrimination of different tissues by proton density alone is poor since the concentration of protons in soft biological tissues does not vary greatly. Instead, since protons in different tissues often have different spin relaxation behaviors, detection of these can improve tissue discrimination. As described elsewhere, both the application of radiofrequency (RF) pulses at specific frequencies and the switching of magnetic field gradients can be used to distinguish different proton relaxation rates in a biological sample. In principle, there are two main classes of relaxation effects which are called spin–lattice or longitudinal relaxation time (T1) and spin–spin or transverse relaxation time (T2), with the units of time usually expressed in seconds or milliseconds (s or ms). Often these relaxation times are converted to relaxation rates (R1 and R2) by inverting T1 and T2, resulting in units of inverse time, for example, s^{-1}. Furthermore, spin–spin relaxation is represented by two quantities: R2 and R2*, where R2* ≥ R2. While R2 is measured by using spin echoes, R2* is measured with gradient echoes. In addition, while R2 represents an intrinsic property of tissue, R2* is affected not only by tissue properties but also by non-uniformities in the local magnetic field. As such, R2* is very sensitive to the presence of iron, particularly in the form of magnetite. The difference between R2* and R2 is represented by R2′ and is often referred to as the non-recoverable spin–spin relaxation.

9.4.3   PARAMAGNETIC CONTRAST AGENTS

Innate relaxation behaviors of different tissues are often not sufficient to discriminate diseased from normal states within a given organ system (e.g., liver, heart); hence, paramagnetic MRI contrast agents have been developed. These agents provide additional discrimination where they accumulate differentially in a particular tissue. Paramagnetic agents primarily shorten T1 relaxation time of the protons (i.e., increase R1), with extent of shortening dependent on tissue concentration of the paramagnetic contrast agent. It is important to note that paramagnetic contrast agents, in general, do not alter the magnetic field at the tissue level. Rather, through dipole–dipole interactions, they directly increase the R1 relaxation rate of those water molecules that come in close contact with the paramagnetic centers.[64] In addition, paramagnetic agents have little effect on spin–spin relaxation; at all times, T2 ≤ T1. Although, when concentration of the paramagnetic contrast agent becomes high enough, T1 can be reduced to less than the tissue T2 value, at which point T2 will approximate T1. The relaxivity (R1) of paramagnetic contrast agents for a typical MRI in medical imaging (on 1.5T and 3T systems) is about 4 $(mM\ sec)^{-1}$,[65] and the sensitivity of detection is approximately 0.1 mM. To achieve this requires injection of approximately 0.1 mmole contrast agent per kg body weight. This relatively poor sensitivity means that tracking cells labeled with a paramagnetic contrast agent is limited, given the prohibitive amount of contrast agent that must be incorporated by the cell even for specifically designed gadolinium contrast agents with increased relaxivity.

9.4.4   FERROMAGNETIC CONTRAST AGENTS

Ferromagnetic contrast agents primarily affect spin–spin relaxation times and their T2 relaxivity is generally much greater than the T1 relaxivity of paramagnetic agents. For example, SPIO particles were one of the first iron particles to be introduced. Their R2 relaxivity is so high, they can be loaded into cells and tracked after transplantation with MRI. For example, if mammalian cells labeled with approximately 60 pg Fe/cell are injected in a mouse brain, then individual cells can be detected using 1.5T MRI with a specialized RF coil and an enhanced magnetic field gradient for small animal imaging.[66] In this work, Heyn et al. used a 3D Fast Imaging Employing Steady-state Acquisition (FIESTA) pulse sequence to emphasize the difference in spin–spin relaxation times.

SPIO particles have an iron core (magnetite, also called iron (II, III) oxide) of approximately 30 nm in diameter. The reported T2 relaxivity of SPIO particles is 180 $(mM\ s)^{-1}$ at 7T.[48,67] However, when they are incorporated into cells, the T2 relaxivity is considerably reduced.[67] In addition, tracking cells with SPIO labels has the following limitations: a) release of SPIO through apoptosis or exocytosis can label interstitial tissue or phagocytic cells and falsely report the targeted cellular activity, and b) in proliferating cells, the number of SPIO particles per cell drops leading to loss of signal in longitudinal studies.[68]

9.4.5   MR IMAGING OF MAGNETOSOMES

Although SPIO labeled cells give the needed CNR to follow cells immediately after transplantation (refer to section on Ferromagnetic Contrast Agents), one needs an MRI reporter gene to follow the fate of transplanted cells over days and weeks. Meriaux et al. have shown that individual magnetosome particles have a relaxivity similar to SPIO particles.[69] Brewer et al. went a step further and imaged the equivalent of MTB, obtaining an R2 value of 250 (mM s)$^{-1}$, comparable to 178 (mM s)$^{-1}$ for SPIO.[48] However, when particles derived from MTB, called magneto-endosymbionts (ME), were introduced into mammalian cells, the R2 values were reduced to 35 (mM s)$^{-1}$. Similar incorporation of SPIO also reduced relaxivity to approximately the same degree: 62 (mM s)$^{-1}$. Brewer et al. went on to investigate the minimum number of cells that could be observed if ME labeled cells were injected into the mouse brain and imaged by MRI at 7T. They labeled each cell with approximately 400 ME; injected mice with either 100 or 1000 cells; were able to detect both; and concluded that as few as 100 such cells could be tracked. Interestingly, Goldhawk et al. took a different approach.[8] Taking values reported by Benoit et al.,[70] in which MTB were imaged within mouse tumors, and using the amount of iron per MTB as well as R2 values measured *in vivo*, a calculation was performed determining the minimum number of cells needed to change R2 by 1 s^{-1}. That gave the minimum number of cells that could be detected if each cell contained 100 times the amount of iron in MTB. This was based on the diameter of a bacterium, which is about one-tenth the diameter of a mammalian cell. Hence, the volume of a mammalian cell would be approximately 1000 times greater than that of a bacterium. Using this approach, we estimated that a minimum of three cells could be detected in small animals and a minimum of 2600 cells in large animals/humans.[8] These very different approaches give somewhat different results, with incorporation of 400 ME providing a lower sensitivity (approximately 50 cells) than the equivalent number of magnetosomes in 100 ME (approximately three cells) and suggesting that reporter gene expression based on the magnetosome may be more sensitive by a factor of approximately 70 (50 times 4 all divided by 3). Of course, this outcome depends on generating sufficient magnetosome-like particles per cell using key magnetosome genes with essential roles in iron biomineralization to achieve the proposed iron per cell.

Lee et al.[26] investigated the dependence of R2 on interecho time in mammalian cells expressing a single MTB gene: *magA*. Compared with control cells, they found that MagA-expressing cells had greater iron content and that increases in R2 with increasing interecho times were larger. As it is known that R2 relaxivity of iron oxide particles depends on echo spacing,[71,72] this strongly suggests that iron particles were formed from a single MTB iron-handling gene. Lee et al. further suggested how these interecho effects could be used to uniquely image magnetosome-like particles by acquiring image-based R2 measurements with two different interecho times and, from the difference, assessing the iron-related contribution to R2.[26]

Cho et al. further demonstrated that mammalian cells expressing MagA from an inducible promoter could be successfully implemented as an MRI genetic reporter.[18] While significant effects on T2- and T2*-weighted images were reported, the technology has not been widely adapted, perhaps because the contrast was low[73] and relatively large amounts of tissue had to be selected to get sufficient signal to noise.

9.5 CONCLUSION

Prospects for the future development of molecular MRI place the spotlight on synthetic biology to build magnetosome-like nanoparticles in mammalian cells that enhance gene-based iron-labeling and MRI contrast for reporter gene expression. Results obtained from the expression of single MTB genes in mammalian hosts indicate the compatibility of these bacterial iron-handling proteins in a variety of cell types. However, no one MTB or magnetosome gene alone has provided optimal iron contrast for non-invasive imaging. A more faithful adaptation of MTB strategy for iron biomineral formation requires assembling multiple magnetosome proteins on a designated membrane. This role of biomineral compartmentalization in the design and regulation of biogenic iron nanoparticles has generally been under appreciated, perhaps because it adds complexity to the structure. Nevertheless, with such complexity comes multiple opportunities for regulating the magnetosome-like compartment, offering versatility in the genetic programming of reporter gene expression and in specification of the resulting iron biomineral. For MRI, there is potential for creating distinct MR signatures based on the complement of magnetosome genes expressed, the nature of their regulation by transcription factors, and the morphology of the encoded iron biomineral.

REFERENCES

1. Kali, A., Cokic, I., Tang, R., Dohnalkova, A., Kovarik, L., Yang, H., Kumar, A., Prato, F., Wood, J., Underhill, D., Marban, E., and Dharmakumar, R. (2016) Persistent microvascular obstruction After myocardial infarction culminates in the confluence of ferric iron oxide crystals, proinflammatory burden, and adverse remodeling, *Circ Cardiovasc Imaging 9*(11), e004996.
2. Dassanayake, P. (2019) Monocyte MRI Relaxation Rates are Regulated by Extracellular Iron and Hepcidin, Western University, London, Canada, Scholarship@Western Electronic Thesis and Dissertation Repository.
3. Delaby, C., Pilard, N., Goncalves, A., Beaumont, C., and Canonne-Hergaux, F. (2005) Presence of the iron exporter ferroportin at the plasma membrane of macrophages is enhanced by iron loading and down-regulated by hepcidin, *Blood 106*(12), 3979–3984.
4. Wilk, B., Wisenberg, G., Dharmakumar, R., Thiessen, J., Goldhawk, D., and Prato, F. (2019) Hybrid PET/MR imaging in myocardial inflammation post-myocardial infarction, *J Nucl Cardiol*, doi: 10.1007/s12350-019-01973-9.
5. Ramanujan, R. (2009) Magnetic particles for biomedical applications. In *Biomedical Materials* Narayan, R. (Ed.), pp. 477–491, Springer Science Media, Switzerland.
6. Goldhawk, D., Rohani, R., Sengupta, A., Gelman, N., and Prato, F. (2012) Using the magnetosome to model effective gene-based contrast for magnetic resonance imaging, *Wires Nanomed Nanobiotechnol 4*(4), 378–388.
7. Nyström, N., Hamilton, A., Xia, W., Liu, S., Scholl, T., and Ronald, J. (2019) Longitudinal visualization of viable cancer cell intratumoral distribution in mouse models using Oatp1a1-enhanced magnetic resonance imaging, *Invest Radiol 54*(5), 302–311.
8. Goldhawk, D., Gelman, N., Thompson, R., and Prato, F. (2017) Forming magnetosome-like nanoparticles in mammalian cells for molecular MRI. In *Design and Applications of Nanoparticles in Biomedical Imaging* (Bulte, J., and Modo, M., Eds.), pp. 187–203, Springer International Publishing, Switzerland.
9. Hentze, M., Muckenthaler, M., Galy, B., and Camaschella, C. (2010) Two to tango: Regulation of mammalian iron metabolism, *Cell 142*(1), 24–38.
10. Thomas, A., Morani, A., Liu, P., Weadock, W., Hussain, H., and Elsayes, K. (2019) Iron-containing abdominal pathologies: Exploiting magnetic susceptibility artifact on dual-echo gradient-echo magnetic resonance imaging, *J Comput Assist Tomogr 43*(2), 165–175.
11. Kim, J. A., Åberg, C., Salvati, A., and Dawson, K. A. (2012) Role of cell cycle on the cellular uptake and dilution of nanoparticles in a cell population, *Nat Nanotechnol 7*(1), 62–68.
12. Li, M., Wang, Y., Liu, M., and Lan, X. (2018) Multimodality reporter gene imaging: Construction strategies and application, *Theranostics 8*(11), 2954–2973.
13. Terrovitis, J., Stuber, M., Youssef, A., Preece, S., Leppo, M., Kizana, E., Schar, M., Gerstenblith, G., Weiss, R. G., Marban, E., and Abraham, M. R. (2008) Magnetic resonance imaging overestimates ferumoxide-labeled stem cell survival after transplantation in the heart, *Circulation 117*(12), 1555–1562.
14. Meikle, S. R., Kench, P., Kassiou, M., and Banati, R. B. (2005) Small animal SPECT and its place in the matrix of molecular imaging technologies, *Phys Med Biol 50*(22), R45–R61.
15. Youn, H., and Chung, J. K. (2013) Reporter gene imaging, *AJR Am J Roentgenol 201*(2), W206–W214.
16. Bourdeau, R. W., Lee-Gosselin, A., Lakshmanan, A., Farhadi, A., Kumar, S. R., Nety, S. P., and Shapiro, M. G. (2018) Acoustic reporter genes for noninvasive imaging of microorganisms in mammalian hosts, *Nature 553*(7686), 86–90.
17. Farhadi, A., Ho, G. H., Sawyer, D. P., Bourdeau, R. W., and Shapiro, M. G. (2019) Ultrasound imaging of gene expression in mammalian cells, *Science 365*(6460), 1469–1475.
18. Cho, I., Moran, S., Paudya, R., Piotrowska-Nitsche, K., Cheng, P.-H., Zhang, X., Mao, H., and Chan, A. (2014) Longitudinal monitoring of stem cell grafts in vivo using magnetic resonance imaging with inducible Maga as a genetic reporter, *Theranostics 4*(10), 972–989.
19. Guan, X., Yang, B., Xie, M., Kumar Ban, D., Zhao, X., Lal, R., and Zhang, F. (2019) MRI reporter gene MagA suppresses transferrin receptor and maps Fe2+ dependent lung cancer, *Nanomedicine 21*, 102064.
20. Liu, L., Alizadeh, K., Donnelly, S., Dassanayake, P., Hou, T., McGirr, R., Thompson, R., Prato, F., Gelman, N., Hoffman, L., and Goldhawk, D. (2019) MagA expression attenuates iron export activity in undifferentiated multipotent P19 cells, *PLoS One 14*(6), e0217842.

21. Rohani, R., Figueredo, R., Bureau, Y., Koropatnick, J., Foster, P., Thompson, R. T., Prato, F. S., and Goldhawk, D. E. (2014) Imaging tumor growth non-invasively using expression of MagA or modified ferritin subunits to augment intracellular contrast for repetitive MRI, *Mol Imaging Biol 16*(1), 63–73.

22. Zhang, X.-Y., Robledo, B., Harris, S., and Hu, X. (2014) A bacterial gene, mms6, as a new reporter gene for magnetic resonance imaging of mammalian cells, *Mol Imaging 13*, 1–12.

23. Magistretti, P., and Allaman, I. (2015) A cellular perspective on brain energy metabolism and functional imaging, *Neuron 86*(4), 883–901.

24. Yang, H., Oksuz, I., Dey, D., Sykes, J., Klein, M., Butler, J., Kovacs, M., Sobczyk, O., Cokic, I., Slomka, P., Bi, X., Li, D., Tighiouart, M., Tsaftaris, S., Prato, F., Fisher, J., and Dharmakumar, R. (2019) Accurate needle-free assessment of myocardial oxygenation for ischemic heart disease in canines using magnetic resonance imaging, *Sci Transl Med 11*(494), eaat4407.

25. Anazodo, U., Finger, E., Kwan, B., Pavlosky, W., Warrington, J., Günther, M., Prato, F., Thiessen, J., and St Lawrence, K. (2017) Using simultaneous PET/MRI to compare the accuracy of diagnosing fronto-temporal dementia by arterial spin labelling MRI and FDG-PET, *NeuroImage Clin 17*, 405–414.

26. Lee, C., Thompson, R., Prato, F., Goldhawk, D., and Gelman, N. (2015) Investigating the relationship between transverse relaxation rate (R2) and interecho time in MagA-expressing iron-labeled cells, *Mol Imaging 14*(12), 551–560.

27. Araujo, A., Abreu, F., Tavares Silva, K., Bazylinski, D., and Ulysses Lins, U. (2015) Magnetotactic bacteria as potential sources of bioproducts, *Mar Drugs 13*(1), 389–430.

28. Blakemore, R. P., Frankel, R. B., and Kalmijn, A. J. (1980) South-seeking magnetotactic bacteria in the Southern Hemisphere, *Nature 286*(5771), 384–385.

29. Leão, P., Teixeira, L. C. R. S., Cypriano, J., Farina, M., Abreu, F., Bazylinski, D. A., and Lins, U. (2016) North-seeking magnetotactic Gammaproteobacteria in the southern hemisphere, *Appl Environ Microbiol 82*(18), 5595–5602.

30. Alphandéry, E. (2014) Applications of magnetosomes synthesized by magnetotactic bacteria in medicine, *Front Bioeng Biotechnol 2*, 5.

31. Bazylinski, D. A., and Schübbe, S. (2007) Controlled biomineralization by and applications of magnetotactic bacteria. *Adv Appl Microbiol 62*, 21–62, Academic Press.

32. Vargas, G., Cypriano, J., Correa, T., Leao, P., Bazylinski, D. A., and Abreu, F. (2018) Applications of magnetotactic bacteria, magnetosomes and magnetosome crystals in biotechnology and nanotechnology: Mini-review, *Molecules (Basel, Switzerland) 23*(10), doi:10.3390/molecules23102438.

33. Faivre, D. (2015) Formation of magnetic nanoparticle chains in bacterial systems, *MRS Bull 40*(6), 509–515.

34. Klumpp, S., Lefèvre, C., Bennet, M., and Faivre, D. (2018) Swimming with magnets: From biological organisms to synthetic devices, *Phys Rep 789*, 1–54.

35. Chen, C., Chen, L., Yi, Y., Chen, C., Wu, L.-F., and Song, T. (2016) Killing of *Staphylococcus aureus* via magnetic hyperthermia mediated by magnetotactic bacteria, *Appl Environ Microbiol 82*(7), 2219–2226.

36. Martel, S. (2017) Targeting active cancer cells with smart bullets, *Ther Deliv 8*(5), 301–312.

37. Yoshino, T., Matsunaga, T., and Tanaka, T. (2018) Bioengineering and biotechnological applications of bacterial magnetic particles. In *Biological Magnetic Materials and Applications* (Matsunaga, T., Tanaka, T., and Kisailus, D., Eds.), pp. 77–93, Springer Singapore, Singapore.

38. Bain, J., Legge, C. J., Beattie, D. L., Sahota, A., Dirks, C., Lovett, J. R., and Staniland, S. S. (2019) A biomimetic magnetosome: Formation of iron oxide within carboxylic acid terminated polymersomes, *Nanoscale 11*(24), 11617–11625.

39. Bakhshi, P. K., Bain, J., Gul, M. O., Stride, E., Edirisinghe, M., and Staniland, S. S. (2016) Manufacturing man-made magnetosomes: High-throughput in situ synthesis of biomimetic magnetite loaded nanovesicles, *Macromol Biosci 16*(11), 1555–1561.

40. Le Nagard, L., Zhu, X., Yuan, H., Benzerara, K., Bazylinski, D. A., Fradin, C., Besson, A., Swaraj, S., Stanescu, S., Belkhou, R., and Hitchcock, A. P. (2019) Magnetite magnetosome biomineralization in Magnetospirillum magneticum strain AMB-1: A time course study, *Chem Geol 530*, 119348.

41. Yoshino, T., Shimada, T., Ito, Y., Honda, T., Maeda, Y., Matsunaga, T., and Tanaka, T. (2018) Biosynthesis of thermoresponsive magnetic nanoparticles by magnetosome display system, *Bioconjug Chem 29*(5), 1756–1762.

42. Lefevre, C. T., Trubitsyn, D., Abreu, F., Kolinko, S., Jogler, C., de Almeida, L. G., de Vasconcelos, A. T., Kube, M., Reinhardt, R., Lins, U., Pignol, D., Schuler, D., Bazylinski, D. A., and Ginet, N. (2013) Comparative genomic analysis of magnetotactic bacteria from the Deltaproteobacteria provides new insights into magnetite and greigite magnetosome genes required for magnetotaxis, *Environ Microbiol 15*(10), 2712–2735.

43. Murat, D., Quinlan, A., Vali, H., and Komeili, A. (2010) Comprehensive genetic dissection of the magnetosome gene island reveals the step-wise assembly of a prokaryotic organelle, *Proc Natl Acad Sci USA* *107*(12), 5593–5598.

44. Uebe, R., and Schuler, D. (2016) Magnetosome biogenesis in magnetotactic bacteria, *Nat Rev Microbiol* *14*(10), 621–637.

45. Nudelman, H., and Zarivach, R. (2014) Structure prediction of magnetosome-associated proteins, *Front Microbiol 5*, article 9.

46. Quinlan, A., Murat, D., Vali, H., and Komeili, A. (2011) The HtrA/DegP family protease MamE is a bifunctional protein with roles in magnetosome protein localization and magnetite biomineralization, *Mol Microbiol 80*(4), 1075–1087.

47. Uebe, R., Keren-Khadmy, N., Zeytuni, N., Katzmann, E., Navon, Y., Davidov, G., Bitton, R., Plitzko, J. M., Schuler, D., and Zarivach, R. (2018) The dual role of MamB in magnetosome membrane assembly and magnetite biomineralization, *Mol Microbiol 107*(4), 542–557.

48. Brewer, K., Spitler, R., Lee, K., Chan, A., Barrozo, J., Wakeel, A., Foote, C., Machtaler, S., Rioux, J., Willmann, J., Chakraborty, P., Rice, B., Contag, C., Bell, C., and Rutt, B. (2018) Characterization of magneto-endosymbionts as MRI cell labeling and tracking agents, *Mol Imaging Biol 20*(1), 65–73.

49. Komeili, A. (2012) Molecular mechanisms of compartmentalization and biomineralization in magnetotactic bacteria, *FEMS Microbiol Rev 36*(1), 232–255.

50. Schüler, D. (2008) Genetics and cell biology of magnetosome formation in magnetotactic bacteria, *FEMS Microbiol Rev 32*(4), 654–672.

51. Raschdorf, O., Forstner, Y., Kolinko, I., Uebe, R., Plitzko, J., and Schüler, D. (2016) Genetic and ultrastructural analysis reveals the key players and initial steps of bacterial magnetosome membrane biogenesis, *PLoS Genet 126*(6), e1006101.

52. Yanatori, I., and Kishi, F. (2019) DMT1 and iron transport, *Free Radic Biol Med 133*, 55–63.

53. Sun, Q., Fradin, C., Thompson, R., Prato, F., and Goldhawk, D. (2019) Developing Magnetic Resonance Reporter Gene Imaging: Co-Expression of Magnetotactic Bacteria Genes *mamI* and *mamL*, In *World Molecular Imaging Congress*, Montreal, Canada.

54. Goldhawk, D., Koropatnick, D., Figueredo, R., Prato, F., Thompson, R., and Gelman, N. (2019) Combined Expression of Contrast Genes in Eukaryotic Cells, European Patent # 3110952.

55. Prato, F., Goldhawk, D., McCreary, C., McGirr, R., Dhanvantari, S., Thompson, R., Thomas, A., and Hill, D. (2017) Magnetosome Gene Expression in Eukaryotic Cells, USPTO Patent # US 9,556,238 B2; MMI.

56. Kolinko, I., Lohße, A., Borg, S., Raschdorf, O., Jogler, C., Tu, Q., Posfai, M., Tompa, E., Plitzko, J., Brachmann, A., Wanner, G., Muller, R., Zhang, Y., and Schuler, D. (2014) Biosynthesis of magnetic nanostructures in a foreign organism by transfer of bacterial magnetosome gene clusters, *Nat Nanotechnol 9*(3), 193–197.

57. Schuler, D., Uhl, R., and Bauerlein, E. (1995) A simple light scattering method to assay magnetism in Magnetospirillum gryphiswaldense, *FEMS Microbiol Ecol 132*(1–2), 139–145.

58. Le Nagard, L., Morillo-Lopez, V., Fradin, C., and Bzylinski, D. (2018) Growing magnetotactic bacteria of the genus *Magnetospirillum*: Strains MSR-1, AMB-1 and MS-1, *JoVE 140*, e58536.

59. Martel, S., and Mohammadi, M. (2016) Switching between magnetotactic and aerotactic displacement controls to enhance the efficacy of MC-1 magneto-aerotactic bacteria as cancer-fighting nanorobots, *Micromachines 7*(6), 97.

60. Torres de Araujo, F., Pires, M., Frankel, R., and Bicudo, C. (1986) Magnetite and magnetotaxis in algae, *Biophys J 50*(2), 375–378.

61. Nakamura, C., Burgess, J. G., Sode, K., and Matsunaga, T. (1995) An iron-regulated gene, *magA*, encoding an iron transport protein of *Magnetospirillum* sp. Strain AMB-1, *J Biol Chem 270*(47), 28392–28396.

62. Elfick, A., Rischitor, G., Mouras, R., Azfer, A., Lungaro, L., Uhlarz, M., Herrmannsdorfer, T., Lucocq, J., Gamal, W., Bagnaninchi, P., Semple, S., and Salter, D. M. (2017) Biosynthesis of magnetic nanoparticles by human mesenchymal stem cells following transfection with the magnetotactic bacterial gene mms6, *Sci Rep 7*, 39755.

63. De Leon-Rodriguez, L., Martins, A., Pinho, M., Rofsky, N., and Sherry, A. (2015) Basic MR relaxation mechanisms and contrast agent design, *J Magn Reson Imaging 42*(3), 545–565.

64. Moran, G., and Prato, F. (2004) Modeling (1H) exchange: An estimate of the error introduced in MRI by assuming the fast exchange limit in bolus tracking, *Magn Reson Med 51*(4), 816–827.

65. Szomolanyi, P., Rohrer, M., Frenzel, T., Noebauer-Huhmann, I., Jost, G., Endrikat, J., Trattnig, S., and Pietsch, H. (2019) Comparison of the relaxivities of macrocyclic gadolinium-based contrast agents in human plasma at 1.5, 3, and 7 T, and blood at 3 T, *Invest Radiol 54*(9), 559–564.

66. Heyn, C., Ronald, J., Mackenzie, L., MacDonald, I., Chambers, A., Rutt, B., and Foster, P. (2006) In vivo magnetic resonance imaging of single cells in mouse brain with optical validation, *Magn Reson Med* 55(1), 23–29.

67. Taylor, A., Herrmann, A., Moss, D., Sée, V., Davies, K., Williams, S., and Murray, P. (2014) Assessing the efficacy of nano- and micro-sized magnetic particles as contrast agents for MRI cell tracking, *PLoS One* 9(6), e100259.

68. Bernau, K., Lewis, C., Petelinsek, A., Reagan, M., Niles, D., Mattis, V., Meyerand, M., Suzuki, M., and Svendsen, C. (2016) In vivo tracking of human neural progenitor cells in the rat brain using magnetic resonance imaging is not enhanced by ferritin expression, *Cell Transplant* 25(3), 575–592.

69. Mériaux, S., Boucher, M., Marty, B., Lalatonne, Y., Prévéral, S., Motte, L., Lefèvre, C., Geffroy, F., Lethimonnier, F., Péan, M., Garcia, D., Adryanczyk-Perrier, G., Pignol, D., and Ginet, N. (2015) Magnetosomes, biogenic magnetic nanomaterials for brain molecular imaging with 17.2 T MRI scanner, *Adv Healthc Mater* 4(7), 1076–1083.

70. Benoit, M., Mayer, D., Barak, Y., Chen, I., Hu, W., Cheng, Z., Wang, S., Spielman, D., Gambhir, S., and Matin, A. (2009) Visualizing implanted tumors in mice with magnetic resonance imaging using magnetotactic bacteria, *Clin Cancer Res* 15(16), 5170–5177.

71. Ghugre, N., Coates, T., Nelson, M., and Wood, J. (2005) Mechanisms of tissue-iron relaxivity: Nuclear magnetic resonance studies of human liver biopsy specimens, *Magn Reson Med* 54(5), 1185–1193.

72. Rozenman, Y., Zou, X., and Kantor, H. (1990) Cardiovascular MR imaging with iron oxide particles: Utility of a superparamagnetic contrast agent and the role of diffusion in signal loss, *Radiology* 175(3), 655–659.

73. Pereira, S., Williams, S., Murray, P., and Taylor, A. (2016) MS-1 magA: Revisiting its efficacy as a reporter gene for MRI, *Mol Imaging* 15, 1–9.

10 Magnetic Control of Biogenic Micro-Mirror

Masakazu Iwasaka

CONTENTS

10.1 Introduction 215
10.2 Optical Detection of Magnetic Orientation of Biogenic Micro/Nanoparticles 216
    10.2.1 Methods and Materials for Optical Detection of Biogenic Micro-Mirrors 216
    10.2.2 Optical Detection of Magnetically Oriented Biogenic Micro/Nanoparticles 216
10.3 Magnetic Control of Biogenic Guanine Crystals 217
    10.3.1 Micro-Mirror Properties of Guanine Crystal Platelets from Fish 218
    10.3.2 Mechanism of Magnetic Orientation of Diamagnetic Micro-Mirrors 220
    10.3.3 Synthesis of Artificial Guanine Crystals 221
10.4 Magnetic Control of Uric Acid Crystals 222
    10.4.1 Methods and Materials for Uric Acid Crystal Experiments 224
    10.4.2 Optical Responses of Uric Acid Crystals under Magnetic Fields 224
10.5 Possible Applications of Biogenic Micro-Mirror for Photonic Cellular Analysis 229
Acknowledgments 230
References 231

10.1 INTRODUCTION

Recently, biomimicry has been used to develop new functional materials. For example, shark skin was investigated to find the specialized structure that reduces friction in a water stream [1]. A considerable number of reports on structural color in butterfly wings indicate many mechanisms that display a strategy for survival [2–7]. Fish also use structural color for survival and communication. The materials utilized for the generation of structural color by individual species were naturally selected during their evolution.

In the case of fish, the control of light reflection from the body is very important during daylight to become less visible to predators. Guanine-based bio-reflectors are the most representative biogenic materials for daylight reflection control. Research on these bio-reflectors has increased since the 1970s [8, 9], and now many researchers are engaged in biogenic reflecting optics. In particular, studies on bio-reflectors are now including micro-mirror device technologies.

In addition to specifically designed optical components in living tissues, most components in living animals and plants include optically functional materials. With respect to biomass utilization, exploring new optical materials, such as cellulose, is a big challenge.

Developing a new optical device technology requires new methods of actuating biological components on an artificial substrate. Specifically, physical manipulation must be established for the development of biogenic micro-mirror systems for industries, including environmental, information technology, and biomedical engineering.

In this chapter, the first section describes a method for detecting the alignment direction of micro-particles floating in water via magnetic orientation. This is followed by a discussion of optical and magnetic properties of biogenic micro-mirror particles.

10.2   OPTICAL DETECTION OF MAGNETIC ORIENTATION
OF BIOGENIC MICRO/NANOPARTICLES

The magnetic orientation of macromolecules under strong static magnetic fields (10 T) was reported in the 1970s, and most studies examined how to control biological materials, organic materials, and cells in a magnetic field [10–24]. A theoretical calculation predicted possible mesoscale alignment of diamagnetic objects under 1-T magnetic fields at room temperature. Recently, alignment control of floating micro-crystals was demonstrated for a field less than 1 T [25–29].

This section focuses on correlations between micro-crystal alignment directions and combinations of magnetic field, incident light, and observation direction. The micro-crystals include uric acid, biogenic guanine, micro-crystal-like cellulose, calcium carbonate, and hexagonal boron nitride particles, which were artificially prepared or derived from biological and commercial products. Real-time observations of the micro-crystal suspensions under magnetic fields were performed. By categorizing the crystal shape anisotropy, the field orientation, and the light-scattering anisotropy, a new method is presented to determine the alignment direction of floating meso-/micro-crystals.

10.2.1   METHODS AND MATERIALS FOR OPTICAL DETECTION OF BIOGENIC MICRO-MIRRORS

The magnetic field experiments used an electromagnet and three charge-coupled device (CCD)-equipped microscopes (VHX-2000 and VH-5000 provided by Keyence Co Ltd., Japan) for imaging, as well as a stereomicroscope (QN42HL by ELMO Co Ltd., Japan). A phase-contrast microscope (Olympus IX-73) was placed in the electromagnet (WS15-40-5K-MS, Hayama Inc., Japan), as shown in Figure 10.1(a). Two superconducting magnets (maximum 14 T, 14t70r, and 5 T, Microstat BT, Oxford Inc., U.K.) were also utilized.

Optical fiber-based measurement systems were used with the superconducting magnet or the electromagnet, depending on the magnetic field intensity required. In the magnetic field region were an optical quartz sample cell, mirrors (if necessary), and optical fibers from a light source and to a cooled CCD spectrophotometer. The magnetic field, incident light, and observation directions had three different modes (MODE-I, II, III) for the evaluation of light scattering anisotropies of the micro-crystals in an aqueous solution, as depicted in Figure 10.2(b). Micro-crystals of cellulose and hexagonal boron nitride (h-BN) were prepared from powder products (Ceolus, from pulp Type-Iβ cellulose, Asahi Kasei co. Ltd., Japan, and boron nitride powder GP, DENKA co Ltd., Japan).

Calcium carbonate and guanine crystals were separated from the algae *Emiliania huxleyi* and goldfish scales, respectively. D-glucose and sucrose powders (Wako Co. Ltd., Japan) were dissolved in water. Re-crystallized uric acid and vesicles were prepared from commercial uric acid and lipids, respectively.

10.2.2   OPTICAL DETECTION OF MAGNETICALLY ORIENTED BIOGENIC MICRO/NANOPARTICLES

Figure 10.2(a) shows vertically oriented micro-crystalline cellulose particles in a 500-mT magnetic field. The micro-crystals were floating in water, and the incident light and magnetic field corresponded to Mode-II. Fiber optic measurements for Mode-I, II, and III were performed as shown in Figure 10.2(b). The bottom plot indicates that the light scattering in Figure 10.2(a) was enhanced at 500 mT, whereas, it decreased for the Mode-I and III combinations.

The detection threshold of magnetic orientation of micro-crystalline cellulose floating in water was measured. As shown in Figure 10.3(a), 150 mT was adequate to detect the magnetic micro-crystals. Low molecular weight molecules exhibited no response. In addition, the plate-like uric acid crystals and the calcium carbonate crystals were oriented perpendicular to the field and exhibited similar light scattering anisotropies in Mode-I, II, III. Mesoscale cellulose nanocrystals from algae also exhibited increases in a Mode-II measurement. The results therefore indicated that micro- to nanoscale crystal orientations can be determined by this light scattering method.

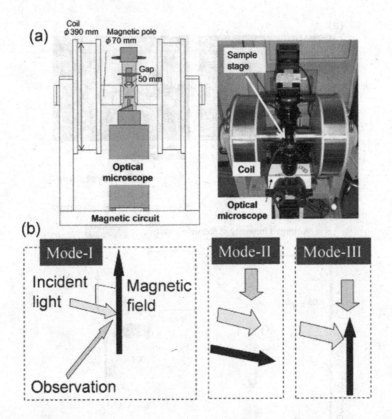

FIGURE 10.1 Experimental setup. (a) Phase-contrast microscope in an electromagnet. (b) Three modes of combining magnetic fields, incident light, and observation of light-scattering anisotropy.

Next, we examined micro-crystals oriented parallel to the magnetic field. Biogenic guanine crystals [25–28], and hexagonal boron nitride (h-BN) crystal plates oriented their broadest face plane parallel to the magnetic field. Figure 10.4 shows a macroscopic view of h-BN plates floating in water with and without magnetic fields. In contrast to micro-crystalline cellulose, the h-BN plates enhanced light scattering in Mode-I and inhibited it in Mode-II and III.

Vesicles (liposomes) have been reported to respond to magnetic fields and exhibited a parallel orientation. This was confirmed for a suspension of tubular and spherical floating vesicles. Light scattering was then measured in Mode-I and II, as shown in Figure 10.5. The light scattering anisotropies were the same as those for h-BN and biogenic guanine crystals from goldfish iridophores.

Tables 10.1 and 10.2 summarize the results obtained. In Table 10.1, the changes in light scattering intensity were complementary for Mode-I and II for parallel and perpendicular orienting micro-objects. In Mode-III, both orientations caused inhibition (decreases).

Micro-objects were categorized into magnetic field-parallel and perpendicular orienting groups in Table 10.2, along with the detection threshold of the orientation. At present, the mechanism of the light scattering anisotropies and the orientation of individual objects is thought to be attributed to the back-scattering cross section that can be changed by magnetic orientations.

10.3 MAGNETIC CONTROL OF BIOGENIC GUANINE CRYSTALS

Guanine crystals from fish are the most efficient micro-mirror particles in Tables 10.1 and 10.2 that have a parallel orientation in a magnetic field. Most materials, such as proteins and DNA, orientate under strong magnetic fields. However, the critical field threshold of magneto-mechanical phenomena is still unknown. It was demonstrated that a thin micro-mirror from a high reflectivity

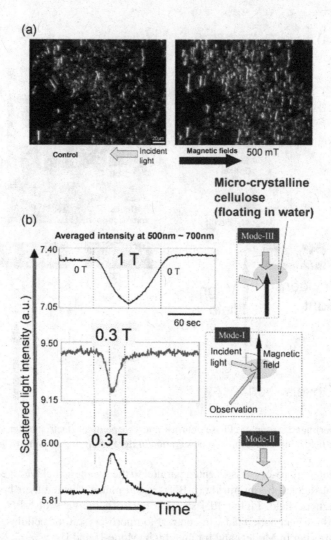

FIGURE 10.2 Magnetically oriented micro-crystalline cellulose particles and light scattering anisotropies. (a) Cellulose oriented perpendicular to the 500-mT magnetic field. (b) Changes in light scattering intensity from the floating cellulose. The incident light and magnetic field orientations in Mode-I, II, and III.

fish scale exhibited a magnetic response at 100 mT. A significant observation is decreased light scattering from guanine crystals under a magnetic field, as well as rapid rotation against the field. Enhancement of light scattering intensity was also observed when the three vectors of light incidence, magnetic field, and observation were orthogonal.

Micro- to nanometer scale thin biogenic plates can be noninvasive, magnetically controlled micro-mirrors for micro-scale light control. Magnetic orientation of the guanine crystal is due to diamagnetic rotation of the entire guanine crystal stack. The noncontact control of guanine micro-mirrors could be used for in situ micro-spectroscopy and cellular analysis.

10.3.1 Micro-Mirror Properties of Guanine Crystal Platelets from Fish

Most fish have a silvery shiny region in their ventral part and eye. The shiny material is a tiny "reflecting platelet." The platelets may direct strong illumination as shown in Figure 10.6. A comparison of light-reflecting properties of reflectors in fish and other species (e.g., cuttlefish) was performed. The reflecting platelet of cuttlefish is a protein, reflectin [30]; whereas, most of the reflecting

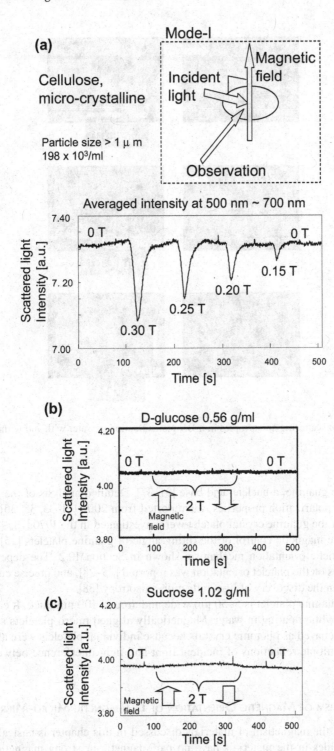

FIGURE 10.3 Light scattering changes in micro-crystalline cellulose, D-glucose, and sucrose (suspended or dissolved in water), by magnetic fields in the Mode-I combination.

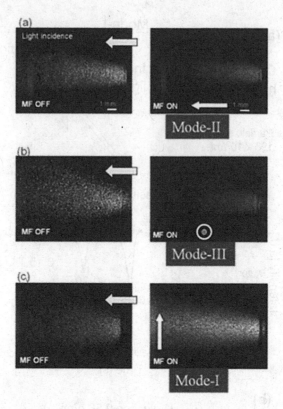

FIGURE 10.4 Light scattering anisotropies in h-BN plates floating in water with and without 400-mT magnetic fields.

platelets in fish are guanine, a nucleic-acid base [31–37]. Detailed analysis of the guanine crystal structure and light polarization properties was reported from 2008 [31–33, 35, 36], while the biological functions of the guanine crystal platelet were investigated in the 1970s [8, 9].

In 2010, possible magnetic control of the tilting angle of guanine platelets [25] was developed based on the magnetic orientation mechanism shown in Section 10.2. The dependence of light reflection intensities on the platelet orientation was reported [25–28], and precise control of the tilting angle resulted in the discovery of light reflection anisotropy [38].

A typical fish guanine platelet is 5–50 μm wide and around 100 nm thick. It can exhibit rapid Brownian motion while floating in water. Magnetically aligned micro-platelets suspended in an aqueous solution behaved as photonic crystals because individual platelets were aligned in parallel. This caused multiple reflections of incident light and light interference between the aligned platelets [39].

10.3.2 Mechanism of Magnetic Orientation of Diamagnetic Micro-Mirrors

Including guanine, the magnetism of materials discussed in this chapter is basically diamagnetic because the molecules in the materials have no paramagnetic or strong magnetic part. At least in the case of orientational behavior, the anisotropy in diamagnetic susceptibility dominates the orientation.

A molecular ring structure can produce a large anisotropy in diamagnetic susceptibility. In particular, guanine has two rings, and a simple estimation indicates that the diamagnetic anisotropic energy of a guanine platelet from fish exceeds the energy of thermal agitation.

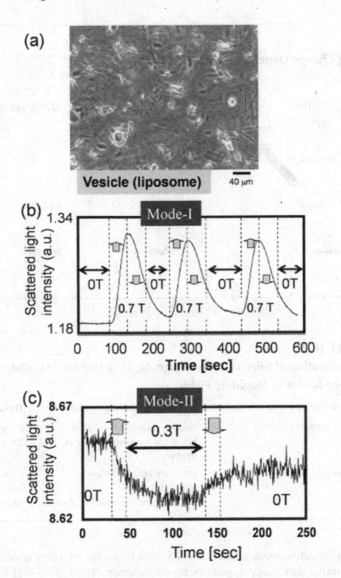

FIGURE 10.5 Light scattering changes from vesicles by a magnetic field in Mode-I and II combinations. (a) Photograph of sample. (b) Scattered light changes during three magnetic field exposures in Mode-I. (c) A response in Mode-II.

The crystal structure of a guanine platelet was investigated, but is conjectured to have hidden underlying structures. A recent analysis of goldfish and Japanese anchovy platelets suggested multiple grating structures that crossed each other at 60 degrees (Figure 10.7).

10.3.3 SYNTHESIS OF ARTIFICIAL GUANINE CRYSTALS

Mimics of fish guanine platelets have potential applications in photonic devices, where control of crystal shape and size is desirable. In bio-crystallization, biogenic guanine crystals may be controlled by other molecules such as proteins, but no such proteins have been identified. A genetic process is most likely controlling the guanine synthesis as well as aligning and stacking the molecules.

Because the guanine is almost insoluble in water, if its bio-crystallization occurs in a re-crystallization that is solubilized, it requires a higher pH.

TABLE 10.1

Light Scattering Change Dependence on Mode-I, II, III

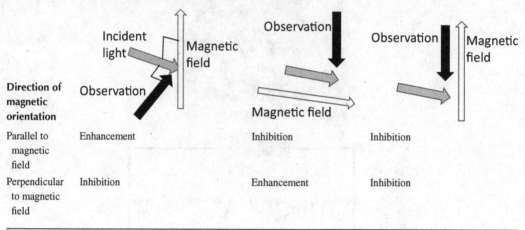

Direction of magnetic orientation			
Parallel to magnetic field	Enhancement	Inhibition	Inhibition
Perpendicular to magnetic field	Inhibition	Enhancement	Inhibition

TABLE 10.2

Classification of Micro-Crystal Magnetic Orientations (Parallel, Perpendicular to Magnetic Field)

Direction of magnetic orientation	Objects (micro-crystals)	Threshold
Parallel to magnetic field	biogenic guanine crystal (fish)	100 mT
	hexagonal boron nitride platelet	200 mT
	lipid vesicle	300 mT
Perpendicular to magnetic field	micro-crystalline cellulose	150 mT
	uric acid crystal	200 mT
	calcium carbonate crystal (*E. huxleyi*)	400 mT

Guanine has a refractive index greater than 1.8, which enables the shiny appearance. Efforts in making crystals similar to the biogenic guanine crystal have been reported [40–42]. Re-crystallization of commercial guanine resulted in the same magnetic orientation of the crystal platelet [40].

A rapid process for producing guanine crystals involves re-crystallization in an aqueous solution with NaOH at pH 13 by adding HCl. Guanine sodium salt crystals were obtained [42]. Figure 10.8 shows the crystals floating in a strong basic solution, where they exhibited intense light reflection that was nearly equal to that of biogenic guanine. Light reflection under a magnetic field (Mode-III parallel orientation) suggested that the synthetic guanine particles were oriented parallel to the field, similar to a biogenic guanine platelet (Figure 10.9).

10.4 MAGNETIC CONTROL OF URIC ACID CRYSTALS

Uric acid crystals can be generated in a human body, whereas, guanine crystals are not found. Synthetic uric acid crystals are magnetically oriented under a field greater than 100 mT [43–45]. The platelets or crystal needles are oriented perpendicular to the field, and light reflection from them was similar to the "magnetic field perpendicular orientation" group in Table 10.1.

Like guanine, uric acid has two rings that induce a diamagnetic anisotropic energy similar to that of guanine. The strength of the magnetic fields in which uric acid or guanine crystals rotate

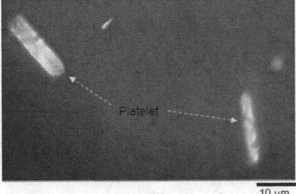

FIGURE 10.6 Guanine platelets from fish under light irradiation from the side.

FIGURE 10.7 Triangle grating structures in guanine platelets from Japanese anchovies.

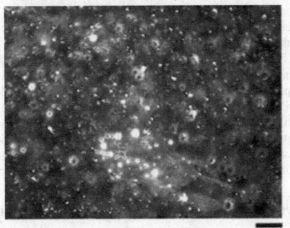

FIGURE 10.8 Reflection image of synthetic guanine sodium salt crystals. The incident light came from left to right.

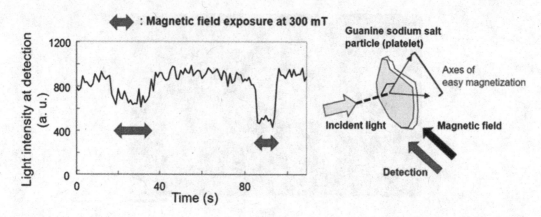

FIGURE 10.9 Time course of light reflection intensity from an aqueous suspension of synthetic guanine sodium salt crystals. The right panel shows the configuration of the incident light, magnetic field, and detection.

can be obtained with a permanent magnet, even when using a commercial product. This suggests new medical applications of magnetic fields for the treatment of gout caused by uric acid crystals.

Uric acid crystals are also utilized as optical materials by insects such as fireflies. The luciferase-luciferin reaction that generates their light emission occurs in a photochemical reaction chamber that is thought to have precise light-controlling units [46, 47]. One unit is a spherical uric acid crystal that can be observed inside the lantern bud of the fireflies [48] and diffuses light from the emission chamber. The intensity of firefly light emission was suppressed and modulated when exposed to magnetic fields up to 14 T [49, 50]. Hence, this section focuses on the correlation between magnetic fields and light reflection in uric acid crystals.

10.4.1 METHODS AND MATERIALS FOR URIC ACID CRYSTAL EXPERIMENTS

The firefly species *Luciola cruciata* was used as the light emission source for the analysis of light transport in uric acid crystals. The white-yellow part of the lantern bag from the firefly was added to distilled water (100 µl), resulting in a white suspension containing biogenic uric acid particles. Synthetic uric acid crystals were prepared by re-crystallization, where needle- or plate-like crystals precipitated from a strong basic solution at 100°C. The crystal particles were suspended in water and then placed in a glass cell. A firefly specimen was placed in the chamber and light that was reflected from the chamber was introduced to a spectrophotometer.

The uric acid crystals and the configuration used for the magnetic field measurements are shown in Figure 10.10. Scanning electron microscope (SEM) images of the materials that were extracted from the white-yellow part of the firefly lantern bag revealed spherical particles (Figure 10.10(a)). Magnified images show needle-shaped objects attached to the surface of the spherical particles, as shown in Figure 10.10(b). An optical microscope image of synthetic monosodium urate (MSU) crystals is shown in Figure 10.10(c). The light intensity from synthetic uric acid crystals was measured with a spectrophotometer (FLAME, Ocean optics Inc., USA) for aqueous suspensions contained in a quartz cuvette, as shown schematically in Figure 10.10(d).

10.4.2 OPTICAL RESPONSES OF URIC ACID CRYSTALS UNDER MAGNETIC FIELDS

Light scattering by biogenic uric acid crystals was measured as the applied magnetic field was swept over 0–5 T and back to 0 T. Scattering at 450 nm, 530 nm, 560 nm, and 632 nm was examined with light-emitting diodes (Figure 10.11(a)). The red (632 nm) and green (530 nm) LEDs covered the 530–632 nm region, and the intensity of 560-nm light was sufficient for detection.

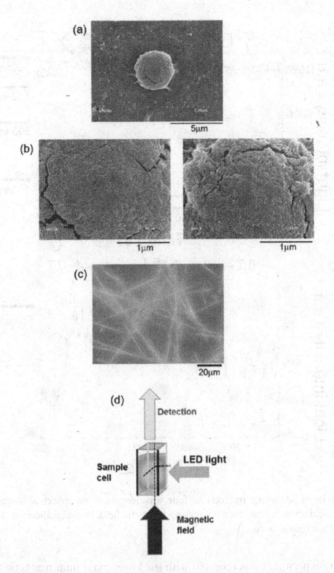

FIGURE 10.10 (a) SEM image of a spherical uric acid crystal particle and (b) magnified surface images of the particle. (c) Optical microscope image of a synthetic monosodium urate (MSU) crystal. (d) Configuration used for the light-scattering measurements of an aqueous suspension containing either firefly uric acid crystals or MSU crystals.

A magnification of changes at 530 nm is shown in Figure 10.11(b). Changes in the light scattering at the other wavelengths were similar. As the field was increased at a rate of 1 T/min, a peak (crest) was observed at approximately 1 T, which then decreased at ~2 T and then continued to increase until 5 T. As the magnetic field decreased from 5 T back to 0 T, a similar but opposite behavior was observed. However, the field at which the peak appeared was different.

The effect of the highest magnetic field strength in the sweep on the 560-nm light intensity shift was examined by sweeping from 0 T to a range of maxima (5 T, 4 T, 3 T, and 2 T) and back to 0 T, as shown in Figure 10.12(a). The magnetic field sweep rate was always +/− 1 T/min.

When the maximum magnetic field strength was 5 T (top panel, Figure 10.12(a)), the light intensity began to increase at 200 s, 5 T was reached at 500 s and was held for 100 s, and then decreased back to 0 T. Three distinct peaks were observed: 1.5 T (sweep-up), 5 T (stable), and 0.7 T

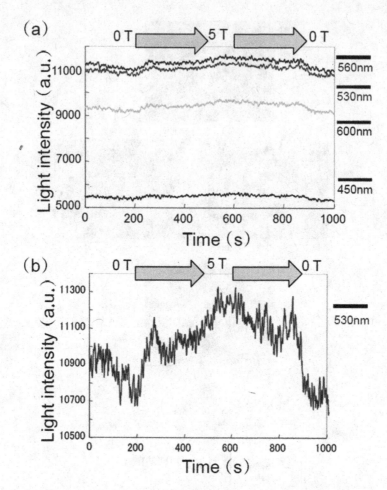

FIGURE 10.11 (a) Light scattering intensity at four wavelengths from spherical biogenic crystals in an increasing magnetic field up to 5 T and then decreasing magnetic field. (b) Magnification of the light intensity change from the 560 nm light source.

(sweep-down). This experiment was repeated with the lower maximum magnetic field strengths (4 T, 3 T, and 2 T), and peaks in the scattered light intensity were obtained. We refer to the magnetic field intensity at which these peaks occurred as "critical fields."

The field strengths at which the first and last scattered light peaks occurred were plotted as a function of the maximum field strength reached in the sweep (Figure 10.12(b)). The plot revealed that the critical field values for the sweep-down were lower than those for the sweep-up. This behavior was likely related to the response time of the magnetic orientation and its relaxation.

A possible mechanism for the light scattering changes involves magnetic deformation of the spherical uric acid crystals (Figure 10.13). Diamagnetic rotation of platelet-like uric acid crystals has been previously reported [38], where a two-stage rotation occurred in the magnetic field. The peaks in the field sweeps indicated that two magneto-mechanical events occurred, changing the light scattering pattern twice. It is possible that the second magnetic field event affected the spherical biogenic crystals after the light intensity decreased at 3 T. The delay in the appearance of the peak in the decreasing field sweep might have been caused by the time required for the magnetically deformed structures to relax.

SEM images in Figure 10.10(b) indicated that the spheres were composed of needle-like crystals that were aligned circumferentially. The spherical particles were examined with a high-resolution

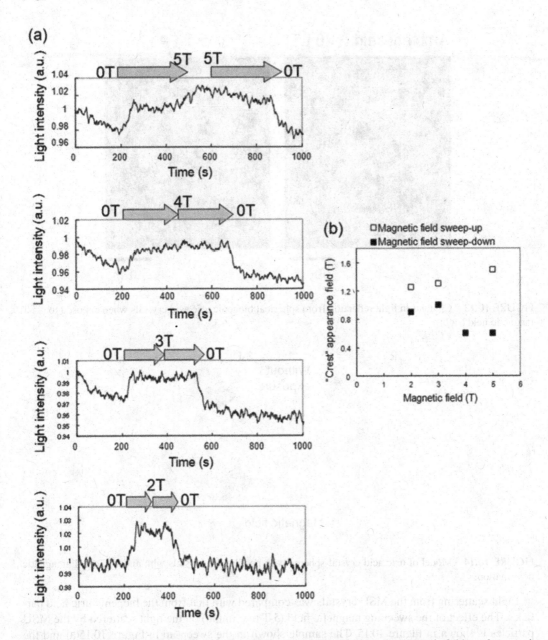

FIGURE 10.12 (a) 560-nm light scattering by spherical biogenic crystals as the magnetic field was increased up to 2, 3, 4, or 5 T and then decreased. (b) Correlation between the magnetic fields at which the most intense peaks in the scattered light appeared (critical fields) for each different magnetic field sweep. The maximum peaks in the scattered light were the first and last peaks in the increasing and decreasing sweeps, respectively.

microscope in real-time in the presence and absence of the 5-T magnetic field (Figure 10.13). Light reflection from individual particles was enhanced in the field, which was attributed to rotation or deformation of the particles. For deformation to occur, they must be pliable, which would be possible if they contained water.

A model of uric acid crystal spheres (Figure 10.14) has needle-like crystals oriented perpendicular to the magnetic field. This magnetic orientation is analogous to that of lipid membranes that deform vesicles (liposomes). It is thus plausible that the spherical uric acid crystals became ellipsoidal in the magnetic fields.

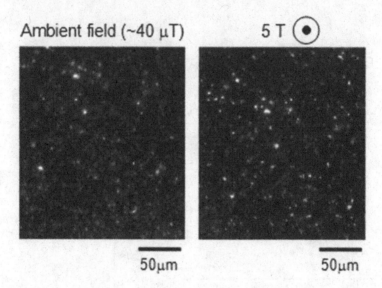

FIGURE 10.13   Changes in light reflection from spherical biogenic uric acid crystals when exposed to a 5-T magnetic field (right).

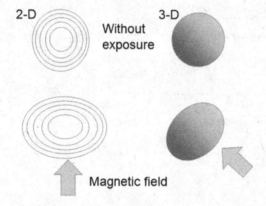

FIGURE 10.14   Model of uric acid crystal spheres with needle-like crystals with and without diamagnetic deformation.

Light scattering from the MSU crystals was compared with that from the biogenic uric acid particles. The effect of the sweeping magnetic field (5-T maximum) on the light scattered by the MSU particles is shown in Figure 10.15. The sample shown in the sweep-up in Figure 10.15(a) and the sweep-down in Figure 10.15(b) exhibited an overall decrease in light scattering, with two peaks in sweeps. Another sample in Figure 10.15(c) exhibited very similar light scattering behavior to that of the biogenic uric acid crystal particles shown in Figures 10.11 and 10.13.

This data showed that the spherical particles within the lantern bag of the fireflies, which are reported to be an assembly of uric acid crystals, exhibited changes in light scattering during increasing and decreasing magnetic field sweeps. This behavior was also observed for the synthetic MSU crystals. It is possible that the partial degradation of the spherical, biogenic particles enhanced the magnetic response. However, the behavior of individual particles in the magnetic fields (Figure 10.13) indicated that changes in the light intensity were caused by the particles.

It is possible that the biogenic uric acid crystals in the firefly lantern produced a light reflection anisotropy when exposed to magnetic fields. The work here indicated that the design and fabrication of novel bio-mimic optical devices may be possible via magnetic control of crystal particles.

10.5 POSSIBLE APPLICATIONS OF BIOGENIC MICRO-MIRROR FOR PHOTONIC CELLULAR ANALYSIS

Biogenic micro-mirror particles have potential applications in industries and for measurements and imaging techniques targeting biological materials and cells in aqueous media. Measurement of the diamagnetic rotation of MSU crystals with near-infrared-light may open a new method of gout diagnosis [44].

Because the biogenic guanine platelets exhibited very novel and unusual behavior, and a detailed reflection mechanism is required, it is important to search for more applications of biogenic guanine platelets in the life sciences, such as bio-imaging.

For example, illumination of cells in the vicinity of guanine platelets was demonstrated in Figure 10.16. In that optical configuration, incident light came from the side of the chamber containing

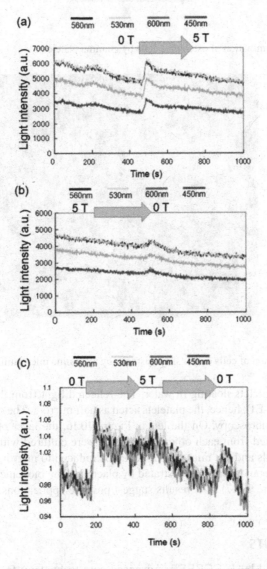

FIGURE 10.15   (a) Light scattering from synthetic needle-like crystals (MSU) at four wavelengths as the magnetic field was increased to 5 T and then (b) decreased. (c) Light scattering (normalized) from synthetic needle-like crystals (MSU) at four wavelengths as the field was increased to 5 T and then decreased.

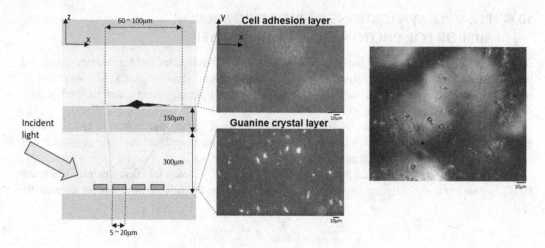

FIGURE 10.16   Light illumination of osteoblast cells by guanine platelets.

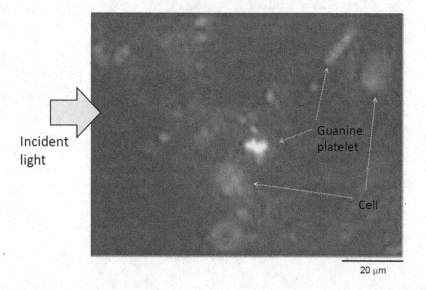

FIGURE 10.17   Illumination of cells over a short distance by a guanine micro-mirror on the cell layer.

biogenic fish guanine platelets floating in water. The reflected light from the platelets illuminated osteoblast cells (MC3T3-E1); hence, the platelets acted as half mirrors. The cells were imaged much like in phase-contrast microscopy. On the left in Figure 10.16, the layer of guanine platelets and the cell layer are separated from each other. When cells were cultured with guanine platelets, the distance between the cells and the micro-mirrors can be reduced to micro-meters. Light illumination of cells at a short distance was demonstrated by placing a tiny biogenic reflecting platelet on a cultured cell layer (Figure 10.17). The results suggest possible applications of cellular imaging by the reflecting particles.

ACKNOWLEDGMENTS

This work was supported by JST-CREST "Advanced core technology for creation and practical utilization of innovative properties and functions based upon optics and photonics (Grant number: JPMJCR16N1)." The author declares no competing interests.

REFERENCES

1. A. G. Domel, M. Saadat, J. C. Weaver, H. Haj-Hariri, K. Bertoldi and G. V. Lauder, Shark skin-inspired designs that improve aerodynamic performance, *J. R. Soc. Interface*, 15(139), 0828 (2017).
2. P. Vukusic and J. R. Sambles, Photonic structures in biology, *Nature*, 424(6950), 852–855 (2003).
3. M. Srinivasarao, Nano-optics in the biological world: Beetles, butterflies, birds, and moths, *Chem. Rev.*, 99(7), 1935–1961 (1999).
4. A. R. Parker and N. Martini, Structural colour in animals - Simple to complex optics, *Opt. Laser Technol.*, 38(4–6), 315–322 (2006).
5. D. Zhu, S. Kinoshita, D. S. Cai and J. B. Cole, Investigation of structural colors in Morpho butterflies using the nonstandard-finite-difference time-domain method: Effects of alternately stacked shelves and ridge density, *Phys. Rev. E*, 80(5 Pt 1), 051924 (2009).
6. S. Berthier, E. Charron and A. Da Silva, Determination of the cuticle index of the scales of the iridescent butterfly Morpho Menelaus, *Opt. Commun.*, 228(4–6), 349–356 (2003).
7. L. Plattner, Optical properties of the scales of Morpho Rhetenor butterflies: Theoretical and experimental investigation of the back-scattering of light in the visible spectrum, *J. R. Soc. Interface*, 1(1), 49–59 (2004).
8. E. J. Denton and M. F. Land, Mechanism of reflexion in silvery layers of fish and cephalopods, *Proc. R. Soc. Lond. B*, 178(1050), 43–61 (1971).
9. P. J. Herring, Reflective systems in aquatic animals. *Comp., Biochem. Physiol.*, 109A, 513–546 (1994).
10. P. Schmelcher, L. S. Cederbaum and U. Kappes, *Molecules in Magnetic Fields: Fundamental Aspects, Conceptual Trends in Quantum Chemistry*, Springer Netherlands, 1–51 (1994).
11. L. Pauling and C. D. Coryell, The magnetic properties and structure of hemoglobin, oxyhemoglobin and carbonmonoxyhemoglobin, *Proc. Natl Acad. Sci. U.S.A.*, 22(4), 210–216 (1936).
12. T. Higashi, A. Yamagishi, T. Takeuchi, N. Kawaguchi, S.Sagawa, S. Onishi and M. Date, Orientation of erythrocytes in a strong static magnetic field, *Blood*, 82(4), 1328–1334 (1993).
13. J. Torbet, J. M. Freyssinet and G. Hudry-Clergeon, Oriented fibrin gels formed by polymerization in strong magnetic fields, *Nature*, 289(5793), 91–93 (1981).
14. J. Torbet and M. C. Ronziere, Magnetic alignment of collagen during self-assembly, *Biochem. J.*, 219(3), 1057–1059 (1984).
15. J. Torbet, Solution behavior of DNA studied with magnetically induced birefringence, *Methods Enzymol.*, 211, 518–532 (1992).
16. J. Torbet, M. Malbouyres, N. Builles, V. Justin, M. Roulet, O. Damour, A. Oldberg, F. Ruggiero and D. J. Hulmes, Orthogonal scaffold of magnetically aligned collagen lamellae for corneal stroma reconstruction, *Biomaterials*, 28(29), 4268–4276 (2007).
17. S. Ueno, M. Iwasaka and H. Tsuda, Effects of magnetic fields on fibrin polymerization and fibrinolysis, *IEEE Trans. Magn.*, 29(6), 3352–3354 (1993).
18. T. Kimura, Y. Sato, F. Kimura, M. Iwasaka and S. Ueno, Micropatterning of cells using modulated magnetic fields, *Langmuir*, 21(3), 830–832 (2005).
19. M. Iwasaka and S. Ueno, Detection of intracellular macromolecule behavior under strong magnetic fields by linearly polarized light, *Bioelectromagnetics*, 24(8), 564–570 (2003).
20. M. F. Testorf, P. Åke Öberg, M. Iwasaka and S. Ueno, Melanophore aggregation in strong static magnetic fields, *Bioelectromagnetics*, 23(6), 444–449 (2002).
21. T. Suda and S. Ueno, Magnetic orientation of red blood cell membranes, *IEEE Trans. Magn.*, 30(6), 4713–4715 (1994).
22. K. Suzuki, T. Toyota, K. Sato, M. Iwasaka, S. Ueno and T. Sugawara, Characteristic curved structure derived from collagen-containing tubular giant vesicles under static magnetic field, *Chem. Phys. Lett.*, 440(4–6), 286–290 (2007).
23. M. Chabre, Diamagnetic anisotropy and orientation of α helix in frog rhodopsin and meta II intermediate, *Proc. Natl Acad. Sci. U.S.A.*, 75(11), 5471–5474 (1978).
24. A. P. Ramirez, R. C. Haddon, O. Zhou, R. M. Fleming, J. Zhang, S. M. McClure and R. E. Smalley, Magnetic susceptibility of molecular carbon nanotubes and fullerite, *Science*, 265(5168), 84–86 (1994).
25. M. Iwasaka, Effects of static magnetic fields on light scattering in red chromatophore of goldfish scale, *J. Appl. Phys.*, 107, 09B314 (2010).
26. M. Iwasaka, Y. Miyashita, M. Kudo, S. Kurita and N. Owada, Effect of 10-T magnetic fields on structural colors in guanine crystals of fish scales, *J. Appl. Phys.*, 111, 07B316 (2013).
27. M. Iwasaka, Y. Miyashita, Y. Mizukawa, K. Suzuki, T. Toyota and T. Sugawara, Biaxial alignment control of guanine crystals by diamagnetic orientation, *Appl. Phys. Express*, 6(3), 037002 (2013).

28. M. Iwasaka and Y. Mizukawa, Light reflection control in biogenic micro-mirror by diamagnetic orientation, *Langmuir*, 29(13), 4328–4334 (2013).

29. Y. Mizukawa, Y.Miyashita, M. Satoh, Y. Shiraiwa and M. Iwasaka, Light intensity modulation by coccoliths of *Emiliania huxleyi* as a micro-photo-regulator, *Sci. Rep.*, 5, 13577 (2015).

30. W. J. Crookes, L.-L. Ding, Q. L. Huang, J. R. Kimbell, J. Horwitz, M. J. McFall-Ngai, Reflectins: The unusual proteins of squid reflective tissues, *Science*, 303(5655), 235 (2004).

31. A. Levy-Lior, E. Shimoni, O. Schwartz, E. Gavish-Regev, D. Oron, G. Oxford, S. Weiner and L. Addadi, Guanine based biogenic photonic crystal arrays in fish and spiders, *Adv. Funct. Mater.*, 20(2), 320–329 (2010).

32. J. Teyssier, S. V. Saenko, D. van der Marel and M. C. Milinkovitch, Photonic crystals cause active colour change in chameleons, *Nat. Commun.*, 6, 6368 (2015).

33. B. A. Palmer, G. J. Taylor, V. Brumfeld, D. Gur, M. Shemesh, N. Elad, A. Osherov, D. Oron, S. Weiner and L. Addadi, The image-forming mirror in the eye of the scallop, *Science*, 358(6367), 1172–1175 (2017).

34. J. Chae and S. Nishida, Integumental ultrastructure and color patterns in the iridescent copepods of the family Sapphirinidae (Copepoda: Poecilostomatoida), *Mar. Biol.*, 119(2), 205–210 (1994).

35. D. Gur, B. Leshem, M. Pierantoni, V. Farstey, D. Oron, S. Weiner and L. Addadi, Structural basis for the brilliant colors of the sapphirinid copepods, *J. Am. Chem. Soc.*, 13726(26), 8408–8411 (2015).

36. A. Levy-Lior, B. Pokroy, B. Levavi-Sivan, L. Leiserowitz, S. Weiner and L. Addadi, Biogenic guanine crystals from the skin of fish may be designed to enhance light reflectance, *Cryst. Growth Des.*, 8(2), 507–511 (2008).

37. T. M. Jordan, J. C. Partridge and N. W. Roberts, Non-polarizing broadband multilayer reflectors in fish, *Nat. Photon.*, 260(11), 759–763 (2012).

38. M. Iwasaka, Y. Mizukawa and N. W. Roberts, Magnetic control of the light reflection anisotropy in a biogenic guanine microcrystal platelet, *Langmuir*, 32(1), 180–187 (2015).

39. M. Iwasaka and H. Asada, Floating photonic crystals utilizing magnetically aligned biogenic guanine platelets, *Sci. Rep.*, 8(1), 16940 (2018).

40. A. Mootha, K. Suzuki, T. Kimura, M. Kurahashi, E. Muneyama, M. Iwasaka and H. Asada, Refinement of synthetic guanine crystals for fast diamagnetic rotation, *AIP Adv.*, 9(3), 035340 (2019).

41. Y. Oaki, S. Kaneko and H. Imai, Morphology and orientation control of guanine crystals: A biogenic architecture and its structure mimetics, *J. Mat. Chem.*, 22(42), 22686 (2012).

42. D. Gur, M. Pierantoni, N. E. Dov, A. Hirsh, Y. Feldman, S. Weiner and L. Addadi, Guanine crystallization in aqueous solutions enables control over crystal size and polymorphism, *Cryst. Growth Des.*, 16(9), 4975–4980 (2016).

43. Y. Takeuchi, Y. Miyashita, Y. Mizukawa and M. Iwasaka, Two-stage magnetic orientation of uric acid crystals as gout initiators, *Appl. Phys. Lett.*, 104(2), 024109 (2014).

44. Y. Takeuchi and M. Iwasaka, Detection of monosodium urate crystals for gout diagnosis using magnetic fields and near-infrared light, *IEEE Trans. Magn.*, 52(7), 2529061 (2016).

45. Y. Takeuchi, M. Sekiya, A. Hamasaki, M. Iwasaka and M. Matsuda, Magnetic orientational properties of monosodium urate crystals, *IEEE Trans. Magn.*, 53(11), 2726525 (2017).

46. A. B. Lall, H. H. Seliger, W. H. Biggley and J. E. Lloyd, Ecology of colors of firefly bioluminescence, *Science*, 210(4469), 560–562 (1980).

47. A. B. Lall and K. M. Worthy, Action spectra of the female's response in the firefly *Photinus pyralis* (Coleoptera: Lampyridae): Evidence for an achromatic detection of the bioluminescent optical signal, *J. Insect. Physiol.*, 46(6), 965–968 (2000).

48. K.-S. Goh, H.-S. Sheu, T.-E. Hua, M.-H. Kang and C.-W. Li, Uric acid spherulites in the reflector layer of firefly light organ, *PLoS One*, 8(2), e56406 (2013).

49. M. Iwasaka and S. Ueno, Bioluminescence under static magnetic fields, *J. Appl. Phys.*, 83(11), 6456–6458 (1998).

50. M. Iwasaka, Y. Miyashita, A. G. Barua, S. Kurita and N. Owada, Changes in the bioluminescence of firefly under pulsed and static magnetic fields, *J. Appl. Phys.*, 109(7), 07B303 (2011).

11 Non-Invasive Techniques in Brain Activity Measurement Using Light or Static Magnetic Fields Passing Through the Brain

Osamu Hiwaki

CONTENTS

11.1   Introduction 233
11.2   Non-Invasive Measurement of Brain Activity Using Light 234
      11.2.1   Characteristics of Biological Tissue to Light...234
      11.2.2   Near-Infrared Spectroscopy...236
      11.2.3   Fast Optical Signal...238
      11.2.4   Brain Activity Measurement Using Light Penetrating the Brain 239
11.3   Non-Invasive Measurement of Brain Activity Using a Static Magnetic Field...................240
      11.3.1   Brain Activity Measurement Using a Static Magnetic Field Penetrating
            the Brain...240
      11.3.2   Brain Activity Measurement Using a Static Magnetic Field Passing
            Through the Human Brain...243
11.4   Summary..246
References..247

11.1 INTRODUCTION

The development of non-invasive techniques to allow the measurement of brain activity is imperative for revealing how the human brain functions. Conventional techniques for the non-invasive measurement of brain activity such as functional magnetic resonance imaging (fMRI), near-infrared spectroscopy (NIRS), magnetoencephalography (MEG), and electroencephalography (EEG) are burdened by critical limitations in spatial or temporal resolution. NIRS, which uses near-infrared light, has been developed to measure cerebral activity conveniently. The signals detected using NIRS originate from changes in the concentration of oxyhemoglobin and deoxy-hemoglobin due to vascular changes in the cortex. However, following an electrophysiological response, the hemodynamic response is delayed by several seconds. Furthermore, in conventional NIRS, the distance between the light source and the detector located on the scalp ranges from 20 to 40 mm. As the light diffuses in all directions inside the head, both before and after passing through the brain tissue, the distribution of the light is rather complex in conventional NIRS. This means that it is difficult to achieve high spatial resolution by conventional NIRS. In this chapter, we introduce a non-invasive brain function measurement using light including NIRS at first. In the latter part of this chapter, the techniques which enable non-invasive measurement of brain activity accompanied by high spatial and temporal resolution using near-infrared light or a static magnetic field are discussed.

11.2 NON-INVASIVE MEASUREMENT OF BRAIN ACTIVITY USING LIGHT

11.2.1 CHARACTERISTICS OF BIOLOGICAL TISSUE TO LIGHT

Light is scattered and goes in various directions in biological tissue. The nature of the medium with respect to light can be characterized by absorption and scattering. Water and glass are hardly scattered against ultraviolet rays, visible light, and infrared rays. They have the property of absorption only, which is small depending on the wavelength. A glass plate and water with a smooth surface are transparent without scattering or absorption in the wavelength range of visible light as shown in Figure 11.1(a). Colored glass appears colored to absorb light at wavelengths other than that color. However, polished glass, crushed glass fragments, and small glass balls where the surface is not smooth are not transparent. This is not because light is absorbed. In these cases, because of the roughness of the surface and internal heterogeneity, the light is randomly reflected at the surface, and the direction of the light changes inside the glass as shown in Figure 11.1(b). This phenomenon is scattering. Although the energy of light is attenuated by the absorption, there is no attenuation of energy by summing the energy of the scattered light.

Light can be regarded as energy, and the biological tissue basically has the characteristics of scattering and absorption to light. The characteristics of scattering and absorption of the biological tissue depend largely on the characteristics of light. When the electromagnetic wave is regarded as light, the wavelength is classified into the light with various wavelengths from the cosmic ray with a wavelength shorter than 10^{-7} μm to gamma rays, X-rays, ultraviolet rays, visible light, infrared light, microwave, shortwave, medium wave, and long wave with a wavelength in the range of 10^{10} μm (10 km).

Electromagnetic waves with a wavelength shorter than X-rays have strong energy and are hardly scattered in biological tissue. X-rays and gamma rays are known to ionize biological tissue for its strong energy and they damage the living body. X-ray computed tomography (CT) has been developed to draw an anatomical tomographic image of the body using X-rays going straight into the living body. Positron emission tomography (PET) and single-photon emission computerized tomography (SPECT) draw a tomographic image of physiological information of the body by using gamma rays.

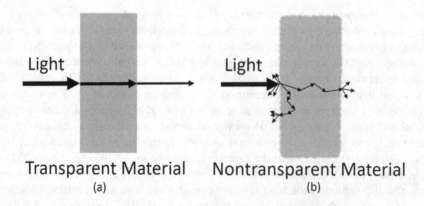

FIGURE 11.1 (a) Light in transparent material. Scattering and absorption of light is small in clear water and in glass with a flat surface. (b) Light in non-transparent material. The light is reflected randomly on the coarse surface and undergoes a zigzag path in the material due to scattering.

On the other hand, light and electromagnetic waves with a wavelength longer than ultraviolet rays are scattered in living tissue. It is characterized by not only scattering, but also the phenomenon associated with the propagation of waves such as reflection, refraction, and diffraction. Transmittance of the biological body of the electromagnetic wave is exponentially reduced in proportion to the frequency. Therefore, infrared light with a wavelength in the range from several micrometers to about 10 micrometers reaches only a few micrometers from the body surface. Microwaves with a wavelength of about 10 cm can be transmitted through tissue thickness of 10 cm. If the component of the transmitted microwave going straight is measured, it is possible to develop a microwave CT drawing of a tomographic image about the characteristics of the living body. Radio waves with a longer wavelength are more permeable to biological tissue, and radio waves generated inside the human body can be accurately measured outside the human body. Magnetic resonance imaging (MRI) utilizes this property.

Thus, the radio wave with a very long wavelength has permeability in the living body, and a signal passed through the human body can be detected. Ultraviolet, visible, and infrared light are not well permeable in the living body because of strong scattering or absorption. In particular, ultraviolet and far-infrared light absorbed in the biological tissue penetrate only the depth of the micrometer order. However, the absorption of biological tissue is weak for red light in visible light and near-infrared light, and the depth of penetration is of the millimeter order. The main cause is that hemoglobin in blood and water have absorption spectrum to visible light and near-infrared light as shown in Figure 11.2. Hemoglobin and water have low absorption at more than 700 nm and less than 900 nm respectively. Therefore, the range from 700 nm to 900 nm of light wavelength is called optical windows in biological tissues and fluids. The depth of the penetration of light in this wavelength range is longer than the light of other wavelengths. Absorption of light is lowered and scattering of light is dominant in this window. Scattering of light in visible and near-infrared light other than this window makes it difficult to quantitatively estimate the characteristics of light in biological tissues.

Visible light other than a red color light cannot penetrate the body because it is absorbed by blood and cells. Bone does not appear as a shadow by transmitting red light to the body, as bones do not scatter light in the same way as soft tissues. A reason for bone being visible by X-rays is that bone

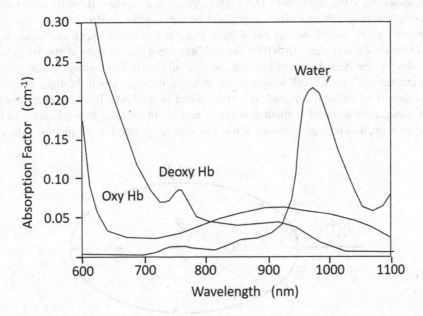

FIGURE 11.2 Absorption characteristics of oxy-hemoglobin (oxy-Hb), deoxy-hemoglobin (deoxy-Hb), and water according to the light wavelength.

absorbs X-rays more strongly than soft tissue. Moreover, the shape of the bone can be seen cleanly by X-rays because it goes straight without being scattered in the body as described above. If the bone absorbs visible light more strongly than soft tissue, it may look like a dark shadow by red light. Blood is a biological material that absorbs visible light, especially venous blood with little oxygen, resulting in the vein appearing to be a black line in transmitted visible light; nonetheless, the line is not a clear shape due to scattering. Scattering of light by biological tissue is similar to light scattering by irregularly shaped glass particles. In a microscopic view, scattering of light in biological tissue is a complex of the refraction at the cell membrane and scattering by mitochondria and other organelles. In the observation of biological tissue of a larger size than sub-mm, it is important to understand the macro scattering phenomenon composed with micro scattering. Generally, macro scattering phenomenon is represented by the scattering coefficient μ_s and the phase function $p(\theta)$. Phase function $p(\theta)$ of scattering light in biological tissue represents forward scattering as shown in Figure 11.3. Anisotropic scattering parameter g which is the average of cosine components of $p(\theta)$ is used as a parameter representing the direction of scattering. Anisotropic scattering parameter g takes a value from -1 to 1: $+1$ as pure forward scattering, 0 as isotropic scattering, and -1 as pure backscattering. It can be approximated as isotropic scattering when non-isotropic scattering is repeated in the medium. The equivalent scattering coefficient in that case can be represented as $\mu_s' = (1-g)\,\mu_s$. On the other hand, the absorption phenomenon can be expressed only by the absorption coefficient μ_a. The equivalent scattering coefficient μ_s' of the red color and near-infrared light is stronger in one to two orders than the absorption coefficient μ_a.

11.2.2 Near-Infrared Spectroscopy

In 1977, Jöbsis first measured non-invasively the oxygenation status of the heart and brain of animals using near-infrared light (Jöbsis 1977). NIRS has been promoted as a method for monitoring blood flow and oxygen metabolism in biological tissues. Furthermore, in the 1990s, the method using hemoglobin (Hb) changes in cerebral blood flow linked to nerve activity was developed as a new imaging method of brain function (functional NIRS, fNIRS) (Ferrari and Quaresima 2012).

Biomaterials with characteristic absorbers for near-infrared light (wavelengths 700 to 1500 nm) are hemoglobin (Hb), myoglobin (Mb), and cytochrome oxidase (CytOx) in mitochondria. Absorption spectrum of Hb and Mb is dependent on their oxidation/deoxidation state. Oxidized copper ions in CytOx. absorb near-infrared light, and the absorption spectrum varies depending on the oxidation/reduction state. In NIRS, the oxidation/deoxidation state of the Hb is measured non-invasively by the Beer–Lambert law (absorbance of a certain wavelength is proportional to the concentration and optical path length of the material through which the light is transmitted) using the intensity of the near-infrared light transmitted in the body. Beer–Lambert law holds in a uniform transparent material without scattering, and it can be used approximately in non-uniform substances such as biological tissue. When the light is scattered in the medium, the modified

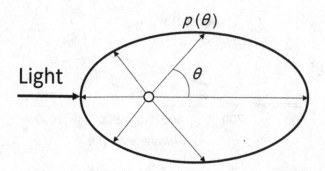

FIGURE 11.3   Phase function $p(\theta)$ of scattering light in biological tissue.

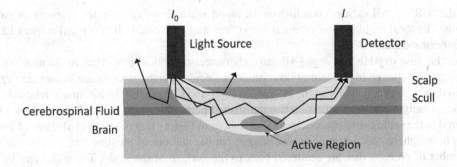

FIGURE 11.4 Light propagation in the head. The modified Beer–Lambert law accounts for tissue scattering by using the mean path length traveled by photons through the biological tissue as a best estimate for the actual photon path lengths.

Lambert–Beer law is used (Figure 11.4) (Delpy 1988). The absorbance (A) is defined with the source intensity I_0 and transmitted intensity I by

$$A = -\log\frac{I}{I_0} = \varepsilon CL + S \tag{11.1}$$

Here, ε is molar absorption coefficient. C is the concentration of the light absorber. L is the optical path length. S is a constant representing the attenuation of light due to scattering. The optical path length L in a scattered medium is the average path length of scattered light in the medium. In NIRS measurement, the concentration changes of oxidized Hb (oxy-Hb), deoxidized Hb (deoxy-Hb), and the total Hb (t-Hb), which is the sum of both, are measured. It is impossible to measure the optical path length L in the scattered medium. It is important to understand in the detected signal how the light propagated through the living body. Light propagation patterns in the human head have been studied by computer simulation. Due to the presence of cerebrospinal fluid in the head, there is no significant difference in the depth of light while the distance between the light source and the detector is 30–50 mm (Fukui et al. 2003). When the distance between the light source and the detector is 30 mm, the region deeper than 25 mm from the scalp is not detected (Hoshi et al. 2005). In other words, the area measured with NIRS is the cerebral cortex under the skull, rather than the deep region of the brain.

The relationship between local neural activity and subsequent changes in cerebral blood flow is called neurovascular coupling. It is possible to know the activity of the brain by the measurement of metabolic changes. Local neural activity in the brain leads to large increases in local blood flow. These increases in blood flow seemingly overcompensate for local oxygen consumption, leading to increases in the local concentration of both oxy-Hb and t-Hb and decreases in deoxy-Hb (Fox et al. 1988), and NIRS measurements often show an increase in oxy-Hb and t-Hb and a decrease in deoxy-Hb in the active region. The changes of t-Hb and deoxy-Hb may not always show in such a way due to the amount of change in cerebral blood flow. For example, when the change in cerebral blood flow is small, oxy-Hb and deoxy-Hb are changed oppositely and t-Hb is not changed. In addition, deoxy-Hb changes not only with the oxygenation state of venous blood but also with the amount of blood, so deoxy-Hb may increase when the increase in cerebral blood flow is larger than the decrease of deoxy-Hb originated from the oxidization of blood flow. On the other hand, the direction of oxy-Hb change always coincides with that of cerebral blood flow, so oxy-Hb in NIRS measurement is a good indicator of changes in local cerebral blood flow (Hoshi et al. 2001). It is often argued from which blood vessels NIRS signals are originated. In simple terms, the signal originates from venous blood which has a large proportion in the brain. However, many arterioles run into the brain vertically from the brain surface. It is considered that the change in the concentration of Hb in the artery and capillary cannot be ignored. Therefore, it is considered that the contribution of each intravascular

Hb to the NIRS signal varies depending on the blood vessel distribution in the propagation path of the light. The local regulation of cerebral blood flow and metabolism by neuronal activity has yet to be understood clearly.

Thus, because oxy-Hb and deoxy-Hb have characteristic optical properties in the near-infrared light range, the change in concentration of these molecules during neurovascular coupling can be measured using NIRS (Chance et al. 1998; Villringer and Chance 1997). Because a relatively predictable quantity of photons follows a banana-shaped path back to the scalp, these photons can be measured at the scalp using photodetectors, as shown in Figure 11.4 (Gratton et al. 1994). Changes in the chromophore concentrations cause changes in the number of photons that are absorbed and the number of photons that are scattered back to the surface of the scalp. These changes in light intensity measured at the surface of the scalp are estimated by a modified Beer–Lambert law as mentioned above. Changes in the relative concentrations of these chromophores can be calculated by measuring absorption changes at two different wavelengths, one of which is more reactive to oxy-Hb, the other to deoxy-Hb. It is possible to assess brain activity non-invasively by this principle of NIRS (Chance et al. 1993; Gratton et al. 1995; Hoshi and Tamura 1993; Kato et al. 1993; Villringer et al. 1993). Usually, a NIRS apparatus is comprised of a light source that is coupled to the head via light-emitting diodes (LEDs) through optical fiber bundles and with a light detector that receives the light after it has interacted with the tissue. Because light is scattered in the head, the detector placed about 30 mm away from the light source can detect light passed through the tissue. The NIRS signal can be detected as hemodynamic changes within the top 2–3 mm of the cortex and extends laterally 1 cm to either side, perpendicular to the axis of source-detector spacing (Chance et al. 1988). Using this NIRS technique, studies of brain activity have been conducted, including motor activity, visual activation, auditory stimulation, and the performance of cognitive tasks (Villringer and Chance 1997).

11.2.3 Fast Optical Signal

The metabolic and hemodynamic signals in NIRS are relatively slow and do not provide information on the characteristics of neural dynamics.

At the cellular level, electrical activation of nerves also causes fast intrinsic optical changes in scattering and birefringence that are largely independent of wavelength and are associated with action potentials and postsynaptic potentials (Cohen 1973; Hochman 1997; Obrig and Villringer 2003; Sable et al. 2007; Tasaki 1999). Some studies have revealed optical changes in the neuronal membrane. The voltage dependent changes in birefringence in the membrane of the nerve axon was found (Cohen 1973)., and it is suggested that the observed effect was due to a thinning of the membrane. A similar signal, which was practically simultaneous to the action potential, was observed (Stepnoski et al. 1991). It was found that thermal (heat production and absorption), mechanical (swelling and shrinking), and optical (birefringence) changes in the axon were simultaneous and related to a sudden swelling of the gel layer of the membrane due to an increase in water content (Tasaki 1999). This gel layer is superficial and about 0.5 μm thick and an integral part of the membrane of the axon. Fast intrinsic birefringence signals associated with neural activation were recorded from isolated nerves relative to simultaneous measurements of the light-scattering response (Carter et al. 2004; Yao et al. 2005). Optical change located at the nerve terminal was also found. It was investigated in the intact neurohypophysis of mice and reported large and rapid decreases in light scattering, which accompanied the secretion by nerve terminals (Salzberg et al. 1985). It indicated other possible effects, such as cell swelling, changes in the cytoskeleton, and changes in the geometry of microtubules, which change the light scattering of neuronal tissue (Sable et al. 2007). Studies on the exposed cortex in animals were also performed. Fast optical changes associated with evoked neural activation in the dorsal brainstem of anesthetized rats were measured (Rector et al. 2001).

The measurement of fast intrinsic optical responses is a promising method that may provide a useful alternative to electrophysiological techniques for dynamic measurements of neural activation.

Optical techniques offer a number of technical advantages; measurements are fast, cost-effective, and non-invasive, with high spatial and temporal resolution. Although discovered earlier than the hemodynamic optical signal observed in NIRS, the so-called fast optical signal (FOS) remains controversial when attempting to record it non-invasively through the scalp. There have been several attempts to record FOS non-invasively in human subjects (Gratton et al. 1995; Gratton and Fabiani 2003; Gratton et al. 1997; Steinbrink et al. 2000; Wolf et al. 2002). However, the results of these studies are still controversial. Several research laboratories have therefore tried to detect the neuronal signal non-invasively in human subjects. The first attempt was published in 1995 (Gratton et al. 1995). Using phase measurement with a frequency-domain optical instrument, Gratton et al. reported short-latency and well-localized event-related optical signal (EROS) recorded from the human occipital cortex during visual stimulation (Gratton et al. 1997). Since then, several reports replicating the initial results were published by the same group, including a recording of the fast optical signal from the auditory cortex during auditory processing and motor cortex during somatosensory stimulation (Gratton and Fabiani 2003; Maclin et al. 2004; Steinbrink et al. 2000). Steinbrink et al. were the first group to measure fast optical signals using a continuous-wave (CW) optical instrument and intensity measurements, rather than photon delay (Steinbrink et al. 2000), but subsequent results reported by this group have been less consistent.

The ability to measure the actual depolarization state of neuronal tissue non-invasively provides obvious advantages. Electroencephalography (EEG) and magnetoencephalography (MEG) can be measured with time resolution in milliseconds, but with the superior spatial resolution lacking. There are also a number of limitations to the non-invasive use of the EROS signal in humans. A primary disadvantage of the measurement of the fast optical signal is the low signal-to-noise ratio (SNR). Basic sensory and motor movements such as tactile stimulation and finger-tapping require between 500 and 1000 trials to obtain a reliable signal (Franceschini and Boas 2004). The low SNR may cause current difficulties with experimental replication. Another constraint is that these methods require a more expensive and cumbersome laser-based light source (versus an LED-based light source), which are not portable, and the potential risk of inadvertent damage to the eyes is increased relative to the systems available for measuring hemodynamic responses. LED-based near-infrared sources pose very little, if any, risk upon eye exposure (Bozkurt and Onaral 2004). In spite of these current limitations, the fast optical signal continues to be an important area of investigation because it offers valuable observations in neuroimaging: the direct measurement of neuronal activity with millisecond time resolution and superior spatial resolution.

11.2.4 BRAIN ACTIVITY MEASUREMENT USING LIGHT PENETRATING THE BRAIN

We have developed a non-invasive technique in the measurement of neural activity with a superior time and spatial resolution using near-infrared light, which is different from NIRS or EROS mentioned above. The use of a light source and detector located on the scalp, such as in conventional NIRS or EROS, cannot achieve enough of a clear signal or measurement because of a poor yield of light back to the head surface via the cerebral cortex following a banana-shaped path as shown in Figure 11.4. In the experiment using isolated nerves, it was shown that transmitted light leads to a clearer nerve signal than reflected light (Carter et al. 2004). This result suggests that it is considered to be more effective to place the light source and detector so that light crosses the brain to measure the brain signal. Therefore, we contrived a method where light crossed the brain. That is, a near-infrared light LED was placed on the palate in the mouth and the detector was placed on the scalp, which we term as optoencephalography (OEG) (Hiwaki and Miyaguchi 2018). The effectiveness and validity of this method was verified with common marmosets. We measured the OEG responses following the stimulation of the median nerve of the left wrist. For OEG recording, an LED emitting 850 nm near-infrared light was fixed on the palate in the mouth. The LED was embedded in polyethylene resin, shaped in the form of a mouthguard, and fitted onto the palate of the upper jaw as shown in Figure 11.5. The near-infrared light penetrating the head was detected

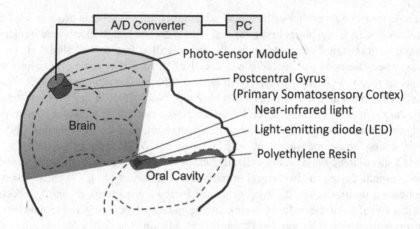

FIGURE 11.5 Schematic drawing of the optoencephalography (OEG) measurement system. Near-infrared light was emitted from a light-emitting diode (LED) located on the palate of the upper jaw. The LED was embedded in the polyethylene resin shaped like a mouthguard. The near-infrared light penetrating the brain was detected on the scalp using a photo-sensor module. The optical signal measured with the photo-sensor module was sampled with an A/D converter. The photo-sensor was placed on the scalp and aimed at the hand region of the postcentral gyrus of the cortex (corresponding to the S1 primary somatosensory cortex).

on the scalp using a photo-sensor module which was located at (R7, 0); 7 mm right from vertex on the scalp. The OEG signal was expressed as the average of 300 values of $\Delta I/I$; the average optical signal during a 125 ms prestimulus period was defined as I, and the difference between the measured optical signal and I was defined as ΔI. In order to evaluate the validity of the OEG signals, somatosensory evoked potential (SSEP) recording using EEG apparatus was also conducted. In the SSEP recording, the center point of a sintered Ag/AgCl recording electrode of 10 mm diameter was placed at (R7, 0), i.e., at the same position as the photo-sensor in the OEG measurement. The OEG response and SSEP measured at (R7, 0) are shown in Figure 11.6(a) and Figure 11.6(b) respectively. The cross-correlation coefficient between the OEG response and SSEP was 0.72, which denotes high similarity between the OEG and SSEP. We measured the OEG responses at an additional four points surrounding the position (R7, 0) in order to evaluate the spatial resolution of OEG. The positions for OEG measurement were: vertex (0, 0); 14 mm right from vertex (R14, 0); 7 mm right and 7 mm anterior from vertex (R7, A7); and 7 mm right and 7 mm posterior from vertex (R7, P7), in addition to (R7, 0) as shown in Figure 11.7(a). The OEG responses measured at four additional points around position (R7, 0) are shown in Figure 11.7(b). The amplitudes of the OEG responses measured at four points at a distance of 7 mm from (R7, 0) were less than half from the original measurement at (R7, 0). Especially, the OEG response measured at the vortex was small.

Our results demonstrate that the technique using OEG measurement enables non-invasive measurement of a localized and fast brain signal, which is tightly coupled to the electrophysiological activity of the brain.

11.3 NON-INVASIVE MEASUREMENT OF BRAIN ACTIVITY USING A STATIC MAGNETIC FIELD

11.3.1 BRAIN ACTIVITY MEASUREMENT USING A STATIC MAGNETIC FIELD PENETRATING THE BRAIN

It is well known that a static magnetic field can penetrate the human body with little or no attenuation because the permeability of biological tissue is essentially the same as that of air. If the OEG signal we observed originated in the mechanical (i.e., swelling and shrinking) changes of the

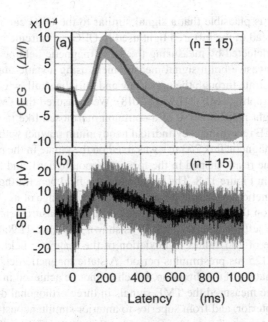

FIGURE 11.6 (a) Somatosensory evoked optoencephalography (OEG) response. (b) Somatosensory evoked potential (SSEP). The bold lines represent the averages of the OEG responses and SSEPs. The gray lines indicate the standard deviations at each sampling time.

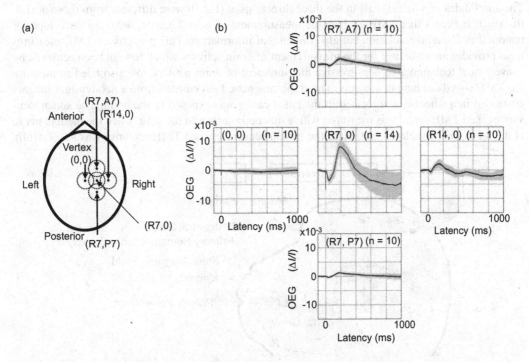

FIGURE 11.7 (a) Locations of photo-sensors to measure optoencephalography (OEG) responses. (b) Observed OEG responses. The locations of photo-sensors were vertex (0, 0); 14 mm right from vertex (R14, 0); 7 mm right and 7 mm anterior from vertex (R7, A7); and 7 mm right and 7 mm posterior from vertex (R7, P7), in addition to (R7, 0). Dotted circles in the locations of photo-sensors represent the window area of the photo-sensor. In the OEG responses, the black lines represent the averages of the OEG responses and the gray lines indicate the standard deviations at each sampling time.

neurons in the cortex, it is plausible that a signal similar to the OEG can be observed by using a static magnetic field instead of the infrared light used in OEG measurement. We hypothesized that the brain signal can be detected by measuring the static magnetic field penetrating the brain. We tried to conduct a non-invasive brain signal measurement using a static magnetic field (instead of the light in OEG) to penetrate through the upper jaw and into the skull, which we term as transcranial magnetoencephalography (TME) (Hiwaki 2018). We measured the TME responses following the stimulation of the right median nerve of a common marmoset, like in the OEG measurement described above. For TME recording, a cylindrical neodymium magnet with 2 mm diameter, 1 mm height, and the surface magnetic flux 263 mT was fixed on the palate in the mouth. The magnet was embedded in polyethylene resin, shaped in the form of a mouthguard, and fitted onto the palate of the upper jaw, as shown in Figure 11.8. The static magnetic field penetrating the head was detected on the scalp using a magnetic sensor which was sensitive in the range of nT. We measured the TME responses at two points on the scalp. The positions for TME measurement were: 8 mm left from vertex (L8, 0) and 4 mm left from vertex (L4, 0) as shown in Figure 11.9(a). The TME signal was expressed as the average of 150 values of deviation of the magnetic field (ΔB) from the average magnetic field during a 125 ms prestimulus period. A static magnetic field can be measured as a vector containing information of the direction, which cannot be achieved in the OEG measurement using light. Therefore, we measured the TME signals in three orthogonal directions: from right to left, from posterior to anterior, and from superior to anterior simultaneously. Figure 11.9(b) shows the TME responses in the three directions measured at (L8, 0) and (L4, 0). The waveforms of these somatosensory evoked TME responses, which have a negative peak and a positive peak, are similar to the somatosensory evoked OEG signals shown in Figure 11.6 and 11.7. This indicates that TME signals originate from the same phenomena accompanied with neural activity as OEG signals. The amplitudes of TME signals in the three directions at (L8, 0) were different from those at (L4, 0), which indicates that TME enables the measurement of neural activity with not only superior temporal and spatial resolution but also directional information. This property of TME measurement provides an advantage for the measurement of brain activity, which has not been achieved by conventional techniques in non-invasive measurement of brain activity. We also tried to measure the TME signals of human subjects. The static magnetic field emitted from a neodymium magnet contained in a silicone tube located in the nasal cavity was exposed to the brain. The somatosensory evoked TME signal was measured with a magnetic sensor on the scalp when the median nerve at the wrist was stimulated. Although the somatosensory evoked TME response was successfully

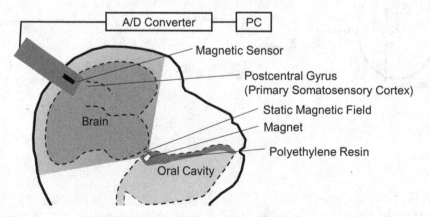

FIGURE 11.8 Schematic drawing of the transcranial magnetoencephalography (TME) measurement system. A static magnetic field was emitted from a magnet located on the palate of the upper jaw. The magnet was embedded in the polyethylene resin shaped like a mouthguard. The static magnetic field penetrating the brain was detected on the scalp using a magnetic sensor. The magnetic signal measured with the magnetic sensor was sampled with an A/D converter.

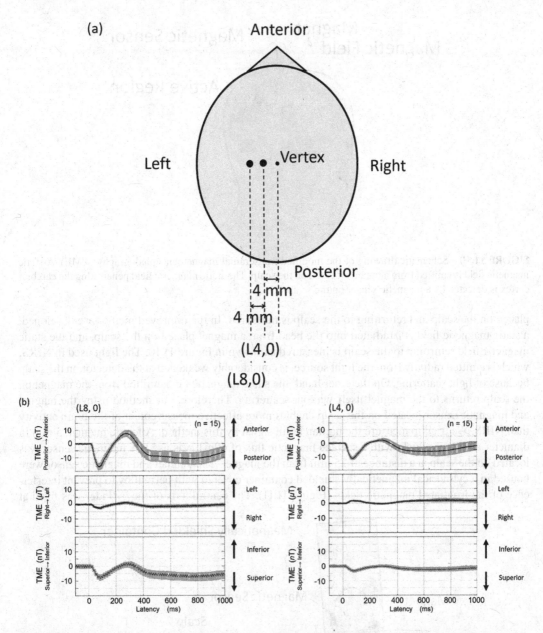

FIGURE 11.9 (a) Locations of magnetic sensors to measure magnetoencephalography (TME) responses. (b) Observed TME responses. The locations of magnetic sensors are 8 mm left from vertex (L8, 0); 4 mm left from vertex (L4, 0). In the TME responses, the bold lines represent the averages of the TME responses and the gray lines indicate the standard deviations at each sampling time.

observed, the signal was ambiguous because of the longer distance between the magnet and the sensor compared with the measurement with common marmosets.

11.3.2 BRAIN ACTIVITY MEASUREMENT USING A STATIC MAGNETIC FIELD PASSING THROUGH THE HUMAN BRAIN

We improved the configuration of the magnet and magnetic sensor of the TME system to measure clearer brain signals (Hiwaki 2019). In NIRS, near-infrared light transmitted from a light source

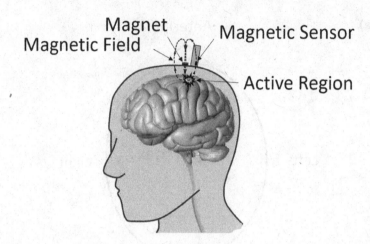

FIGURE 11.10 Schematic drawing of the modified transcranial magnetoencephalography (TME). A static magnetic field is emitted from a magnet located on the scalp. The static magnetic field penetrating the cerebral cortex is detected by a magnetic sensor on the scalp.

placed on the scalp and returning to the scalp is measured. In the improved method we developed, a static magnetic field is irradiated into the head from a magnet placed on the scalp, and the static magnetic field returning to the scalp is measured as shown in Figure 11.10. The light used in NIRS, which is emitted radially from the light source, is considerably weakened at the detector on the scalp because of light scattering. On the other hand, the static magnetic field emitted from the magnet on the scalp returns to the magnet itself without scattering. Therefore, our method using the magnet and magnetic sensor located on the scalp enables more effective measurement of the brain activity than NIRS. We tried to measure the human brain signal by this method. A ferrite magnet, 5 mm in diameter and 2 mm thick, with a surface magnetic flux of 60mT was used. A magnetic sensor was located on the scalp at a distance of 25 mm from the magnet. The magnet and magnetic sensor were housed in a cylindrical magnetically shielded container covered with permalloy to prevent interference of the irradiated magnetic fields (Figure 11.11). The central axis of the cylinder was fixed at

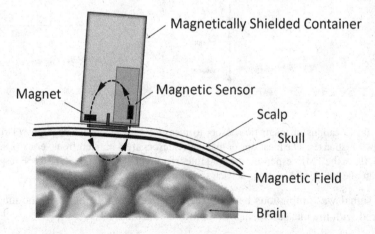

FIGURE 11.11 Transcranial magnetoencephalography (TME) measurement system. A ferrite magnet with a diameter of 5 mm and a thickness of 2 mm was placed on the scalp. A magnetic field, irradiated from the magnet, passing through the cerebral cortex under the skull, was measured by a magnetic sensor at a distance of 25 mm from the magnet on the scalp. In order to prevent interference of the irradiated magnetic fields next to each other, both the magnet and the magnetic sensor were shielded in a cylindrical container with a diameter of 33 mm and a height of 54 mm covered with permalloy sheets. The magnet was put N-pole down on the surface of the head. We made a 16-channel TME system composed of 16 pairs of the magnet and the magnetic sensor.

the center point between the magnet and the magnetic sensor. The cylinder was allowed to rotate around the central axis. The N-pole of the magnet was placed facing the scalp, and the tangential direction from the magnet to the magnetic sensor was defined as the direction of the magnetic field. We have developed a 16-channel TME measurement system consisting of 16 pairs of the magnet and magnetic sensor. Using this system, the effectiveness and validity of this method was verified by measuring the somatic sensory evoked TME responses in human subjects. The electric current stimulation consisted of a monophasic square wave pulse (200 μs in duration) applied every 1.2 s just above the motor threshold to elicit a slight but visible twitch of the thumb. The somatosensory evoked TME signals were obtained by an average of 300 responses. The TME responses were measured at 16 points with the 16-channel TME system as shown in Figure 11.12 (a). The direction of

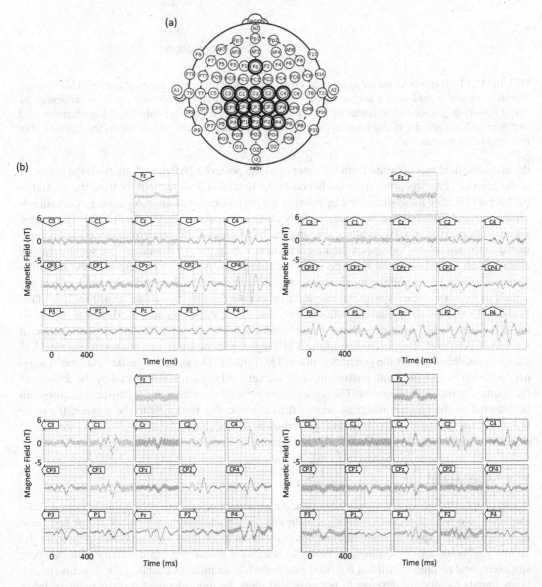

FIGURE 11.12 (a) Locations of the magnet and magnetic sensors to measure transcranial magnetoencephalography (TME) responses. (b) Observed TME responses. The arrows indicate the directions of the magnetic fields. The largest signal was observed at CP4 locally, overlying the hand area of the somatosensory cortex contralateral to the stimulated side when the magnetic field was from the anterior to the posterior (AP direction). The TME signals were changed with the direction of the static magnetic field.

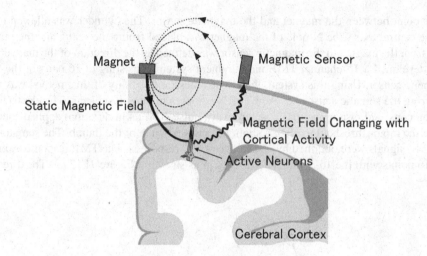

FIGURE 11.13 Hypothesis on the generation of the transcranial magnetoencephalography (TME) signal. The static magnetic field from a magnet on the scalp, passing through the cerebral cortex, is influenced by neural activity (e.g., mechanical [swelling and shrinking] change) in the cerebral cortex. The magnetic field changing with the activity of the neurons can be detected by the magnetic sensor due to the magnetic flux returning to the magnet.

the magnetic field was adjusted from the anterior to the posterior (AP direction), from the posterior to the anterior (PA direction), from the lateral to the medial (LM direction), or from the medial to the lateral (ML direction) directions by rotating the cylindrical container. As a result, the somato-sensory evoked TME signals were successfully observed as shown in Figure 11.12(b). The largest signal was observed at CP4 locally, overlying the hand area of the somatosensory cortex contra-lateral to the stimulated side when the magnetic field was in the AP direction. The somatosensory evoked TME response at CP4 with the magnetic field in the AP direction began to deviate with a latency of about 30 ms and the maximum amplitude was about 6 nT, whereas the amplitude of the signals at the same location with the magnetic fields in other directions were less than 2 nT. That is, the TME signals were changed with the direction of the static magnetic field. It is verified that the technique of TME enables vectorial measurement of brain activity with superior spatiotemporal resolution. Although the mechanism of the TME signal generation is still unknown, Figure 11.13 shows a plausible cause of the generation of the TME signal. The static magnetic field from a mag-net on the surface of the head, passing through the cerebral cortex, is influenced by the activity of the neurons in the cerebral cortex. The signal changing with the activity of the cortical neurons can be detected by the magnetometer by way of the magnetic flux returning to the magnet. It can be presumed that the static magnetic field is perturbed by the mechanical (i.e., swelling and shrinking) changes of the neurons accompanied with neural activity in the cortex.

11.4 SUMMARY

In this chapter, non-invasive techniques in the measurement of brain activity using light or a static magnetic field were discussed. The ideal technique which enables measurement of fast and focal brain signals was explored. The combinations of different modalities were proposed to increase the accuracy of brain activity estimation. However, a method measuring brain activity with excellent spatiotemporal resolution still has not been established. If techniques enabling the detection of fast brain signals focally are developed, not only will there be more detailed investigations of brain function, but also optimal brain–machine interface will be realized. Such techniques, including our methods in the non-invasive measurement of brain activity discussed in this chapter, have the potential to open a new horizon in the understanding and application of human brain function.

REFERENCES

Bozkurt A, Onaral B. 2004. Safety assessment of near infrared light emitting diodes for diffuse optical measurements. *Biomed Eng Online* 3(1):9.

Carter KM, George JS, Rector DM. 2004. Simultaneous birefringence and scattered light measurements reveal anatomical features in isolated crustacean nerve. *J Neurosci Methods* 135(1–2):9–16.

Chance B, Anday E, Nioka S, Zhou S, Hong L, Worden K, Li C, Murray T, Ovetsky Y, Pidikiti D, Thomas R. 1998. A novel method for fast imaging of brain function, non-invasively, with light. *Opt Express* 2(10):411–23.

Chance B, Leigh JS, Miyake H, Smith DS, Nioka S, Greenfeld R, Finander M, Kaufmann K, Levy W, Young M. 1988. Comparison of time-resolved and -unresolved measurements of deoxyhemoglobin in brain. *Proc Natl Acad Sci U S A* 85(14):4971–5.

Chance B, Zhuang Z, UnAh C, Alter C, Lipton L. 1993. Cognition-activated low-frequency modulation of light absorption in human brain. *Proc Natl Acad Sci U S A* 90(8):3770–4.

Cohen LB. 1973. Changes in neuron structure during action potential propagation and synaptic transmission. *Physiol Rev* 53(2):373–418.

Delpy DT. 1988. Developments in oxygen monitoring. *J Biomed Eng* 10(6):533–40.

Ferrari M, Quaresima V. 2012. A brief review on the history of human functional near-infrared spectroscopy (fNIRS) development and fields of application. *Neuroimage* 63(2):921–35.

Fox PT, Raichle ME, Mintun MA, Dence C. 1988. Nonoxidative glucose consumption during focal physiologic neural activity. *Science* 241(4864):462–4.

Franceschini MA, Boas DA. 2004. Noninvasive measurement of neuronal activity with near-infrared optical imaging. *Neuroimage* 21(1):372–86.

Fukui Y, Ajichi Y, Okada E. 2003. Monte Carlo prediction of near-infrared light propagation in realistic adult and neonatal head models. *Appl Opt* 42(16):2881–7.

Gratton G, Corballis PM, Cho E, Fabiani M, Hood DC. 1995. Shades of gray matter: Noninvasive optical images of human brain responses during visual stimulation. *Psychophysiology* 32(5):505–9.

Gratton G, Fabiani M. 2003. The event-related optical signal (EROS) in visual cortex: Replicability, consistency, localization, and resolution. *Psychophysiology* 40(4):561–71.

Gratton G, Fabiani M, Corballis PM, Hood DC, Goodman-Wood MR, Hirsch J, Kim K, Friedman D, Gratton E. 1997. Fast and localized event-related optical signals (EROS) in the human occipital cortex: Comparisons with the visual evoked potential and fMRI. *Neuroimage* 6(3):168–80.

Gratton G, Maier JS, Fabiani M, Mantulin WW, Gratton E. 1994. Feasibility of intracranial near-infrared optical scanning. *Psychophysiology* 31(2):211–5.

Hiwaki O. 2018. Noninvasive measurement technique for dynamic brain signals by magnetic field penetrating the brain. *The 41st Annual Meeting of the Japan Neuroscience Society*, 2P-379.

Hiwaki O. 2019. Novel technique for noninvasive measurement of dynamic brain signals using static magnetic fields passing through the brain. *The 9th International IEEE EMBS Conference on Neural Engineering*, FrPO.132.

Hiwaki O, Miyaguchi H. 2018. Noninvasive measurement of dynamic brain signals using light penetrating the brain. *PLoS One* 13(1):e0192095.

Hochman DW. 1997. Intrinsic optical changes in neuronal tissue: Basic mechanisms. *Neurosurg Clin N Am* 8(3):393–412.

Hoshi Y, Kobayashi N, Tamura M. 2001. Interpretation of near-infrared spectroscopy signals: A study with a newly developed perfused rat brain model. *J Appl Physiol* 90(5):1657–62.

Hoshi Y, Shimada M, Sato C, Iguchi Y. 2005. Reevaluation of near-infrared light propagation in the adult human head: Implications for functional near-infrared spectroscopy. *J Biomed Opt* 10(6):064032.

Hoshi Y, Tamura M. 1993. Dynamic multichannel near-infrared optical imaging of human brain activity. *J Appl Physiol 1985* 75(4):1842–6.

Jöbsis FF. 1977. Noninvasive, infrared monitoring of cerebral and myocardial oxygen sufficiency and circulatory parameters. *Science* 198(4323):1264–7.

Kato T, Kamei A, Takashima S, Ozaki T. 1993. Human visual cortical function during photic stimulation monitoring by means of near-infrared spectroscopy. *J Cereb Blood Flow Metab* 13(3):516–20.

Maclin EL, Low KA, Sable JJ, Fabiani M, Gratton G. 2004. The event-related optical signal to electrical stimulation of the median nerve. *Neuroimage* 21(4):1798–804.

Obrig H, Villringer A. 2003. Beyond the visible—Imaging the human brain with light. *J Cereb Blood Flow Metab* 23(1):1–18.

Rector DM, Rogers RF, Schwaber JS, Harper RM, George JS. 2001. Scattered-light imaging in vivo tracks fast and slow processes of neurophysiological activation. *Neuroimage* 14(5):977–94.

Sable JJ, Rector DM, Gratton G. 2007. Optical neurophysiology based on animal models. *IEEE Eng Med Biol Mag* 26(4):17–24.

Salzberg BM, Obaid AL, Gainer H. 1985. Large and rapid changes in light scattering accompany secretion by nerve terminals in the mammalian neurohypophysis. *J Gen Physiol* 86(3):395–411.

Steinbrink J, Kohl M, Obrig H, Curio G, Syré F, Thomas F, Wabnitz H, Rinneberg H, Villringer A. 2000. Somatosensory evoked fast optical intensity changes detected non-invasively in the adult human head. *Neurosci Lett* 291(2):105–8.

Stepnoski RA, LaPorta A, Raccuia-Behling F, Blonder GE, Slusher RE, Kleinfeld D. 1991. Noninvasive detection of changes in membrane potential in cultured neurons by light scattering. *Proc Natl Acad Sci U S A* 88(21):9382–6.

Tasaki I. 1999. Rapid structural changes in nerve fibers and cells associated with their excitation processes. *Jpn J Physiol* 49(2):125–38.

Villringer A, Chance B. 1997. Non-invasive optical spectroscopy and imaging of human brain function. *Trends Neurosci* 20(10):435–42.

Villringer A, Planck J, Hock C, Schleinkofer L, Dirnagl U. 1993. Near infrared spectroscopy (NIRS): A new tool to study hemodynamic changes during activation of brain function in human adults. *Neurosci Lett* 154(1–2):101–4.

Wolf M, Wolf U, Choi JH, Gupta R, Safonova LP, Paunescu LA, Michalos A, Gratton E. 2002. Functional frequency-domain near-infrared spectroscopy detects fast neuronal signal in the motor cortex. *Neuroimage* 17(4):1868–75.

Yao XC, Foust A, Rector DM, Barrowes B, George JS. 2005. Cross-polarized reflected light measurement of fast optical responses associated with neural activation. *Biophys J* 88(6):4170–7.

Index

Absorbance *(A)*, 237
Acidic-pH-activatable probes, 25, 27
Activatable probes, 8, 16–17
   aminopeptidase-targeted activatable fluorescence
      probes, 28–31
   carboxypeptidase-targeted, 30, 31
   enzyme pre-targeting and, 24, 26
   fluorescence switching mechanisms, 17–18
   PeT-based fluorescence imaging probes, 18–20
   spirocyclization-based fluorescence imaging probes,
      19, 21–25
   targeted to glycosidases, 25, 28
   tumor-targeting antibody and acidic-pH-activatable
      probes, 25, 27
AD, *see* Axial diffusivity
ADC, *see* Apparent diffusion coefficient
ADEPT, *see* Antibody-directed enzyme prodrug therapy
Always-on probes, 8, 16, 17
Amide proton transfer (APT) imaging
   amide proton transfer ratio (APTR), 103
   brain tumors, 105
      differential diagnosis, 107–108
      diffuse gliomas, 105–107
      discrimination between treatment-related changes
         and viable tumors, 108
      meningiomas, 108
   breast, chest, and abdominal lesions, 109–110
   effect, quantifying, 104–105
   head and neck lesions, 109
   method, 104
   in oncology, 110–111
Amide proton transfer ratio (APTR), 103
Aminopeptidase-targeted activatable fluorescence probes,
      28–31
Anisotropic diffusion, 130–131
Antibody-directed enzyme prodrug therapy (ADEPT), 24
Apparent diffusion coefficient (ADC), 129–130
APT imaging, *see* Amide proton transfer imaging
APTR, *see* Amide proton transfer ratio
APT-weighted signal intensity (APTWSI), 104–111
Arterial spin labeling, 98
Artificial guanine crystals, 221–224
Automatic magnetic immunostaining system, 195–196
Axial diffusivity (AD), 132
Axillary lymph nodes, breast cancer, 185

Beer–Lambert law, 236, 238
β-galatosidase, 22–24, 26, 28
Bio-crystallization, 221
Biogenic guanine crystals, 217
   artificial guanine crystals, synthesis of, 221–224
   diamagnetic micro-mirrors, magnetic orientation of,
      220–221, 223
   platelets from fish, micro-mirror properties of, 218,
      220, 223
Biogenic micro-mirrors, 9–10
   magnetically oriented (*see* Magnetically oriented
      biogenic micro/nanoparticles)

magnetic control (*see* Magnetic control)
   optical detection, 216, 217
   for photonic cellular analysis, 229–230
Bioimaging, activatable fluorescence probes for
   PeT-based fluorescence imaging probes, 18–20
   spirocyclization-based fluorescence imaging probes,
      19, 21–25
Bloch equation, 81
Bloch–Torrey equation, 125
Blood oxygenation level dependent (BOLD) effects, 5, 95
Boltzmann distribution, 79
Brain activity measurement, *see* Non-invasive techniques,
      brain activity measurement
Breast, chest, and abdominal lesions, 109–110
Brownian relaxation, 157
Brownian relaxation time, 159
b-tensor, 144

Carbon-hydrogen (CH)-stretching region, 51
Carboxypeptidase-targeted activatable fluorescence
      probes, 30, 31
CARS microscopy, *see* Coherent anti-Stokes Raman
      scattering microscopy
Central limit theorem, diffusion magnetic resonance
      imaging, 124
Chemical exchange saturation transfer (CEST) imaging,
      9, 101
   APT imaging (*see* Amide proton transfer imaging)
   endogenous, 103
   exogenous, 113–115
   of GAGs, 113
   of glycogen, 113
   of pH, 111–113
   principle, 102
   Z-spectrum, 102, 103
C-H stretching vibrational mode, 51
Coherent anti-Stokes Raman scattering (CARS)
      microscopy, 9, 38
   applications, 49–51
   hyperspectral (*see* Hyperspectral CARS microscopy)
   number of photons, 57–58
   as optical phase modulation, 56–57
   quantum-mechanical picture, 61–63
   schematics, 40–41
   signal-to-noise ratios, 57
   two-color pulse sources, 42–43
Coherent Raman scattering (CRS)
   dielectric permittivity, time- and frequency-domain
      responses of, 55–56
   microscopy, 9, 38
      image-forming system, 64–66
      optical resolution, 66–68
   principles, 39–40
   quantum-mechanical picture
      CARS, 61–63
      density matrix, 58
      energy diagrams, 59
      Hamiltonian, 58

nonlinear polarization density, 60–61
SRL, 63–64
third-order perturbation term, 60
two-color optical field, driving force, 54–55
Computed tomography (CT), 4
Constrained spherical deconvolution (CSD), 140
Critical fields, 226
CRS, *see* Coherent Raman scattering
CSD, *see* Constrained spherical deconvolution
Cumulant expansion, 133

Diamagnetic CEST (diaCEST), 114
Dielectric constant, 2
Dielectric permittivity, time- and frequency-domain
responses of, 55–56
Diffusion, imaging of, 97
Diffusion kurtosis imaging (DKI), 122, 133–134
Diffusion magnetic resonance imaging (dMRI)
anisotropic diffusion, 130–131
biophysical models of white matter, 134–136
Bloch–Torrey equation, 125
central limit theorem, 124
diffusion kurtosis imaging (DKI), 133–134
diffusion pattern, 121
diffusion propagator, 123–124
diffusion tensor imaging (DTI)
applications and limitations, 133
axial diffusivity (AD), 132
b-matrix, 131
fractional anisotropy (FA), 132, 133
linear least squares (LLS), 132
mean diffusivity (MD), 132
radial diffusivity (RD), 132
tractography, 139
diffusion time dependence, 144
diffusion time effect, 127–129
diffusion-weighted image, 129–130
Fick's law, 122–123
free, hindered, and restricted diffusion, 124
limitations, 137–139
multiple diffusion encoding, 144–146
neurite orientation dispersion and density imaging
(NODDI), 136–137
q-space, 126–127
random walk, 124
Stejskal–Tanner pulsed gradient spin–echo sequence,
125–126
white matter tract integrity (WMTI), 137
Diffusion propagator, 123–124
orientation information of, 139–140
Diffusion spectrum imaging (DSI), 139–140
Diffusion tensor imaging (DTI)
applications and limitations, 133
axial diffusivity (AD), 132
b-matrix, 131
diffusion kurtosis imaging (DKI), 133–135
fractional anisotropy (FA), 132, 133
linear least squares (LLS), 132
mean diffusivity (MD), 132
radial diffusivity (RD), 132
tractography, 139
Diffusion time dependence, 144
DKI, *see* Diffusion kurtosis imaging
dMRI, *see* Diffusion magnetic resonance imaging

DSI, *see* Diffusion spectrum imaging
DTI, *see* Diffusion tensor imaging

EEG, *see* Electroencephalogram; Electroencephalography
Electroencephalogram (EEG), 5
Electroencephalography (EEG), 239
somatosensory evoked potential (SSEP) recording, 240
Electromagnetic fields/waves
classification, 3
ionizing, 4
light velocity, 2
non-ionizing, 3
propagation velocity, 2
wavelength, 2
Endogenous CEST, 103
ESCCs, *see* Esophageal squamous cell carcinoma
Esophageal squamous cell carcinoma (ESCCs), 30
Exogenous CEST, 113–115

FA, *see* Fractional anisotropy
FDG, *see* Fluorodeoxyglucose
Ferromagnetic contrast agents, 208
FFL, *see* Field-free line
FFP, *see* Field-free point
Fiber tracking/tractography, 139
Fick's law, 122–123
Field-dependent relaxation time, 160
Field-free line (FFL), 7, 155
generation and scanning, 173–175
Field-free point (FFP), 7, 155, 156, 160
DC gradient field with, 170
generation and scanning, 172–173
MNP magnetization, 161, 162
Fingerprint region, 51
Fluorescence switching mechanisms, 17–18
Fluorescent magnetic nanoparticles, 194
Fluorodeoxyglucose (FDG), 114
fMRI, *see* Functional magnetic resonance imaging
fNIRS, *see* Functional NIRS
Folate receptor-α (FR-α), 16
Förster resonance energy transfer (FRET), 17–18
Fourier-transformation, hyperspectral CARS microscopy,
43, 44
Fractional anisotropy (FA), 132, 133
FRET, *see* Förster resonance energy transfer
FT-CARS method, 44
Functional magnetic resonance imaging (fMRI), 5
applications, 96–97
neurovascular coupling, 95–96
Functional NIRS (fNIRS), 7
Funk–Radon transform, 140

GAG, *see* Glycosaminoglycan
gagCEST, 113
γ-ray radiation, 4
GFP, *see* Green fluorescent protein
gGlu-HMRG, 30
Gliomas, APT imaging
genetic mutation prediction, 106–107
malignancy grade prediction, 105–106
Glucose-enhanced CEST imaging (glucoCEST), 114
glycoCEST, 113
Glycogen, 113
Glycosaminoglycan (GAG), 113

Glycosidases, activatable probes, 25, 28
Green fluorescent protein (GFP), 7
Guanine-based bio-reflectors, 10, 215
Guanine crystals platelets, from fish
    artificial guanine crystals, synthesis of, 221–224
    micro-mirror properties of, 218, 220, 223
    osteoblast cells, light illumination of, 229, 230
Guanine sodium salt crystals, 222
Gyromagnetic ratio, 77

Hall effect sensor, 189–191
Harmonic signals, 163–165
Head and neck lesions, 109
Hematoxylineosin (HE) staining, 194
HIFU treatment, *see* High-intensity focused ultrasound
        treatment
Highest Occupied Molecular Orbital (HOMO)
        level, 19
High-intensity focused ultrasound (HIFU) treatment, 98
HMDER derivatives, *see* Hydroxymethyl diethylrhodol
        derivatives
HMDER-βGal, 22, 23, 26
HMRG, *see* Hydroxymethyl rhodamine green
HOMO level, *see* Highest Occupied Molecular Orbital
        level
Hydroxymethyl diethylrhodol (HMDER) derivatives, 22,
        23
Hydroxymethyl rhodamine green (HMRG), 28, 29, 30
Hyperspectral CARS microscopy
    Fourier-transformation, 43, 44
    multiplex/broadband method, 43, 44
    multiplex CARS microscopy/microspectroscopy,
        45–46
    Raman spectrum retrieval, 44–45
    spectral focusing method, 43, 44
    vibrational bands and assignments, 46, 47
    wavelength scanning method, 43
Hyperspectral SRS microscopy
    modulation-frequency-division multiplexing
        technique, 48
    spectral focusing method, 46, 48
    wavelength multiplexing method, 46, 48
    wavelength scanning method, 46, 48
    wavelength-tuning based hyperspectral SRS
        microscopy, 48–49

IDH mutation, *see* Isocitrate dehydrogenase mutation
Image-forming system, 64–66
Image processing, 98–99
Immunostaining, 194
    intraoperative, 196–197
    magnetic, 194–196
Intra-axonal diffusivity, 135
Intramolecular spirocyclization, 18, 19, 21
Intraoperative immunostaining, 196–197
Isocitrate dehydrogenase (IDH) mutation, 107

Kerr effect, 56
k space, 89–90

Larmor frequency, 79, 80
    electromagnetic wave, 76
Lattice, 80
Linear least squares (LLS), 132

lipoCEST, 114
LLS, *see* Linear least squares

Macroscopic magnetization, 80
magA, 209
Magnetically oriented biogenic micro/nanoparticles,
        optical detection
    light scattering anisotropies
        biogenic guanine crystals and hexagonal boron
            nitride (h-BN) crystal plates, 217, 220
        micro-crystalline cellulose particles and, 216, 218
    light scattering change
        dependence, 217, 222
        in micro-crystalline cellulose, D-glucose, and
            sucrose, 216, 219
        from vesicles, 217, 221
    micro-crystal magnetic orientations, classification of,
        217, 222
Magnetic beads, 194–195
Magnetic control
    biogenic guanine crystals, 217
        artificial guanine crystals, synthesis of, 221–224
        diamagnetic micro-mirrors, magnetic orientation,
            220–221, 223
        platelets from fish, micro-mirror properties of, 218,
            220, 223
    uric acid crystals
        luciferase-luciferin reaction, 224
        magnetic field perpendicular orientation, 222
        under magnetic fields, optical responses, 224–229
        methods and materials, 224, 225
        synthetic, 222
Magnetic immunostaining
    automatic magnetic immunostaining system, 195–196
    cancer metastasis, intraoperative diagnosis of, 194
    intraoperative immunostaining, 196–197
    permanent magnet design, 194–195
Magnetic marker, 156
Magnetic moment effect, 166
Magnetic nanoparticles (MNPs), 6–7, 155
    AC and DC gradient fields, 160–162
    biocompatible, 156
    biofunctionalized, 156
    Brownian relaxation, 157
    effective relaxation time, 157
    elementary particle, 156
    fluorescent, 194
    frequency response, 159–160
    imaging technique, 162–164
    magnetic core, 156
    magnetic marker, 156
    magnetization, 161, 164, 168
        DC, 158–159
        dynamic, 157
        hysteresis, 169–170
        saturation, 157, 160
    Néel relaxation, 157
    sensing for SLN biopsy (*see* Sentinel lymph node
        biopsy)
Magnetic orientation
    of guanine crystal, 218
    of living systems and biogenic micromirrors, 9–10
Magnetic particle imaging (MPI), 6–7; *see also* Magnetic
        nanoparticles

hardware
  drive coil, 175–176
  excitation field, 171–172
  FFL generation and scanning, 173–175
  FFP generation and scanning, 172–173
  one-dimensional MPI system, 170
  pickup (receiver) coil, 172
imaging technique
  imaging system, 179–181
  signal voltage and inversion problem, 176–178
  system function matrix, 178–179
operating principle, 155–163
point spread function
  AC *vs.* DC fields, 167–169
  amplitude effect, AC field, 166–167
  harmonic signals, 163–165
  magnetic moment effect, 166
  magnetization hysteresis, 169–170
schematic description, 155, 156
Magnetic permeability, 2
Magnetic resonance frequency, *see* Larmor frequency
Magnetic resonance imaging (MRI), 5
advances and applications
  diffusion, 97
  electric properties of tissue, 98
  flow and pressure, 97–98
  image processing, 98–99
  MRI-guided surgery and treatment, 98
Bloch equation, 81–86
  numerical analysis, 91–95
CEST (*see* Chemical exchange saturation transfer
    imaging)
contrast, 90–91
functional
  applications, 96–97
  neurovascular coupling, 95–96
gene-based iron-labeling, 202
hardware
  permanent magnets, 75, 76
  superconducting magnets, 75–77
  types, 75
iron-related measures, 201
magnetization
  excitation, using RF pulse, 86–87
  macroscopic, 80
  motion of, 81–82, 87
  relaxation of, 80–81
molecular imaging, 9
molecular probing, 101
non-invasive detection, 201
nuclear magnetic moment, 77–80
pulse sequence
  gradient echo imaging, 87–88
  image reconstruction, 89–90
  k space, 89–90
  phase changes, 88
reporter gene expression detection
  ferromagnetic contrast agents, 208
  magnetosome, 207–209
  paramagnetic contrast agents, 208
  reporter, 203
  signal, principles of, 207–208
Magnetic resonance spectroscopy (MRS), 101, 112
Magnetization

excitation, using RF pulse, 86–87
macroscopic, 80
magnetic nanoparticles, 161, 164, 168
  DC, 158–159
  dynamic, 157
  hysteresis, 169–170
  saturation, 157, 160
motion of, 81–82, 87
relaxation of, 80–81
saturation, 97
Magnetoencephalography (MEG), 5–6, 239
Magnetosome
  biosynthesis, 204
  iron biomineral, 206–207
  with MRI, 207–209
  stepwise assembly, 204–205
  vesicles, 205–206
Magnetosome membrane (Mam) proteins, 205–206
Magnetotactic bacteria (MTB)
  amount of iron in, 209
  magnetic properties, 203–204
  magnetosomes, 204–207
Mam proteins, *see* Magnetosome membrane proteins
Maple syrup urine disease (MSUD), 130
Master oscillator fiber amplifier (MOFA) laser source, 51
MD, *see* Mean diffusivity
MDE, *see* Multiple diffusion encoding
Mean diffusivity (MD), 132
MEG, *see* Magnetoencephalography
Meningiomas, APT imaging, 108
MNPs, *see* Magnetic nanoparticles
Modulation-frequency-division multiplexing
    technique, 48
MOFA laser source, *see* Master oscillator fiber amplifier
    laser source
Molecular and cellular imaging, advances in, 7–10
Molecular imaging, cancer cells
  non-invasive molecular imaging, 15–16
  optical fluorescence imaging
    advantages, 16
    probes (*see* Optical fluorescence imaging probes)
Molecular vibrational imaging, coherent Raman
    scattering, *see* Coherent Raman scattering
MPI, *see* Magnetic particle imaging
MRI, *see* Magnetic resonance imaging
MRS, *see* Magnetic resonance spectroscopy
MSUD, *see* Maple syrup urine disease
MTB, *see* Magnetotactic bacteria
Multiple diffusion encoding (MDE), 144–146
Multiplex/broadband method, 43, 44
Multiplex CARS microscopy/microspectroscopy, 45–46

Near-infrared spectroscopy (NIRS), 1, 96
  brain activity measurement, 233
    absorbance *(A)*, 237
    biomaterials with characteristic absorbers, 236
    deoxidized Hb (deoxy-Hb), 237, 238
    neurovascular coupling, 237
    optical path length *(L)*, 237
    oxidized Hb (oxy-Hb), 237, 238
    total Hb (t-Hb), 237
  imaging, 7
Néel relaxation, 157
Néel relaxation time, 159

Neurite orientation dispersion and density imaging (NODDI), 136–137
Neurovascular coupling, 95–96, 237, 238
NIRS, *see* Near-infrared spectroscopy
NNLS method, *see* Nonlinear nonnegative least square method
NODDI, *see* Neurite orientation dispersion and density imaging
NOE, *see* Nuclear Overhauser effect
Non-invasive techniques, brain activity measurement
    conventional techniques, 233
    using light
        biological tissue characteristics, 234–236
        fast optical signal, 238–239
        near-infrared spectroscopy (NIRS), 233, 236–238
        penetrating the brain, 239–241
    using static magnetic field
        passing through the human brain, 243–246
        penetrating the brain, 240–243
Nonlinear nonnegative least square (NNLS) method, 163
Non-magnetic titanium retractors, 192
Non-specific magnetization transfer (MT) effect, 104, 105
Nuclear Overhauser effect (NOE), 104, 105

ODF, *see* Orientation distribution function
OEG, *see* Optoencephalography
OGSE, *see* Oscillating gradient spin echo
OPA, *see* Optical parametric amplifier
OPOs, *see* Optical parametric oscillators
Optical detection, of biogenic micro-mirrors
    magnetically oriented biogenic micro/nanoparticles, 216–222
    methods and materials, 216, 217
Optical fiber-based measurement systems, 216
    Mode-I, II, and III, 216, 217
Optical fluorescence and cancer therapy, 8
Optical fluorescence imaging probes
    activatable probes, 16–17
        aminopeptidase-targeted activatable fluorescence probes, 28–31
        carboxypeptidase-targeted, 30, 31
        enzyme pre-targeting and, 24, 26
        fluorescence switching mechanisms, 17–18
        PeT-based fluorescence imaging probes, 18–20
        spirocyclization-based fluorescence imaging probes, 19, 21–25
        targeted to glycosidases, 25, 28
        tumor-targeting antibody and acidic-pH-activatable probes, 25, 27
    always-on probes, 16, 17
Optical parametric amplifier (OPA), 42
Optical parametric oscillators (OPOs), 42
Optical path length (*L*), 237
Optical resolution, CRS microscopy, 66–68
Optical signal, 238–239
Optical transfer function (OTF), 67–68
Optoencephalography (OEG), 239–240
    photo-sensors, locations of, 240, 241
    somatosensory evoked OEG response, 240–242
Optogenetics, neuronal circuit dynamics, 8
Orientation distribution function (ODF), 136
Oscillating gradient spin echo (OGSE), 128
OTF, *see* Optical transfer function

paraCEST agents, 114
Paramagnetic contrast agents, 208
PCNSLs, *see* Primary central nervous system malignant lymphomas
Permanent magnets, 75, 76
Permittivity, *see* Dielectric constant
PeT, *see* Photoinduced electron transfer
PET, *see* Positron emission tomography
PeT-based fluorescence imaging probes, 18–20
pH
    acidic-pH-activatable probes, 25, 27
    CEST imaging of, 111–113
    physiological, 22, 28
Photoinduced electron transfer (PeT), 18
Photonic cellular analysis, 229–230
pH-weighted imaging, hyperacute stroke, 112
Point spread function (PSF), 162
    AC *vs.* DC fields, 167–169
    amplitude effect, AC field, 166–167
    harmonic signals, 163–165
    magnetic moment effect, 166
    magnetization hysteresis, 169–170
Positron emission tomography (PET), 1, 4
Primary central nervous system malignant lymphomas (PCNSLs), 107
Prostate-specific membrane antigen (PSMA), 30, 31
PSF, *see* Point spread function
PSMA, *see* Prostate-specific membrane antigen

q-space imaging (QSI), 127

Radial diffusivity (RD), 132
Radioisotope (RI) method, 185
Radon transformation, 4
Raman scattering, 8–9, 38
RD, *see* Radial diffusivity
Re-crystallization, 222, 224
Reflecting platelet, cuttlefish, 218, 223
Refractive index modulation, 56
Relaxation of magnetization, 80–81
Reporter gene expression
    housekeeping factors, 202
    imaging reporter, 202–203
    magnetic resonance imaging (MRI)
        ferromagnetic contrast agents, 208
        magnetosome, 207–209
        paramagnetic contrast agents, 208
        reporter, 203
        signal, principles of, 207–208
    magnetotactic bacteria, magnetic properties, 203–204
    transcription factor (TF) regulation, 202
Rhodamine 110, 21
RI method, *see* Radioisotope method

Self-quenching, 17
Sentinel lymph node biopsy (SLNB)
    clinical trials, 193–194
    current state and challenges, 186–187
    dye method, 188
    magnetic immunostaining
        automatic magnetic immunostaining system, 195–196
        cancer metastasis, intraoperative diagnosis of, 194

intraoperative immunostaining, 196–197
permanent magnet design, 194–195
magnetic probe
    axillary sentinel lymph nodes identification,
        189, 191
    development and improvement, 189–190
    Hall effect sensor, 189–191
    magnetic detection principles, 190
    magnetic null field, formation of, 191, 192
    · non-magnetic titanium retractors, 192
    principle, 189
radioisotope (RI) method, 185
using magnetic nanoparticles, 189
using magnetic techniques, 188–189
Sentinel lymph nodes (SLNs), 185–186
    and cancer metastasis, 186, 187
Silent region, 51–52
Single-photon emission computed tomography (SPECT), 4
SLNB, *see* Sentinel lymph node biopsy
SLNs, *see* Sentinel lymph nodes
Somatosensory evoked potential (SSEP) recording,
        240, 241
SPECT, *see* Single-photon emission computed tomography
Spectral focusing method
    hyperspectral CARS microscopy, 43, 44
    hyperspectral SRS microscopy, 46, 48
Spin–lattice relaxation, 80
Spin–spin relaxation, 80
Spin–spin relaxation times, 208
SPIO particles, *see* Superparamagnetic iron oxide particles
Spirocyclization-based fluorescence imaging probes, 19,
        21–25
Spontaneous Raman scattering, 38–39
SQUID, *see* Superconducting quantum interference device
SRL, *see* Stimulated Raman loss
SRS, *see* Stimulated Raman scattering
SSEP recording, *see* Somatosensory evoked potential
        recording
Stejskal–Tanner pulsed gradient spin–echo sequence, 125–126
Stimulated Raman loss (SRL), 39, 59, 60, 63–64
Stimulated Raman scattering (SRS)
    microscopy, 9, 38
        applications, 51–53
        hyperspectral (*see* Hyperspectral SRS microscopy)
        schematics, 40–42
        two-color pulse sources, 42–43
    number of photons, 57–58
    as optical phase modulation, 56–57
    signal-to-noise ratios, 57

Superconducting magnets, 75–77
Superconducting quantum interference device
        (SQUID), 4
Superparamagnetic iron oxide (SPIO) particles, 201,
        208–209
System function matrix, 178–179

T_1 relaxation time, 80, 81, 90, 91
T_2 relaxation time, 81, 90, 91
Telaxation time, 157
TG, *see* TokyoGreen
TME, *see* Transcranial magnetoencephalography
TMS, *see* Transcranial magnetic stimulation
TokyoGreen (TG), 19, 20
Tractogram, 139, 143
Tractography
    applications, 140–142
    DTI, 139
    higher-order methods, 139–140
    limitations, 143–144
Transcranial magnetic stimulation (TMS), 6
Transcranial magnetoencephalography (TME),
        242–246
Tumor-targeting antibody, 25, 27
Two-color optical field, driving force of, 54–55

Ultramultiplex CARS spectroscopic imaging, 51
Uric acid crystals
    luciferase-luciferin reaction, 224
    magnetic field perpendicular orientation, 222
    under magnetic fields, optical responses, 224–229
    methods and materials, 224, 225
    synthetic, 222

Vacuum ultraviolet ray, 3
Vibrational spectroscopic imaging, *see* Hyperspectral
        CARS microscopy

Wavelength multiplexing method, 46, 48
Wavelength scanning method
    hyperspectral CARS microscopy, 43
    hyperspectral SRS microscopy, 46, 48
Wavelength-tuning based hyperspectral SRS microscopy,
        48–49
White matter tract integrity (WMTI), 137

Zeeman level, 79
Zeeman splitting, 79, 80
Z-spectrum, 105

Printed in the United States
by Baker & Taylor Publisher Services